3

Springer Series on Fluorescence

Methods and Applications

Series Editor:
O. S. Wolfbeis

Fluorescence Spectroscopy in Biology

Advanced Methods and their Applications to Membranes, Proteins, DNA, and Cells

Volume Editors: M. Hof · R. Hutterer · V. Fidler

Springer

About this series:

Fluorescence spectroscopy, fluorescence imaging and fluorescent probes are indispensible tools in numerous fields of modern medicine and science, including molecular biology, biophysics, biochemistry, clinical diagnosis and analytical and environmental chemistry. Applications stretch from spectroscopy and sensor technology to microscopy and imaging, to single molecule detection, to the development of novel fluorescent probes, and to proteomics and genomics. The Springer Series on Fluorescence aims at publishing state-of-the-art articles that can serve as invaluable tools for both practitioners and researchers being active in this highly interdisciplinary field. The carefully edited collection of papers in each volume will give continuous inspiration for new research and will point to exciting new trends.

Library of Congress Control Number: 2004114543

ISSN 1617-1306
ISBN 3-540-22338-X Springer Berlin Heidelberg New York

Springer is a part of Springer Science+Business Media
springeronline.com

Printed in Germany

Cover-design: design and production GmbH
Dataconversion: Fotosatz-Service Köhler GmbH, Würzburg

Printed on acid-free paper 52/3020xv – 5 4 3 2 1 0

Series Editor

Professor Dr. Otto S. Wolfbeis
University of Regensburg
Institute of Analytical Chemistry, Chemo- and Biosensors
Universitätsstraße 31
93040 Regensburg
Germany
e-mail: otto.wolfbeis@chemie.uni-regensburg.de

Volume Editors

Professor M. Hof
J. Heyrovský Institute of Physical Chemistry
Academy of Sciences of the Czech Republic
and Centre for Complex Molecular Systems and Biomolecules
Dolejškova 3
18223 Prague 8
Czech Republic
e-mail: hof@jh-inst.cas.cz

Professor R. Hutterer
Institute of Analytical Chemistry, Chemo- and Biosensors
Universitätsstraße 31
93040 Regensburg
Germany
e-mail: rudolf.hutterer@chemie.uni-regensburg.de

Professor V. Fidler
Faculty of Nuclear Sciences and Physical Engineering
Czech Technical University in Prague
Břehová 7
11519 Praha 1
Czech Republic
e-mail: fidler@troja.fjfi.cvut.cz

Preface

Take any combination of the following features: supramolecular structures with a specific fluorescent probe localized as you would like; nanoscale spatial resolution; tailor-made molecular and/or solid-state fluorescing nanostructures; user-friendly and/or high- throughput fluorescence techniques; the ability to do whatever you wish with just one single (supra)molecule; utilization of non-linear optical processes; and, last but not least, physical understanding of the processes resulting in a (biological) functionality at the single molecule level. What you will then have is some recent progress in physics, chemistry, and the life sciences leading to the development of a new tool for research and application. This was amply demonstrated at the 8th Conference on Methods and Applications of Fluorescence: Probes, Imaging, and Spectroscopy held in Prague, the Czech Republic on August 24th–28th, 2003. This formed a crossroad of ideas from a variety of natural science and technical research fields and biomedical applications in particular.

This volume – the third book in the Springer-Verlag Series on Fluorescence – reviews some of the most characteristic topics of the multidisciplinary area of fluorescence applications in life sciences either presendted directly at th 8th MAF Conference or considered to be a cruical development in the field.

In the initial contribution in Part 1 – Basics and Advanced Approaches, the editors explain the basics of fluorescence and illustrate the relationship between some modern fluorescence techniques and classical approaches. The second contrigution by B. Valeur, with his many years of personal experience, helps the fluorescence spectroscopist to answer teh perennial question of whether to use pulse or phase modulation fluorescence detection. A technically demanding but promising new approach for extracting distance information from fluorescence kinetics data is presented by ist innovator L. Johansson in the third contribution. The three subsequent contributions also have the pioneers of each new approach among their authors: D. Birch – nanotomography, M. Hof – solvent relaxation used micro-polarity and fluidity probing, and N. Thompson – total internal reflection fluorescence microscopy. The last contribution in Part 1, written by J. Enderlein, is devoted to single molecule spectroscopy using a quantitative approach to data analysis in this important new experimental field. Part 2 – Fluorescence in Biological Membranes – addresses a hot topic in membrane research, i.e., the formation of microdomains. G. Duportail summarizes the recent results in the study of lipid rafts using fluorescence quenching and L. Bagatolli demonstrates the use of fluorescence microscopy in the charcterization of domain formation.

Part 3 consisting of contributions ten and eleven deals with advanced fluorescene kinetics analysis in protein sciences. G. Krishnamoorthy's chapter shows what we can learn with time-resolved fluorescence about protein dynamics and folding. Y. Mély combines time-resolved fluorescence with FCS to elucidate the mechnaism of interaction of the HIV-1 nucleocapsid protein with hairpin loop oligonucleotides.

The development of efficient non-viral dug carriers is one of the most urgently needed requirements in the biological sciences. It has become obvious that modern fluorescence is capable of helping in the development of such supramolecular assemblies. Thus the two contributions (I. Blagbrough and M. Langner) in Part 4 are devoted to this field.

The final part of this volume focuses on two new approaches in cell fluorescence microscopy. R. Brock shows how to characterize diffusion in cells by fluorescence correlation spectroscopy. The last two contributions by S. Rosenthal and O. Minet are devoted to photophysics and the use of quantum dots in cell imaging.

Prague, October 2004 Martin Hof, Rudi Hutterer, and Vlastimil Fidler

Contents

Contributors

N. Adjimatera
Department of Pharmacy and Pharmacology, University of Bath, Bath BA2 7AY, UK

J. Azoulay
Laboratoire de Pharmacologie et Physicochimie, UMR 7034 du CNRS, Faculté de Pharmacie, Université Louis Pasteur de Strasbourg, 67401 Illkirch Cedex, France

Luis A. Bagatolli
Memphys – Center for Biomembrane Physics, Department of Biochemistry and Molecular Biology, University of Southern Denmark, Campusvej 55, DK-5230 Odense M, Denmark
e-mail: bagatolli@memphys.sdu.dk

S. Bernacchi
Laboratoire de Pharmacologie et Physicochimie, UMR 7034 du CNRS, Faculté de Pharmacie, Université Louis Pasteur de Strasbourg, 67401 Illkirch Cedex, France

H. Beltz
Laboratoire de Pharmacologie et Physicochimie, UMR 7034 du CNRS, Faculté de Pharmacie, Université Louis Pasteur de Strasbourg, 67401 Illkirch Cedex, France

J. Beuthan
Charité – Universitätsmedizin Berlin/Campus Benjamin Franklin, Institut für Medizinische Physik und Lasermedizin, Fabeckstraße 60–62, 14195 Berlin, Germany

D. J. S. Birch
University of Strathclyde, Department of Physics, John Anderson Building, 107 Rottenrow, Glasgow G4 0NG, UK

I. S. Blagbrough
Department of Pharmacy and Pharmacology, University of Bath, Bath BA2 7AY, UK
e-mail: prsisb@bath.ac.uk

R. Brock
Group of Cellular Signal Transduction, Institute for Cell Biology, University of Tübingen, Auf der Morgenstelle 15, 72076 Tübingen, Germany
e-mail: roland.brock@uni-tuebingen.de

K. Bryl
Department of Physics and Biophysics, University of Warmia and Mazury, 10-719 Olsztyn, Poland
e-mail: kris@uwm.edu.pl

J. Clamme
Laboratoire de Pharmacologie et Physicochimie, UMR 7034 du CNRS, Faculté de Pharmacie, Université Louis Pasteur de Strasbourg, 67401 Illkirch Cedex, France

J. Darlix
LaboRétro, Unité de Virologie Humaine INSERM, Ecole Normale Supérieure de Lyon, 46 allée d'Italie, 69364 Lyon, France

C. Dressler
Charité – Universitätsmedizin Berlin/Campus Benjamin Franklin, Institut für Medizinische Physik und Lasermedizin, Fabeckstraße 60–62, 14195 Berlin, Germany

G. Duportail
Laboratoire de Pharmacologie et Physicochimie, UMR 7034 du CNRS, Faculté de Pharmacie, Université Louis Pasteur de Strasbourg, 67401 Illkirch Cedex, France
e-mail: duportai@aspirine.u-strasbg.fr

J. Enderlein
Institute for Biological Information Processing 1, Forschungszentrum Jülich, 52425 Jülich, Germany
e-mail: j.enderlein@fz-juelich.de

D. Ficheux
IBCP, 7, passage du Vercors, 69367 Lyon Cedex 07, France

M. Isaksson
Department of Chemistry, Biophysical Chemistry, Umeå University, S-901 87 Umeå, Sweden

L. B.-Å. Johansson
Department of Chemistry, Biophysical Chemistry, Umeå University, S-901 87 Umeå, Sweden
e-mail: lennart.johansson@chem.umu.se

S. Kalinin
Department of Chemistry, Biophysical Chemistry, Umeå University, S-901 87 Umeå, Sweden

G. Krishnamoorthy
Dept. of Chemical Sciences, Tata Institute of Fundamental Research, Homi Bhabha Road, Mumbai 400 005, India

M. Langner
Laboratory for Biophysics of Macromolecular Aggregates, Institute of Physics, Wrocław University of Technology, Wyb. Wyspiańskiego 27, 50-370 Wrocław, Poland, and Academic Centre for Biotechnology of Lipid Aggregates, ul. Przybyszewskiego 63/77, 51–148 Wrocław, Poland
e-mail: marek.langner@pwr.wroc.pl

Y. Mély
Laboratoire de Pharmacologie et Physicochimie, UMR 7034 du CNRS, Faculté de Pharmacie, Université Louis Pasteur de Strasbourg, 67401 Illkirch Cedex, France
e-mail: mely@aspirine.u-strasbg.fr

O. Minet
Charité – Universitätsmedizin Berlin/Campus Benjamin Franklin, Institut für Medizinische Physik und Lasermedizin, Fabeckstraße 60–62, 14195 Berlin, Germany
e-mail: minet@zedat.fu-berlin.de

A. P. Neal
Department of Pharmacy and Pharmacology, University of Bath, Bath BA2 7AY, UK

I. K. Pero
Department of Chemistry, Campus Box 3290, University of North Carolina at Chapel Hill, Chapel Hill, NC 27599-3290, USA

E. Piemont
Laboratoire de Pharmacologie et Physicochimie, UMR 7034 du CNRS, Faculté de Pharmacie, Université Louis Pasteur de Strasbourg, 67401 Illkirch Cedex, France

S. J. ROSENTHAL
Department of Chemistry, Vanterbilt University, Nashville, Tennessee 37235, USA
e-mail: sandra.j.rosenthal@vanderbilt.edu

O. J. ROLINSKI
University of Strathclyde, Department of Physics, John Anderson Building, 107 Rottenrow, Glasgow G4 0NG, UK
e-mail: o.j.rolinski@strath.ac.uk

B. ROQUES
Départment de Pharmacochimie Moléculaire et Structurale, INSERM U266 Faculté de Pharmacie, 4 Avenue de l'Observatoire, 75270 Paris Cedex 06, France

A. SAXENA
Départment of Chemical Sciences, Tata Institute of Fundamental Research, Homi Bhabha Road, Mubai 400 005, India

E. SCHAUB
Laboratoire de Pharmacologie et Physicochimie, UMR 7034 du CNRS, Faculté de Pharmacie, Université Louis Pasteur de Strasbourg, 67401 Illkirch Cedex, France

J. SÝKORA
J. Heyrovský Institute of Physical Chemistry, Academy of Sciences of the Czech Republic, and Centre for Complex Molecular Systems and Biomolecules, Dolejškova 3, 18223 Prague 8, Czech Republic

N. L. THOMPSON
Department of Chemistry, Campus Box 3290, University of North Carolina at Chapel Hill, Chapel Hill, NC 27599-3290, USA
e-mail: nlt@unc.edu

J. B. UDGAONKAR
National Centre for Biological Sciences, TIFR, UAS-GKVK Campus, Bangalore 560 065, India

B. VALEUR
CNRS UMR 8531, Laboratoire de Chimie générale, CNAM, 292 rue Saint Martin, 75141 Paris Cedex 03, France and Laboratoire PPSM, ENS-Cachan, 61 avenue du Président Wilson, 94235 Cachan Cedex, France

M. R. WARNEMENT
Department of Chemistry, Vanderbilt University, Nashville, Tennessee 37235, USA

Part 1

Fluorescence Spectroscopy: Basics and Advanced Approaches

1 Basics of Fluorescence Spectroscopy in Biosciences

M. Hof, V. Fidler and R. Hutterer

Keywords: Fluorescence polarization; Time-correlated single photon counting; Fluorescence energy transfer; Fluorescence quenching; Solvent relaxation; Fluorescence correlation spectroscopy

Abbreviations

BODIPY	Derivatives of 4-bora-3a,4a-diaza-*s*-indacene
DPH	1,6-Diphenyl-1,3,5-hexatriene
ET	Energy transfer
2D FLIM	2-Dimensional fluorescence lifetime imaging
FRAP	Fluorescence recovery after photobleaching
FRET	Fluorescence resonance energy transfer
NBD	Derivatives of 7-nitrobenz-2-oxa-1,3-diazol-4-yl
NFOM	Near-field optical microscopy
SNOM	Scanning near-field optical microscopy
SR	Solvent relaxation
STM/AFM	Scanning tunnelling microscopy/atomic force microscopy
SUV	Small unilamellar vesicles
TCSPC	Time-correlated single photon counting
TIRF	Total internal reflection fluorescence
TIR-FRAP	Total internal reflection fluorescence recovery after photobleaching
TMA-DPH	1-(4-trimethylammonium-phenyl)-6-phenyl-1,3,5-hexatriene
TRES	Time-resolved emission spectra

1.1 Introduction

After the pioneering works of Kasha, Vavilov, Perrin, Jabłoński, Weber, Stokes and Förster (including the appearance of the first book of the latter on fluorescence of organic molecules in 1951 [1]), fluorescence spectroscopy became a widely used scientific tool in biochemistry, biophysics and material sciences. In recent years, however, several new applications based on fluorescence have been developed, promoting fluorescence spectroscopy from a primarily scientific to a more routine method. The phenomenon of fluorescence is, for example, exploited in simple analytical assays in environmental science and clinical chemistry, in cell identification and sorting in flow cytometry, and in imaging of single cells in medicine. Though there is a rapid growth in the number of routine applications of fluorescence, the principles remain the same. This contribution aims at a condensed but comprehensive description of the principles and selected applications of fluorescence spectroscopy. Standard approaches like the detection of anisotropy, quenching and solvent shifts will be discussed in some detail in this chapter, while more advanced techniques will only be mentioned briefly. For the more detailed description of these advanced techniques the reader is referred to the following chapters in this book written by experts in the respective fields.

The fluorescence of a molecule is the light emitted spontaneously due to transitions from excited singlet states (usually S_1) to various vibrational levels of the electronic ground state, i.e. $S_{1,0}{\rightarrow}S_{0,v}$. It can be characterized by several parameters. The most important among them are the fluorescence intensity at a given wavelength, $F(\lambda)$, the emission spectrum (i.e. dependence of emission intensity on the emission wavelength), quantum yield (Φ; see Sect. 1.4.1), lifetime (τ; see Sect. 1.4.1) and polarization (P; see Sect. 1.3). These parameters can be monitored in a steady-state or time-resolved manner. They carry information about both the photophysical properties of the fluorescing molecule and the chemical and physical nature of its microsurroundings. The following section will specify such parameters and describe how they are influenced by intra- and intermolecular processes.

1.2 Fluorescence and its Measurement

1.2.1 Molecular Electronic Relaxation

Schematic representation of spontaneous molecular relaxation processes that follow any excitation of a molecule to a higher electronic excited state (e.g. by absorption of a photon) is depicted in Fig. 1.1 in the form of a Jabłoński diagram.

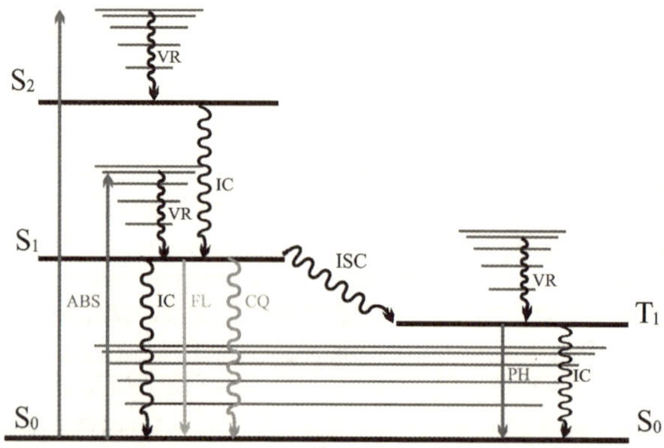

Fig. 1.1. Jabłoński diagram illustrating the creation and fate of a molecular excited singlet state, including absorption (ABS), fluorescence (FL), phosphorescence (PH), internal conversion (IC), intersystem crossing (ISC), vibrational relaxation (VR) and collisional quenching (CQ). Not included are processes like solvent relaxation, energy transfer and photochemical reactions

Fluorescence emission is clearly one of the several possible, mostly non-radiative processes that compete with each other. Thus, fluorescence intensity, emission wavelength, time behaviour and polarization can be indirectly influenced by every interaction of the fluorescing molecule that can either change the probability of any of the competing relaxation processes (e.g. internal conversion – IC, or intersystem crossing – ISC, in particular), or that can introduce a new relaxation pathway (e.g. photochemical bonding or the simple proximity of a heavy atom or a particular chemical group). Many biochemical and biological applications of the fluorescence are based on these phenomena, such as the widespread usage of a broad variety of fluorescent probes, just to name one.

1.2.2 Detecting Fluorescence

The principal fluorescence measurement arrangement is depicted in Fig. 1.2, where the most important properties (parameters) are listed for both exciting radiation and fluorescence emission. Not all of these parameters are necessarily known or well specified for every spectrofluorometric instrument; any attempt at sophisticated analysis and interpretation of the fluorescence data should be accompanied by a rigorous measurement of all the listed parameters that are relevant to the interpretation. The following examples of fluorescence spectroscopy applications also indicate this aspect of practical fluorescence measurements.

The fluorescence of an object of interest can be detected in various ways. Besides the classical solution fluorescence measurement in different types of cuvette, there are several advanced ways of detecting the fluorescence signal. The use of fibre optics allows measurement of fluorescence even in biological organs in vivo. When looking at cells (see Chaps. 14 and 16 in this book) one can use cell culture plates or flow cytometry in combination with optical microscopy. Selected spots within a cell can be monitored using classical, confocal, or multiphoton microscopy. Advanced techniques of single molecule spectroscopy (Chap. 7), total internal reflection fluorescence microscopy (Chap. 6), fluorescence correlation spectroscopy (Chaps. 12, 13 and 14) and other advanced techniques are described elsewhere in this book. Two trends in recent developments of fluorescence techniques, often combined within one instrument, should be mentioned here: (1) high spatial resolution (extremely small volume probed or a combination of local fluorescent probe FRET with SNOM techniques as in [2]); and (2) high time resolution, performed simultaneously [3]. An illustration of such instrumental development is the space-resolved TCSPC detector used by [4] for 2D FLIM with 500 nm spatial and 100 ps time resolution. New technology combining, for example, NFOM or STM/AFM with high-resolution photon timing, when each detected photon is tagged with all other information related to it, allows multi-dimensional fluorescence lifetime and fluorescence correlation spectroscopy to be performed during one measurement [3]. Single molecule fluorescence characterization can thus now be done with unprecedented accuracy and depth (see, e.g., Chap. 7 in this book and recent reviews [3, 5]).

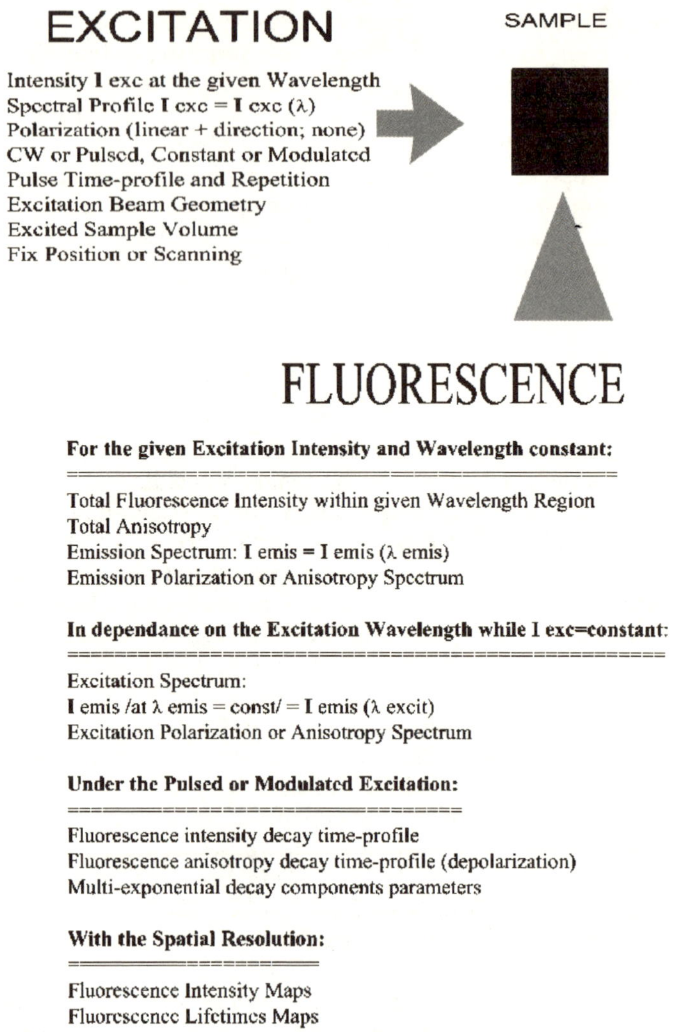

Fig. 1.2. Summary of main variables and read-out parameters of fluorescence experiments

1.2.3 Data Evaluation

For the primary spectroscopic raw data treatment relevant to the technique used for a particular fluorescence measurement (such as correction of spectral intensity for the sensitivity of a detector), we refer to the instrument producer manuals, to basic physics textbooks and to comprehensive books on fluorescence [22–24]. Topics such as fluorescence quantum yield evaluation and steady-state spectra analysis (e.g. decomposition) are also covered by such literature [24]. Mathematically much more complicated is the fluorescence kinetics data treatment necessary for fluorescence lifetime and rotational correlation time calculation. Furthermore, mathematical models differ substantially with the detection technique used: time-correlated single photon counting [e.g. 23] or phase shift measurement [e.g. 22]. Comparison of the two basic fluorescence lifetime measurement techniques is done in detail in Chap. 2 of this book. For data evaluation methods in fluorescence correlation spectroscopy (such as number of particles and diffusion time calculations) see [25] or Chap. 14, for fluorescence recovery after photobleaching (rate and extent of recovery calculations) see [26, 27] and for internal reflection fluorescence parameters see [28] and Chap. 6 of this book. Moreover, there are many comprehensive books on optical spectroscopy covering aspects of techniques and data analysis of fluorescence spectroscopy as well – e.g. [29, 30], to name but two.

1.3 Polarized Fluorescence

Interaction of the exciting light with a molecule can be described as an interaction of the electric field component of the light with the relevant transition (electrical) dipole moment of the molecule. Thus, the absorption of the light quantum is proportional to the cosine of the angle between the two directions, i.e. between the excitation light polarization plane and the transition moment vector. Consequently, excitation by linear polarized light leads to an anisotropic spatial distribution of the excited molecules: those with transition dipole moment parallel to the light polarization can be excited, whereas those in a perpendicular position can not (this phenomenon is called photoselection). The resulting anisotropy can persist even up to the later moment of fluorescence emission, yielding partially polarized emitted light. Such fluorescence polarization will decay faster with higher rotational diffusion of the excited molecule, and it can be diminished further, e.g. by an excitation energy transfer. The rotational diffusion depends on the (micro)viscosity of the environment, and on the size and shape of the excited molecule. This connection represents the basis for applications of fluorescence polarization studies. The depolarization by excitation energy transfer is often an undesirable process. However, it occurs only in concentrated solutions when the average distance between molecules is not much above 5 nm. Thus, this kind of depolarization can be avoided by the use of dilute solutions.

1.3.1 Definition of Polarization and Anisotropy

The direction of light polarization is conventionally specified with reference to a system of laboratory coordinates defined by the propagation directions of the excitation beam and of the fluorescence beam. It is customary to observe the fluorescence beam resolved in directions parallel ($F_{||}$) and perpendicular ($F_{\perp}$) to the direction of the linear polarized excitation light ($E_{||}$). The degree of fluorescence polarization P is defined as

$$P = (F_{||} - F_{\perp})/(F_{||} + F_{\perp}) \tag{1.1}$$

An equivalent parameter used for the description of fluorescence polarization is the anisotropy a:

$$a = (F_{\parallel} - F_{\perp})/(F_{\parallel} + 2F_{\perp}) \qquad (1.2)$$

Though both parameters are equivalent for the description of polarized light, anisotropy is usually preferred because it leads to simpler equations for the time-dependent behaviour. Following a pulse excitation, the fluorescence anisotropy of a spherical particle in a homogeneous isotropic medium decays exponentially, given by

$$a = a_0 \exp(-t/\tau_p) \qquad (1.3)$$

where τ_p is the rotational correlation time of a sphere and a_0 is the anisotropy at t=0. The anisotropy stays constant at the initial value a_0 if the fluorophore is fixed in space. Thus, it can be experimentally determined by measuring the steady-state anisotropy of the dye in a rigid and homogeneous medium like vitrified solutions. The value of a_0 depends on the angle between the absorption and emission transition moments of the dye, β. Since the orientation of absorption and emission transition moments is characteristic for each corresponding electronic transition, the angle β is a constant for every given pair of electronic transitions of a dye. As explained earlier, fluorescence usually arises from a single transition. Thus, a_0 is supposed to be invariant to the emission wavelength. However, the solvent relaxation (Sect. 1.5) occurring on a nanosecond timescale can result in a time-dependent shift of the emitting S_1 state energy and lead to a decrease of anisotropy across the emission spectrum. Since the excitation spectrum might be composed of several absorption bands with different transition moments, the fluorescence anisotropy might change with the exciting light wavelength. Thus, polarization excitation spectra can be used to identify partially overlapping electronic transitions. Using linear polarized light under one-photon excitation conditions (for multi-photon excitation see [6]) a_0 for a randomly orientated molecule is

$$a_0 = 0.6 \cos^2 \beta - 0.2 \qquad (1.4)$$

For colinear absorption and emission transition dipole moments, the theoretical initial anisotropy value a_0 is equal to 0.4.

1.3.2
Steady-State Fluorescence Anisotropy

In low-viscosity solvents the rotational relaxation of low molecular weight compounds occurs on the picosecond timescale [7]. Since in this case the rotation is much faster than the fluorescence (typically with nanosecond decay time), the steady-state emission is depolarized. If a fluorophore rotational motion is on the same timescale as its fluorescence decay time, steady-state fluorescence polarization is observed. In the simplest case, i.e. for a spherical-rotor-like molecule with a single-exponential fluorescence intensity decay (τ), the expected steady-state fluorescence anisotropy is given by

$$a = a_0/[1 + (\tau/\tau_p)] \tag{1.5}$$

The rotational correlation time of a sphere τ_p is given by

$$\tau_p = \eta\, V/RT \tag{1.6}$$

where η is the viscosity, T the temperature, R the gas constant and V the volume of the rotating unit. It is important to note that these equations only hold for spherically symmetrical molecules. Corresponding expressions for spherically unsymmetrical and ellipsoidal molecules can be found in the literature [8–11]. By combining Eqs. (1.5) and (1.6) it can be seen that a plot of $1/a$ versus T/η should be linear, with an intercept equal to $1/a_0$ and with a slope/intercept that is directly proportional to τ and indirectly proportional to V. If one of the latter two parameters is known, the other can be calculated from such a plot. An absence of the viscosity dependence indicates that some other depolarizing process dominates. A non-linear plot of $1/a$ versus T/η indicates the existence of more than one rotational mode.

Prior to the availability of time-resolved measurements, such so-called Perrin plots were extensively used to determine the apparent hydrodynamic volume of proteins [12–14]. Since protein association reactions usually affect the rotational correlation time of the protein label, such reactions have been characterized by steady-state anisotropy measurements [15, 16].

1.3.3 Time-Resolved Fluorescence Polarization

As described by Eq. (1.3), the anisotropy of spherical particles in homogeneous isotropic medium decays exponentially. Anisotropy decay, however, can be more complex. The three most important origins of a deviation from mono-exponential decay are as follows.

1.3.3.1 *Non-Spherical Particles in Homogenous Isotropic Medium*

The theory for rotational diffusion of non-spherical particles is complex; the anisotropy decay of such a molecule can be composed of a sum of up to five exponentials [17]. The ellipsoids of revolution represent a smooth and symmetrical shape, which is often used for description of the hydrodynamic properties of proteins. They are three-dimensional bodies generated by rotating an ellipse about one of its characteristic axes. In this case the anisotropy decay displays only three rotational correlation times, which are correlated to the rotational diffusion coefficients $D_{\parallel}$ and $D_{\perp}$. The indexes $\parallel$ and $\perp$ denote the rotation around the main and side axes, respectively [11]. The pre-exponential factors of the three exponentials depend on the angle between the emission transition moment and the main axis of the rotational ellipsoid. In practice, due to the limited time resolution, one rarely resolves more than two exponentials [11].

1.3.3.2
Segmental Mobility of the Chromophore

An important factor is that a typical chromophore is not rigidly fixed to the biopolymer but rotates around the bond linking it to the biopolymer. Consequently, the anisotropy decay kinetics are found to be double or triple exponential, due to the contributions from internal and global rotation of the macromolecule. The same concept applies to the rotational wobble of that portion of a biopolymer that is in proximity to the fluorophore or, in the more defined case, to the rotation of a molecular domain [18].

1.3.3.3
Hindered Rotors: Fluorescent Dyes in Biological Membranes

If isotropic rotors are imbedded in an anisotropic environment, like in phospholipid bilayers, the decay of fluorescence anisotropy can be complex. Let us consider a dye, such as 1-(4-trimethylammoniumphenyl)-6-phenyl-1,3,5-hexatriene (TMA-DPH) or 1,6-diphenyl-1,3,5-hexatriene (DPH), intercalated inside the bilayer. The polarization of its fluorescence depends on its motion dumping exerted by the molecular environment. In the case of a fixed hindrance to rotational relaxation motion, the anisotropy value decreases exponentially, not to zero but to a finite value a_∞, yielding the formula:

$$a = (a_0 - a_\infty) \exp(-t/\tau_p) + a_\infty \quad (1.7)$$

The time-resolved measurement of such membrane probes contains information on the dynamics of the hindered probe rotation, often interpreted as the microviscosity, and about the hindrance of this rotation, usually interpreted as the static packing arrangement of the lipids or the so-called membrane order [19, 20]. Fluorescence polarization studies in membranes, however, exhibit some major limitations: the experimentally determined steady-state and time-resolved anisotropies characterize the motional restrictions of the 'reporter' molecule itself and give therefore only *indirect* information about its environment. The consequence is that the fluorescence of a probe – namely when it is bound covalently to the lipid (like a phosphatidylcholine-fixed DPH) – might report more about this attachment than about the surrounding membrane. The membrane order parameters obtained from freely mobile probes like DPH result from a broad distribution of localizations within the hydrophobic interior, the detailed characterization of which reveals inherent ambiguities [21].

Despite these drawbacks, among the fluorescence techniques employed so far, the determination of fluorescence anisotropy has certainly been the dominating method in studies of biological systems. For a detailed description of the theory and several examples of its application see review articles [11, 20].

1.4 Influence of Fluorescence Quenching

1.4.1 Fluorescence Quantum Yield and Lifetime

In the gas phase or in non-interacting solvents, exactly speaking in the absence of intermolecular photophysical or photochemical processes (see Fig. 1.1), the fluorescence intensity F after a short pulse excitation decays according to the mono-exponential law with an average fluorescence lifetime τ. The rate constant of this fluorescence decay, k ($=1/\tau$), represents the sum of the emission rate constant of the unperturbed fluorophore, k_0 ($=1/\tau_0$), and the rate constants of its internal radiationless processes: internal conversion and intersystem crossing, k_{ic} and k_{isc}, respectively. The radiative lifetime τ_0 can be correlated to the emission transition dipole moment **M** by

$$\tau_0 \approx \text{constant}/k_{ave}^3\, n^2\, |\mathbf{M}|^2, \tag{1.8}$$

where n is the refractive index of the solvent and k_{ave} the wavenumber of the centre of gravity of the fluorescence emission spectrum. The radiative lifetime τ_0 can be considered as a photophysical constant of a chromophore surrounded by a solvent shell with the refractive index n. In the case of planar aromatic systems it appears to be temperature independent [31]. Since the internal conversion and intersystem crossing processes compete with fluorescence for deactivation of the lowest excited singlet state, not all molecules will return to the ground state by fluorescence. The fraction of the excited molecules that do fluoresce is called the quantum yield Φ. In terms of the above defined rate constants and lifetimes, Φ is given by Eq. (1.9):

$$\Phi = k_0/(k_0 + k_{ic} + k_{isc}) = \tau/\tau_0 \tag{1.9}$$

The fluorescence lifetime τ can be determined directly by monitoring the decay curve of fluorescence intensity following a short excitation pulse [23] or by detecting the emission response delay (phase shift) to the intensity modulated excitation light [22]. When a standard steady-state spectrofluorometer is used for the fluorescence quantum yield measurement, Φ is usually determined by fluorescence intensity and spectra comparison with those of standard compounds of known quantum yield [32].

1.4.2 Fluorescence Quenchers

A fluorescence quencher is a compound, the presence of which in the vicinity of a fluorophore leads to a decrease of the fluorescence quantum yield and lifetime of the latter. For example, those molecules or ions can function as a quencher that are added to the solution and introduce new or promote already existing non-radiative deactivation pathways (solute quenching) by molecular contact with the chromophore. Further possibilities are self-quenching by simply another fluorophore molecule of the same type, and quenching by solvent molecules. In any case, the quenching term $k_Q[Q]$ has to be added to Eq. (1.9) yielding

$$\Phi = k_0/(k_0 + k_{ic} + k_{isc} + k_Q\,[Q]) \tag{1.10}$$

where k_Q is the bimolecular quenching constant and $[Q]$ the concentration of the quencher.

1.4.2.1
Solute Quenching

Solute quenching reactions are a very valuable tool for studies of proteins, membranes and other supra- or macromolecular assemblies, providing information about the location of fluorescent groups in the examined molecular structure. A fluorophore that is located on the surface of such a structure will be relatively accessible to a solute quencher (for a list of common quenchers see Table 1.1). A quenching agent will quench the chromophore that is buried in the core of the molecular assembly to a lesser degree. Thus, the quenching experiment can be used to probe topographical features of the examined structure and to detect structural changes that may be caused by addition of external compounds or changed physical conditions. In usual quenching experiments the quencher is added successively to the fluorophore-containing solution. The analysis of the dependence of fluorescence intensity F, quantum yield Φ, or lifetime τ on the quencher concentration yields quantitative information about the accessibility of the chromophore within the macro- or supramolecular structure.

Depending on the chemical nature of both the quenching agent and the chromophore, one has to distinguish between two forms of quenching: dynamic and static quenching. Static quenching results from the formation of a non-fluorescent fluorophore-quencher complex, formed in the fluorophore's ground state. Characteristic for this type of quenching is that increasing quencher concentration decreases the fluorescence intensity or quantum yield but does not affect the fluorescence lifetime. An important feature of static quenching is its decrease with increasing temperature, as the stability of the fluorophore-quencher ground state complexes is generally lower at higher temperatures.

If quenchers act (e.g. through collisions) by competing with the radiative deactivation process (see Eq. (1.10) and Fig. 1.1), the ratio of the quantum yield in the absence, Φ_a, and the presence, Φ, of the quencher will be equal to the ratio of the corresponding lifetimes, τ_a/τ (see Eq. 1.9). The concentration dependence of this so-called dynamic or collisional quenching is described by the Stern-Volmer equation, where the Stern-Volmer constant K_{SV} is equal to $k_Q\tau_a$:

$$\Phi_a/\Phi = \tau_a/\tau = F_a/F = 1 + K_{sv}\,[Q] = 1 + k_Q\tau_a\,[Q]) \tag{1.11}$$

Thus, from a plot of one of these ratios versus the quencher concentration, and by knowing τ_a independently, the bimolecular quenching constant, k_Q, can be determined. From physical considerations, the k_Q magnitude can be expressed as follows:

$$k_Q = 4\gamma\,\pi D\,r\,N' \tag{1.12}$$

where γ is the efficiency of the quenching reaction, D is the sum of the diffusion coefficients of quencher and chromophore, r is the sum of the molecular radii of quencher and chromophore and $N'=6.02\times10^{20}$. The diffusion coefficient for each species i can be expressed by using the Stokes-Einstein relationship:

$$D_i = b\,T/6\pi\,\eta\,r_i \tag{1.13}$$

where b is Boltzmann's constant and η is the viscosity of the solution. It follows that the quenching constant increases with increasing temperature T due to the diffusion control of the dynamic quenching.

Another mechanism of the dynamic fluorescence quenching is connected with the chemical nature of the chromophore and the solute quencher: quenchers containing halogen or heavy atoms increase the intersystem crossing rate (generally induced by a spin-orbital coupling mechanism). Acrylamide quenching of tryptophans in proteins is probably due to the excited state electron transfer from the indole to acrylamide. Paramagnetic species are believed to quench aromatic fluorophores by an electron spin exchange process.

In many instances a fluorophore can be quenched by both dynamic and static quenching simultaneously. The characteristic feature for mixed quenching is that the plot of the concentration dependence of the quantum yield or intensity ratios (see Eq. 1.11) shows an upward curvature. In this case the Stern-Volmer equation has to be modified, resulting in an equation which is second order in [Q]. More details on the theory and applications of solute quenching can be found in an excellent review by Maurice Eftink [59]. An overview of typical fluorophore-quencher pairs is given in Table 1.1. In Chap. 8, quenching methods are used for the detection of so-called rafts in membranes.

1.4.2.2
Solute Quenching in Protein Studies: an Application Example

One of the main aims in biophysical studies of the structure and function of proteins is to identify the protein domains that are responsible for interaction of the entire protein with physiologically relevant binding partners. Proteins usually contain several tryptophan residues, which might be distributed over the different protein domains. Since each of these tryptophan residues is located in a distinct environment, each residue might exhibit a different fluorescence decay profile as well as a different accessibility to quenching molecules. Using picosecond time-resolved fluorescence spectroscopy, the tryptophan fluorescence lifetimes in proteins containing up to three tryptophan residues can be determined with high accuracy [37]. As an example may

Table 1.1. List of selected solute quenchers

Type of fluorophore	Used quenchers	References
Indole	Carboxy groups, chlorinated compounds, dimethylformamide	33–35
Tyrosine	Disulphides	36
Tryptophan	Acrylamide, histidine, succinimide, trifluoroacetamide, iodide, disulphides	37–42
Naphthalene	Halogens, nitroxides	43, 44
Anthracene	Amines, halogens, thiocyanate	45–47
Anthranoyloxy probes	Tetracaine	48
Quinolinium ions and their betaines	Chloride, bromide, iodide	49–51
Pyrene	Halothane	52
Carbazole	Amines, chlorinated compounds, halogens	53–56
Common quencher for almost all dyes	Oxygen	57, 58

serve a picosecond tryptophan study of prothrombin fragment 1 (BF1), which is the 1-156 *N*-terminal peptide of a key blood coagulation protein, prothrombin. It is believed to be the region predominantly responsible for the metal ion and membrane binding properties of prothrombin. An important question to answer has been to what extent the conformations of the two protein domains, the so-called Gla and kringle domains, are altered by the interaction with calcium ions and with negatively charged phospholipid surfaces (see Fig. 1.3).

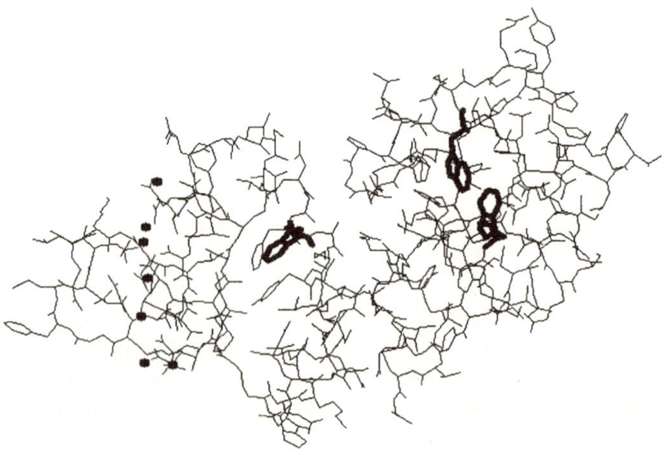

Fig. 1.3. A depiction of the X-ray structure of Ca-BF1. The *right part* of the protein is the kringle domain, where the solvent-inaccessible tryptophan residues Trp90 and Trp126 are located. The Gla domain is the *left part* of the protein, containing the solvent- and quencher-accessible Trp42 and seven calcium ions (*dots*)

Analysis of the fluorescence decay of the three tryptophan residues (Trp42, Trp90, Trp126) in apo-BF1 resulted in a five-exponential decay model, where the five fluorescence lifetimes are wavelength independent. Since structural data show a huge difference in solvent accessibilities for the kringle tryptophans (4×10^{-20} m^2 for Trp90 and Trp126) and the Gla tryptophan (133×10^{-20} m^2 for Trp42), acrylamide quenching studies were performed to assign the five lifetimes to the two types of tryptophan. Acrylamide was added successively up to a concentration of 0.7 M. The Stern-Volmer analysis of the fluorescence decays showed that the five lifetimes are basically due to two different types of tryptophans characterized by two different k_Q values ($0.2\pm0.2\times10^9$ M^{-1} s^{-1} and $1.1\pm0.3\times10^9$ M^{-1} s^{-1} for the kringle and Gla tryptophan components, respectively). Note that the theoretical k_Q-value for a fully exposed polypeptide-tryptophan is about 3×10^9 M^{-1} s^{-1}. The resulting assignment of the lifetime compounds to the two types of tryptophan allowed for a separate investigation of conformational changes in the two protein domains without cleaving BF1 into the isolated Gla (containing Trp42) and kringle domains (containing Trp90 and Trp126) or modifying the protein by site-directed mutagenesis. After the assignment of the lifetimes to the two tryptophan types in BF1, further experiments led to the conclusion that the Gla domain is exclusively responsible for the interaction with calcium ions and negatively charged phospholipids. Moreover, the first experimental evidence for a lipid-specific conformational change in the Gla domain of prothrombin was found, indicating an important role of this domain in the regulation of blood coagulation [60].

1.4.2.3
Solvent Quenching

The influence of solvent molecules on the fluorescence characteristic of a dye solute is certainly one of the most complex issues in fluorescence spectroscopy. Eventually every chromophore shows some dependence of its quantum yield on the chemical structure of the surrounding solvent. This observation is to some extent due to fluorescence quenching by the solvent. One possibility is that the interaction of the chromophore with its solvent shell can promote non-radiative pathways by changing the energy of the S_0, S_1 and T_1 states. Transition probabilities for the internal conversion and intersystem crossing processes are governed by the energy-gap law [61]. This law states that the rate constants k_{ic} and k_{isc} increase exponentially as the energy gap between the corresponding S_1, S_0 and/or T_1 states decreases [61]. Consequently, any change in those energy levels will strongly influence the fluorescence lifetime and quantum yield (see Eq. (1.9)).

Some of the so-called hemicyanine dyes represent special cases for the promotion of non-radiative pathways by increasing solvent polarity [62]. These dyes undergo an intramolecular twist in their excited states. The intramolecular twist leads to an increase of the polarity, and the twisted form of the S_1 state is deactivated very efficiently by fast internal conversion. Increasing solvent polarity promotes the intramolecular twist and, therefore, the non-radiative deactivation by internal conversion [62]. Moreover, evidence has been accumulated that quenching by interaction with solvent molecules can proceed by a vibrational mechanism. It has been speculated that the collision between dye and solvent molecules results in vibrational coupling, favouring efficient internal conversion [63]. In this connection the solvent deuterium effect on the fluorescence lifetime, which has been observed for a variety of chromophores, should be mentioned [64–66]. It has been shown that the quantum yield increases substantially if D_2O is used instead of H_2O as the solvent. Interestingly, this effect appears to be independent of the chemical nature of the fluorophore. It is conceivable that the different energies of the OH versus OD stretching vibrations (3657 and 2670 cm^{-1}, respectively) are responsible for the more effective solvent quenching by H_2O versus D_2O. Regardless of its physical nature, this 'heavy' atom effect in solvent quenching proved to be a very smart tool for characterization of the water accessibility in supra- and macromolecular assemblies [64].

1.4.2.4 *Self-Quenching*

Self-quenching is the quenching of one fluorophore by another one of the same kind. It is a widespread phenomenon in fluorescence, but it requires high concentrations or labelling densities. The general physical description of the self-quenching processes involves a combination of trap-site formation and energy transfer among fluorophores, with a possibility of trap-site migration which results in quenching. Trap sites may be formal fluorophore complexes or aggregates, or they may result from sufficiently high concentrations of fluorophores leading to close proximity of the dye molecules. A mathematical modelling of such processes is given in [67]. Self-quenching experiments are frequently performed by simply monitoring an increase in the fluorescence intensity F due to a decrease in the local dye concentration. One such example is the self-quenching assay for the characterization of leakage of aqueous contents from cells or vesicles as a result of lysis, fusion or physiological permeability. This assay is based on the fact that carboxyfluorescein is >95% self-quenched at concentrations >100 mM [68]. A concentrated solution of this water-soluble dye is encapsulated in liposomes. Upon addition of a fusogen or other permeabilizing agent, the dye release is accompanied by an increase in fluorescence. Other chromophores, the self-quenching properties of which are exploited in biochemical assays, are NBD (derivatives of 7-nitrobenz-2-oxa-1,3-diazol-4-yl) [69, 70], BODIPY (derivatives of 4-bora-3a,4a-diaza-*s*-indacene) [71] and DPH (derivatives of 1,6-diphenyl-1,3,5-hexatriene) [72].

1.4.2.5 *Trivial Quenching*

Trivial quenching arises from attenuation of the exciting beam and/or inability of the fluorescence photon to reach out of the sample, which occurs mainly when other compounds are added that strongly absorb in the UV region. Though the added concentration may be small, they might block the excitation light completely. Another reason for trivial quenching can be the turbidity of the sample. True and trivial quenching, however, are easily differentiated, since in trivial quenching the lifetime and quantum yield remain constant.

1.5 Influence of Solvent Relaxation on Solute Fluorescence

1.5.1 Basics of Solvent Relaxation

Electronic excitation from the ground state S_0 to a higher electronic excited state (such as S_1) is generally accompanied by a change of the permanent dipole moment, $\Delta\mu_c$, of the molecule ($\Delta\mu_c=\mu(S_1)-\mu(S_0)$). Since the timescale of the molecular electronic transition is much shorter than that of nuclear motion, the excitation-induced ultrafast change of the electron density happens virtually under fixed (original) positions and orientations of the surrounding solvent molecules. With the new dipole moment $\mu(S_1)$, the solute-solvent system is no longer in equilibrium. The solvation shell molecules are, thus, forced to adapt to the new situation: they start to reorient themselves into energetically more favourable positions with respect to the excited dye. This dynamic process, starting from the originally created non-equilibrium Franck-Condon state (FC) and leading gradually to a new equilibrium with the solute excited state (R) is called solvent relaxation (SR). This relaxation red-shifts the solute's fluorescence emission spectrum continuously from the emission maximum frequency corresponding to the Franck-Condon state ($\upsilon(0)$ for $t=0$) down to the emission maximum frequency corresponding to the fully relaxed R state ($\upsilon(\infty)$ for $t=\infty$). Since a more polar solvent typically leads to a stronger stabilization of the polar R state, the overall shift $\Delta\upsilon$ ($\Delta\upsilon=\upsilon(0)-\upsilon(\infty)$) increases with increasing solvent polarity for a given change of the solute's dipole moment $\Delta\mu_c$. The detailed mathematical description of this relationship depends on the set of assumptions that each particular dielectric solvation theory formulates [73–78].

The fundamental 'dielectric continuum solvation model' [76–78] predicts a linear proportionality between $\Delta\upsilon$ and a dielectric measure of the solvent polarity for a large variety of solvents [79]. According to this model, changes in $\Delta\upsilon$ directly reflect polarity changes in the dye environment – which can be a major desired piece of information thus accessible through solvent relaxation studies. Another important piece of information that can be obtained from a solvent relaxation investigation follows from the fact that the kinetics are determined by the mobility of the dye environment. The response of solvent molecules to a dye's electronic rearrangement is fastest in the case of water: more than half of its overall solvation response occurs within 55 fs [80]. If the dye is located in a viscous medium, the typical solvent relaxation takes place on a nanosecond timescale [81]. In vitrified solutions, on the other hand, the dye may fluoresce before the solvent relaxation towards the R state is completed [82].

1.5.2
Influence of Solvent Relaxation on Steady-State Spectra

1.5.2.1
Non-Viscous Solvents

At ambient temperature, non-viscous solvents respond to the photoinduced ultrafast change of the solute dipole moment by a fast inertial (librational) motion in the range from 50 to 500 fs. After this initial stage of the solvation response, diffusion of the solvent molecules, occurring typically on pico- to sub-nanosecond timescales, leads to further relaxation towards the R state [79, 80, 83, 84]. Fluorescence decay times, τ, of common chromophores are usually 1 ns or longer. In such a case, most of the fluorescence detected in a steady-state experiment occurs from the equilibrium state R. Based on the above-described relations between $\Delta\upsilon$, the dipole moments of the solute, $\Delta\mu_c$, and the solvent polarity, there are two basic consequences for the spectral position of the steady-state fluorescence spectrum:

1. Increased solvent polarity generally leads to a red shift of the emission spectrum.
2. The larger $\Delta\mu_c$, the more pronounced is the solvent polarity effect on the emission spectrum position.

1.5.2.2
Viscous and Vitrified Solutions

If the dye is located in a viscous medium, the solvent relaxation might take place on the nanosecond timescale. Thus, emission occurs, to a substantial extent, during solvent relaxation, and the steady-state emission spectrum represents a time-average of the emissions from different partially relaxed states. In this case, the maximum of the emission spectrum is no longer directly correlated with the polarity of the solvent. Any increase of temperature leads to a faster solvent reorientation process and, in this case, to a red shift of the emission spectrum peak. Moreover, the emission band maximum for a polar fluorophore placed in motionally restricted media (such as very viscous solvents [85, 86] or membranes [81]) shifts to a longer wavelength when the excitation wavelength approaches the red edge of the absorption band [87]. The observed shift should be maximal when the solvent relaxation is much slower than the fluorescence, and it should be zero if the solvent relaxation is fast enough. Thus, the red-edge excitation shift can serve as an indicator of the mobility of the probe's surroundings [85, 86, 88]. Usually, such a red-edge excitation shift value ranges from several nm up to 40 nm depending on the chosen solute/solvent system. The red-edge excitation shift is an especially useful piece of information for dyes, the absorption and fluorescence maxima of which hold linear correlations with the polarity of low-viscosity solvents [82, 89]. The probed polarity as well as the hypothetical emission maximum of the fully relaxed R state can be estimated from the behaviour of the absorption maximum. In vitrified solutions like sol-gel matrices [82], solvent relaxation becomes much slower than the fluorescence, most of which in that case arises from states close to the initial Franck-Condon state.

1.5.3 Quantitative Characterization of Solvent Relaxation by Time-Resolved Spectroscopy

Although there have been several attempts to simplify the characterization of the solvent relaxation process, the determination of time-resolved emission spectra (TRES) is certainly the most general and most precise way to quantitatively describe the solvent response. The TRES are usually determined by 'spectral reconstruction' [79, 80, 89]. The TRES at a given time t is calculated from the wavelength-dependent time-resolved decays by relative normalization to the steady-state spectrum. By fitting the TRES at different times to the empirical 'log-normal' function, the emission maximum frequencies $\upsilon(t)$ (or $\lambda(t)$) and the total Stokes shift $\Delta\upsilon$ (or $\Delta\lambda$) are usually derived [89]. Since $\upsilon(t)$ contains both information about polarity ($\Delta\upsilon$) and viscosity of the reported environment, the spectral shift $\upsilon(t)$ may be normalized to the total shift $\Delta\upsilon$. The resulting 'correlation functions' $C(t)$ Eq. (1.14) describe the time course of the solvent response and allow for comparison of the solvent relaxation kinetics and, thus, of the relative micro-viscosities, reported from environments of different polarities [79, 80, 89, 90–95]:

$$C(t) = (\upsilon(t) - \upsilon(\infty))/\Delta\upsilon \tag{1.14}$$

Solvent relaxation probes used for the characterization of micro-viscosities and polarities are listed in Table 1.2. They are characterized by a large change

Table 1.2. List of solvent relaxation probes

Dye or chromophore	References
1,8-ANS (1-anilinonaphthalene-8-sulphonate)	96
2,6-ANS (2-anilinonaphthalene-6-sulphonate)	96
2,6-TNS (2-(*p*-toluidinylnaphthalene)-6-sulphonate)	97, 98
NPN (*N*-phenyl-1-naphthylamine)	99
Dansyl lysine (*N*-(5-dimethylaminonaphthalene-1-sulphonyl)-L-lysine)	100
Prodan (6-propionyl-2-(dimethylamino)-naphthalene)	93–95
Laurdan (2-dimethylamino-6-lauroylnaphthalene)	103
Patman (6-palmitoyl-2-[[2-(trimethylammonium)ethyl]methylamino]-naphthalene chloride)	92–95
NBD (7-nitrobenz-2-oxa-1,3-diazol-4-yl)	81
Coumarin 153	79, 82, 89
Nile red	101
Hemicyanine dyes	88, 102
Aminobenzanthrone derivatives	90

in their dipole moment $\Delta\mu_c$ upon electronic excitation. Chapter 5 gives an overview of applications of the solvent relaxation technique in probing micro-polarities and -mobilities in membranes, proteins, DNA, polymers, micelles and ionic liquids.

1.6 Fluorescence Resonance Energy Transfer as a Spectroscopic Ruler

Fluorescence resonance energy transfer (FRET) is a non-radiative transfer of the excitation energy from a donor to an acceptor chromophore that is mediated by a long-range interaction between the emission and absorption transition dipole moments of the donor and the acceptor, respectively. The rate of energy transfer depends on the extent of the spectral overlap of the donor emission and the acceptor absorption spectra, the donor fluorescence quantum yield, the relative orientation of their transition dipole moments, and the distance between donor and acceptor molecules. The distance dependence has resulted in widespread use of FRET to measure distances between donors and acceptors in macromolecular systems. The quality of a donor/acceptor pair is usually characterized by the parameter R_0, which is typically in the range from 2 to 9 nm. It is defined as the distance at which the efficiency of resonance energy transfer is equal to 50%. R_0 can be estimated as follows:

$$R_0\,[\mathrm{nm}] = 979\,(\kappa^2\, n^4\, \Phi_0\, J)^{1/6} \tag{1.15}$$

where n is the refractive index of the medium, Φ_0 is the fluorescence quantum yield of the donor, J the spectral overlap integral and κ an orientation factor. The energy transfer rate k_{ET} is given by

$$k_{ET} = 1/\tau_d\,(R_0/r)^6 \tag{1.16}$$

where τ_d is the decay time of the donor fluorescence in the absence of an acceptor and r is the distance between donor and acceptor. Thus, the rate depends strongly on distance, providing a spectroscopic ruler for determining distances in macromolecular assemblies.

1.6.1 Donor-Acceptor Pairs at Fixed Distances

The magnitude of k_{ET} can be determined from the efficiency of energy transfer, ET, via

$$k_{ET} = 1/\tau_d\,(ET/(1-ET)) \tag{1.17}$$

and ET, in turn, can be established experimentally by measuring the decrease in the intensity F or lifetime τ of the donor in the presence of the acceptor:

$$ET = 1 - F/F_d = 1 - \tau/\tau_d \tag{1.18}$$

Thus, by determining *ET* and knowing R_0, the separation distance r can be calculated. When distances are estimated this way, there is often some concern about the correct value of the orientation factor κ^2, which depends on the relative orientation of the donor emission transition moment and the acceptor absorption transition moment. The value of κ^2 varies from 4 (parallel orientation of the transition moments) to 0 (perpendicular orientation). Often, a value of $\kappa^2=2/3$ is assumed which corresponds to the situation of a rapid, isotropic rotation of the donor and acceptor molecules. Randomly oriented dipoles that remain fixed during the singlet lifetime give $\kappa^2=0.476$. When needed, the value of κ^2 can be estimated by polarization measurements [104]. A comprehensive discussion on the theory and effects of the orientation factor is given in [105].

1.6.2 Donor-Acceptor Pairs at Variable Distances

Let's assume the simplest case: a donor with mono-exponential fluorescence decay τ_d, a fixed donor-acceptor distance r and a dynamically random orientation factor $\kappa^2=0.476$, for which k_{ET} has to be added to Eq. (1.10). The energy transfer in this situation will simply result in a shortened but still mono-exponential decay of the donor τ_d. In homogeneous solution, however, at low donor concentration and without any significant diffusion of the donor and acceptor within the fluorescence lifetime, the donor fluorescence intensity decay is given by [106–109]:

$$F = F_0 \exp(-t/\tau_d) \exp[-\gamma (t/\tau_d)^{\delta}]; \quad \delta = \mathrm{dim}/6 \tag{1.19}$$

For randomly distributed donor and acceptor molecules the value for the dimension *dim* is equal to 3 and γ is

$$\gamma = 4/3\, g\, \pi^{3/2}\, c_a\, R_0 \quad \text{where} \quad g = (3/2\, \kappa^2)^{0.5} \tag{1.20}$$

with c_a as the acceptor concentration. With knowledge of the acceptor concentration and provided that the donor fluorescence decays mono-exponentially in the absence of the acceptor, the R_0 value and the dimension of the medium in which donor and acceptors are imbedded can be determined. Two-dimensional or so-called fractal energy transfer is of interest if the dye molecules are bound to phospholipid membranes [110, 111] or imbedded in silicate networks [112]. One-dimensional energy transfer has been considered for dyes bound to DNA [113]. A detailed and up-to-date review of the physics and theory of long-range resonance energy transfer in molecular systems was published recently by Scholes [114].

1.6.3 Some Applications of Fluorescence Resonance Energy Transfer

An important class of FRET applications is represented by assays for the characterization of fusion of cells or vesicles. Usually, their membranes are labelled either by donor or acceptor molecules. Fusion leads to an intermixing of these membrane labels in the same bilayer, allowing resonance energy transfer to occur. Examples can be found in the literature [115–119]. Another membrane application of energy transfer is the demonstration of lipid asymmetry in human red blood cells [120]. Moreover, energy transfer has proved to be a very useful tool in elucidating the sub-unit structure of oligomeric assemblies in membranes. Examples are studies on the oligomerization of ATPase of sarcoplasmic reticulum in phospholipid vesicles [121], on gramicidin A trans-membrane channels [122], and on the aggregation state of bacteriorhodopsin [123]. Finally, the combination of energy transfer with flow cytometry [124] and its use in immunoassays should be mentioned [125]. Further information on the theory and application of energy transfer can be found, e.g., in [105, 126]. Some advanced applications of FRET will be encountered in Chaps. 3, 4, 10, 11 and 12 of this book.

1.7 Irreversible Photobleaching

'Fluorescence recovery after photobleaching' (FRAP) was introduced as a method to measure the local mobility of fluorescently labelled particles bound to the plasma membrane of living cells [26, 127, 128]. It has been used to study transport phenomena in a wide variety of biological membrane-bound systems, as well as to probe the photobleaching properties of fluorescent molecules [129]. FRAP is based on observing the rate of fluorescence recovery due to the movement of a new fluorescent marker into an area of the membrane which contains the same kind of marker, but which has been rendered non-fluorescent via an intense photobleaching pulse of laser light. The two-dimensional diffusion coefficient of the fluorescent marker is related to both its rate and extent of fluorescence recovery. For a discussion of the photophysical mechanism of photobleaching see reference [130].

In order to create a finite observation area, usually both laser pulses – the single short pulse with rather high intensity leading to photobleaching and the less intensive pulse monitoring the fluorescence recovery – are focused by an epifluorescence or confocal microscope. A very elegant variant is the combination of FRAP with the total internal reflection fluorescence (TIRF) technique (total internal reflection fluorescence recovery after photobleaching; TIR-FRAP [28]) which is covered in greater depth in the contribution of N. Thompson (Chap. 6). Here, a laser beam totally internally reflects at a solid/liquid interface, creating an evanescent field which penetrates only a fraction of the laser beam's wavelength into the liquid domain. Using planar phospholipid bilayers and fluorescently labelled proteins, this method allows the determination of adsorption/desorption rate constants and surface diffusion constants [28, 131, 132]. Figure 1.4 shows a representative TIR-FRAP curve for fluorescein-labelled prothrombin bound to a planar membrane. In this experiment the conditions are chosen in such a way that the recovery curve is characterized by the prothrombin desorption rate. It should be mentioned that, in analogy with other fluorescence microscopy techniques, two- and three-photon absorption might be utilized for FRAP in the near future.

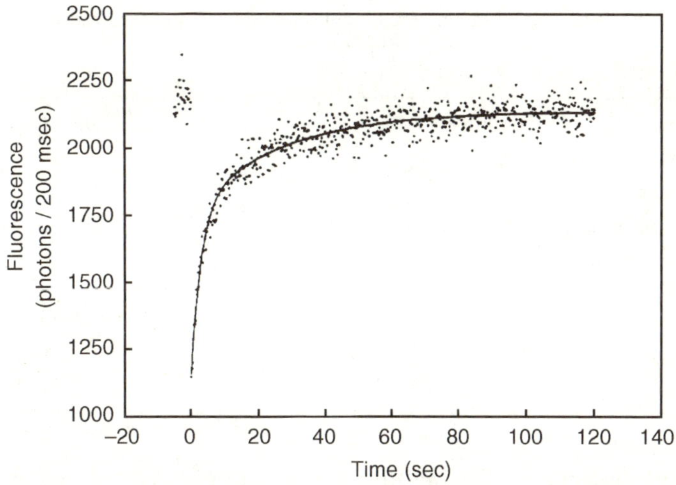

Fig. 1.4. Representative TIR-FRAP curve for fluorescein-labelled prothrombin bound to planar membranes. Shown is a typical recovery curve for the binding of 1 μM prothrombin (labelled with fluorescein) to a planar bilayer. The *dotted points* represent the experimental data and the *line* the best fit, yielding the desorption rate. Note that the fluorescence intensity does not recover fully. This effect is generally observed in photobleaching experiments and is one of the major drawbacks of this method

1.8 Single Molecule Fluorescence

Recent advances in ultra-sensitive instrumentation have allowed the detection of individual atoms and molecules in solids [133, 134], on surfaces [135, 136], and in the condensed phase [137, 138] using laser-induced fluorescence. In particular, single molecule detection in the condensed phase enables scientists to explore new frontiers in many scientific disciplines, such as chemistry, molecular biology, molecular medicine and nanostructure materials. There are several optical methods to study single molecules, the principles and application of which have been reviewed by Nie and Zare [139]. These methods are listed in Table 1.3. A broader coverage of this topic is given by J. Enderlein's contribution in Chap. 7 or in recent reviews [3, 5].

In contrast to the other single molecule techniques listed above, measurements based on fluorescence correlation spectroscopy (FCS) can be performed both routinely and rapidly. Moreover, FCS is applied in many scientific disciplines and the number of applications of this technique is growing rapidly. In a FCS experiment fluorescence fluctuations due to diffusion, chemical reactions or flow are detected and analysed. Usually in FCS, a sharply focused laser beam illuminates a volume element of about 10^{-15} L by using confocal or multi-photon microscopy. This volume is so small that at a given time, it can host only one fluorescent particle out of many under analysis. The illuminated volume is adjustable in 1 μm steps in three dimensions, providing a high spatial resolution. If diffusion is the investigated process, the single

Table 1.3. Methods for studying single molecules using laser-inducedfluorescence

Method for studying single molecules	References
Solid matrices at low temperatures	140, 141
Liquid streams	142–144
Microdroplets	145
Near-field scanning optical microscopy	136, 146–149
Far-field confocal microscopy, including fluorescence correlation spectroscopy	150, 151
Microscopy combined with multi-photon excitation	152–154
Wide-field epi-illumination	155, 156
Evanescent wave excitation	157

fluorescent molecules diffusing through the illuminated volume give rise to bursts of fluorescence light quanta. Each individual burst, resulting from a single molecule, can be registered. The photons are recorded in a time-resolved manner by a highly sensitive single-photon counting device. In the diffusion case, the autocorrelation function of the time course of the fluorescence signal gives information about the number of molecules in the illuminated volume element and their characteristic translational diffusion time. Since the size of the illuminated volume is known, the concentration and diffusion constant of the fluorescent species can be determined.

The principles of FCS and its application in cell research are outlined in Chap. 14. The combination of FCS with evanescent field excitation is described in Chap. 6. Chapter 11 gives an example of a FCS investigation in protein sciences. Recently FCS has become an important method in the characterization of precursors for targeted drug delivery. This issue is treated in Chaps. 12 and 13.

1.9 Optical Sensors Based on Fluorescence

An optical sensor for chemical analysis is part of the detector system that allows for continuous monitoring of a physical parameter or concentration of an analyte. Such sensors can detect changes in optical absorbance, reflectance, fluorescence, chemiluminescence, Raman scattering, refractive index, light polarization and other optical properties. Due to the high sensitivity, selectivity and versatility of fluorescence spectroscopy, optical sensors based on fluorescence are the most highly developed. Typically, the sensing probe (chemically interacting part) is placed on a carrier material, while the analyte can be either in the gas phase or in solution. Interaction between the sensing probe and the analyte leads to a change in the sensor's fluorescence properties. If needed, fibre optics allow one to perform the sensor fluorescence measurement in a remote part of the detector, which is especially useful in clinical applications. Such fluorescence change monitoring, e.g. pH, O_2 pressure, or the concentration of ions in blood, can

Table 1.4. Examples of fluorescence sensing

Analyte	Sensing dye	Sensing mechanism	References
Oxygen	Several Ru complexes (e.g. $[Ru(Ph_2phen)_3]^{2+}$	Collisional quenching	159
Chloride	Sultones (betaines) of quinolinium and acridinium ions	Collisional quenching	51, 160
Calcium	Blue and green fluorescent proteins	Resonance energy transfer	25
pH	Fluoresceins	pH-dependent ionization	161, 162
Glucose	Fluorescein and rhodamine	Resonance energy transfer	163

occur in the intensity, emission spectrum, anisotropy, or lifetime of the sensor probe. The mechanism for the fluorescence change can be collisional quenching, resonance energy transfer, photoinduced electron transfer, or analyte-induced change of the state of the sensing chromophore. Table 1.4 gives some examples of analyte, sensing probe and sensing mechanism combinations.

Since intensity measurements are dependent upon the concentration of the fluorophore, they are often not applicable since they might be inaccurate, e.g. if photobleaching occurs. Moreover, intensity-based systems suffer from other problems including turbidity, limited range of detection, low signal-to-noise ratio and optical losses. Fluorescence lifetime-based sensing, on the other hand, does not suffer from these problems and it seems likely to become widely used in the near future. A comprehensive overview on this topic is given in the chapter 'fluorescence sensing' written by Otto Wolfbeis in reference [158].

Acknowledgements
The authors gratefully acknowledge the financial support of the Ministry of Education, Youth and Sports of the Czech Republic (M.H.: via LN 00A032; V.F.: J04/98:210000022).

References

1. Foerster T (1951) Fluoreszenz organischer Verbindungen. VandenHoech und Ruprecht, Goettingen
2. Shubeita GT, Sekatskii SK, Dietler G, Letokho VS (2002) Appl Phys Lett 80:2625–2627
3. Bohmer M, Enderlein J (2003) Chem Phys Chem 4:793-808
4. Emilliani V, Sanvitto D, Tramier D, Piolot T, Petrasek Z, Kemnitz K, Durieux C, Coppey-Moisan M (2003) Appl Phys Lett 83:2471–2473
5. Moerner WE, Fromm DE (2003) Rev Sci Instrum 74:3597–3619
6. Callis RP (1997) Annu Rev Phys Chem 48:271–297
7. Hansen JE, Rosenthal SJ, Fleming GR (1992) J Phys Chem 96:3034–3029
8. Perrin F (1926) J Phys Radium 7:390
9. Jablonski A (1960) Bull Acad Pol Sci 8:529
10. Kawski A (1993) Crit Rev Anal Chem 23:459–529

11. Steiner RF (1991) In: Lakowicz JR (ed) Topics in fluorescence spectroscopy: principles. Plenum, New York, pp 1–52
12. Ferguson BQ, Yang DCH (1986) Biochemistry 25:529–539
13. Gorovits BM, Horowitz PM (1995) J Biol Chem 270:13057–13062
14. Lim K, Jameson DM, Gentry CA, Herron JN (1995) Biochemistry 34:6975–6984
15. Lukas TJ, Burgess EH, Prendergast FG, Lau W, Watterson DM (1986) Biochemistry 25:1458–1464
16. LeTilly V, Royer C (1993) Biochemistry 31:7753–7758
17. Tao T (1969) Biopolymers 8:609
18. Lipari G, Szabo A (1980) Biophys J 30:489
19. Jaehnig F (1980) Proc Natl Acad Sci USA 76:6361
20. Lentz BR (1989) Chem Phys Lipids 50:171
21. Van der Heide UA, van Ginkel G, Levine YK (1996) Chem Phys Lett 253:118
22. Lakowicz JR, Gryczynski I (1991) In: Lakowicz JR (ed) Topics in fluorescence spectroscopy: techniques. Plenum, New York, pp 293–336
23. O'Connor DV, Philips D (1983) Time-correlated single photon counting. Academic, London
24. Valeur B (2002) Molecular fluorescence. Principles and applications. Wiley-VCH, Weinheim
25. Miyawaki A, Llopis J, Heim R, NcCaffery JM, Adams JA, Ikura M, Tsien RY (1997) Nature 388:882–887
26. Axelrod D, Koppel D, Schlessinger J, Elson E, Webb W (1976) Biophys J 16:1055
27. Lettinga MP, Koenderink GH, Kuipers BWM, Bessels E, Philipse AP (2004) J Chem Phys 120:4517–4529
28. Thompson NL, Burghardt TP, Axelrod D (1981) Biophys J 33:435–454
29. Demtröder W (1996) Laser spectroscopy. Basic concepts and instrumentation, 2nd enlarged edn. Springer, Berlin Heidelberg New York
30. Menzel R (2001) Photonics: linear and nonlinear interactions of laser light with matter. Springer, Berlin Heidelberg New York
31. Lanig H, Hof M, Bringmann G, Schneider FW (1997) Chem Phys Lett 272:478–483
32. Parker CA, Rees WT (1960) Analyst 85:587
33. James DR, Ware WR (1985) J Phys Chem 89:5450–5458
34. Namiki A, Nakashima N, Yoshihara K (1979) J Chem Phys 71:925–930
35. Fucaloro AF, Forster LS, Campbell MK (1984) Photochem Photobiol 39:503–506
36. Swadesh JK, Mui PW, Scheraga HA (1987) Biochemistry 26:5761–5769
37. Hof M, Fleming GR, Fidler V (1996) Proteins 24:485–494
38. Vos R, Engelborghs Y (1994) Photochem Photobiol 60:24–32
39. Eftink MR, Ghiron CA (1984) Biochemistry 23:3891–3899
40. Midoux P, Wahl P, Auchet JC, Munsigny M (1984) Biochim Biophys Acta 801:16–25
41. Lehrer SS (1971) Biochemistry 10:3254–3263
42. Sanyal G, Kim E, Thompson FM, Brady KE (1989) Biochem Biophys Res Commun 165:772–781
43. Behera PK, Mukherjee T, Mishra AK (1995) J Lumin 65:131–136
44. Green SA, Simpson DJ, Zhou G, Ho PS, Blough NV (1990) J Am Chem Soc 112:7337–7346
45. Obyknovennaya IE, Vasileva IM, Cherkasov AS (1986) Opt Spectrosc 60:169–171
46. Bowen EJ, Metcalf WS (1951) Proc R Soc Lond A 206:437–447
47. Mac M, Najbar J, Phillipis D, Smith TA (1992) J Chem Soc Faraday Trans 88:3001–3005
48. Hutterer R, Krämer K, Schneider FW, Hof M (1997) Chem Phys Lipids 90:11–23
49. Chao SC, Tretzel J, Schneider FW (1979) J Am Chem Soc 101:134
50. Urbano E, Offenbacher H, Wolfbeis OS (1984) Anal Chem 56:427–429
51. Wolfbeis OS, Urbano E (1983) Fresenius J Anal Chem 314:577–581
52. Washington K, Sarasua MM, Koehler LS, Koehler KA, Schultz JA, Pedersen LG, Hiskey RG (1984) Photochem Photobiol 40:693–701
53. Bisht PB, Tripathi HB (1993) J Lumin 55:153–158
54. Johnson GE (1980) J Phys Chem 84:2940–2946

55. Daems D, Boens N, Schryver FC (1989) Eur Biophys J 17:25–36
56. Ahmed A, Durocher G (1981) Photochem Photobiol 34:573–578
57. Kikuchi K, Sato C, Watebe M, Ikeda H, Takahashi Y, Miyashi T (1993) J Am Chem Soc 115:5180–5184
58. Subczynski WK, Hyde JS, Kusumi A (1989) Proc Natl Acad Sci USA 86:4474–4478
59. Eftink MR (1991) In: Lakowicz JR (ed) Topics in fluorescence spectroscopy: principles. Plenum, New York, pp 53–126
60. Hof M (1998) Biochim Biophys Acta 1388:143–153
61. Schlag EW, Schneider S, Fischer S (1971) Annu Rev Phys Chem 22:465
62. Fromherz P (1995) J Phys Chem 99:7188–7192
63. Schulman SG (1977) Fluorescence and phosphorescence spectroscopy: physicochemical principles and practice. Pergamon, New York
64. Brand K, Hof M, Schneider FW (1991) Ber Bunsen Phys Chem 95:1511–1514
65. Foerster T, Rokos K (1967) Chem Phys Lett 1:279
66. Sriram R, Hoffman MZ (1982) Chem Phys Lett 85:572
67. Baumann J, Fayer MD (1986) J Chem Phys 85:4087
68. Chen RF, Knutson JR (1988) Anal Biochem 198:119
69. Brown RS, Brennan JD, Krull UJ (1994) J Chem Phys 100:6019–6021
70. Chattopadhyay A (1990) Chem Phys Lipids 53:1
71. Hendrickson HS et al (1999) Anal Biochem 276:27
72. Massenburg D, Lentz BR (1993) Biochemistry 32:9172
73. Bagchi B, Oxtoby SD, Fleming GR (1984) Chem Phys 86:257
74. Rips I, Klafter J, Jortner J (1988) J Chem Phys 89:4288
75. Friedman HL, Raineri FO, Hirata F, Perng PC (1995) J Stat Phys 78:239
76. Bakshiev NG (1964) Opt Spectosc USSR 16:446
77. Mazurenko Yt, Bakshiev NG (1970) Opt Spectosc USSR 28:490
78. Liptay W (1974) In: Lim EC (ed) Excited states, vol 1. Academic, New York, p 129
79. Horng ML, Gardecki JA, Papazyan A, Maroncelli M (1995) J Phys Chem 99:17320
80. Jimenez R, Fleming GR, Kumar PV, Maroncelli M (1994) Nature 369:471
81. Mukherje S, Chattopadhyay A (1995) J Fluoresc 5:237
82. Hof M, Lianos P (1997) Langmuir 13:290
83. Zhang H, Jonkman AM, van der Meulen P, Glasbeek M (1994) Chem Phys Lett 236:587
84. Bingemann D, Ernsting NP (1995) J Chem Phys 102:2691
85. Demchenko AP (1991) In: Lakowicz JR (ed) Topics in fluorescence spectroscopy: biochemical applications. Plenum, New York, p 65
86. Demchenko AP (1994) Biochim Biophys Acta 1209:149
87. Galley WC, Purkey RM (1970) Proc Natl Acad Sci USA 67:1116
88. Hof M, Lianos P, Laschewsky A (1997) Langmuir 13:2181
89. Maroncelli M, Fleming GR (1987) J Chem Phys 86:6221
90. Sýkora J, Mudogo V, Hutterer R, Nepraš M, Vaněrka J, Kapusta P, Fidler V, Hof M (2002) Langmuir 18:9276–9282
91. Hutterer R, Schneider FW, Lanig H, Hof M (1997) Biochim Biophys Acta 1323:195
92. Sheynis T, Sýkora J, Benda A, Kolusheva S, Hof M, Jelinek R (2003) Eur J Biochem 270:4478–4487
93. Hutterer R, Schneider FW, Sprinz H, Hof M (1996) Biophys Chem 61:151
94. Hutterer R, Schneider FW, Hermens WT, Wagenvoord R, Hof M (1998) Biochim Biophys Acta 1414:155–164
95. Sýkora J, Kapusta P, Fidler V, Hof M (2002) Langmuir 18:571–574
96. Slavik J (1982) Biochim Biophys Acta 694:1
97. Lakowicz JR, Thompson RB, Cherek H (1983) Biochim Biophys Acta 815:295
98. Hof M, Hutterer R, Perez N, Ruf H, Schneider FW (1994) Biophys Chem 52:165
99. Matayoshi ED, Kleinfeld AM (1981) Biochim Biophys Acta 644:233
100. Epand RM, Leon BTC (1992) Biochemistry 31:1550
101. Deye JF, Berger TA, Anderson AG (1990) Anal Chem 62:615
102. Fromherz P (1995) J Phys Chem 99:7185

103. Parasassi T, Di Stefano M, Loiero M, Ravagnan G, Gratton E (1994) Biophys J 66:763
104. Dale RE, Eisinger J, Blumberg WE (1979) Biophys J 26:161
105. Van der Meer BW, Coker G, Chen SYS (1994) Resonance energy transfer theory and data. VCH, Weinheim
106. Foerster T (1959) Discuss Faraday Soc 27:7–17
107. Goesele U (1978) Spectrosc Lett 11:445
108. Hauser M, Klein UKA, Goesele U (1976) Z Phys Chem 101:255
109. Klafter J, Blumen A (1984) J Phys Chem 80:875
110. Lianos P, Duportail G (1993) Biophys Chem 48:293–299
111. Loura LMM, Fedorov A, Prieto M (1996) Biophys J 71:1823–1836
112. Schleicher J, Hof M, Schneider FW (1993) Ber Bunsen Phys Chem 97:172–176
113. Mergny JL, Slama-Schwok A, Monteney-Garestier T, Rougee M, Helene C (1991) Photochem Photobiol 53:555–558
114. Scholes GD (2003) Long-range resonance energy transfer in molecular systems. Annu Rev Phys Chem 54:57–87
115. MacDonald RI (1987) J Biol Chem 262:10392–10397
116. Keller PM, Person S, Snipes W (1968) J Cell Sci 28:167–177
117. Gibson GA, Loew LM (1979) Biochem Biophys Res Commun 88:135–140
118. Struck DK, Hoekstra D, Pagano RE (1981) Biochemistry 20:4093–4099
119. Morris SJ, Bradley D (1984) Biochemistry 23:4642–4650
120. Connor J, Schroit AJ (1987) Biochemistry 26:5099–5105
121. Vanderkoi JM, Ierokomas A, Nakamura H, Martonosi A (1977) Biochemistry 16:1262–1267
122. Veatch W, Stryer L (1977) J Mol Biol 113:89–102
123. Hasselbacher CA, Street TL, Dewey TG (1984) Biochemistry 23:6445–6565
124. Tron L, Szoelloesi J, Damjanovich S, Helliwell SH, Arndt-Jovin DJ, Jovin TM (1984) Biophys J 45:939–946
125. Khanna PL, Ullman EF (1980) Anal Biochem 108:156–161
126. Cheung HC (1991) In: Lakowicz JR (ed) Topics in fluorescence spectroscopy: principles. Plenum, New York, pp 127–176
127. Edidin M, Zagyansky M, Lardner T (1976) Science 191:466
128. Peters R, Peters J, Tews K, Bahr W (1974) Biochim Biophys Acta 367:282–294
129. Periasamy N, Bicknese S, Verkman A (1996) Photochem Photobiol 63:265
130. Eggeling C, Widengren J, Rigler R, Seidel CAM (1999) In: Rettig W, Strehmel B, Schrader S, Seifert H (eds) Applied fluorescence in chemistry, biology and medicine. Springer, Berlin Heidelberg New York, pp 193–240
131. Huang ZP, Pearce KH, Thompson NL (1994) Biophys J 67:1754–1766
132. Pearce KH, Hof M, Lentz BR, Thompson NL (1993) J Biol Chem 268:22984–22991
133. Ambrose WP, Moerner WE (1991) Nature 349:225
134. Plakhotnik T, Walser D, Pirotta M, Renn A, Wild UP (1994) Science 265:364
135. Betzig E, Chichester RJ (1993) Science 262:1422
136. Xie XS, Dunn RC (1994) Science 265:361
137. Goodwin PM, Ambrose WP, Keller RA (1996) Acc Chem Res 29:607
138. Barnes MD, Kung CY, Whitten WB, Ramsey JM (1997) Anal Chem 69:2115
139. Nie S, Zare RN (1997) Annu Rev Biophys Biomol Struct 26:567–596
140. Orrie M, Bernard J (1990) Phys Rev Lett 65:2716–2719
141. Moerner WE, Plakhonik T, Irngartinger T, Croci M, Palm V, Wild UPJ (1994) J Phys Chem 98:7382–7389
142. Dovichi NJ, Martin JC, Jett JH, Keller RA (1983) Science 219:845–847
143. Chen D, Dovichi NJ (1996) Anal Chem 68:690–696
144. Shopa SA, Davis LM, Shera EB (1992) J Opt Soc Am B 9:1761–1769
145. Barnes MD, Ramsay WB, Whitten WB (1995) Anal Chem 67:418–423
146. Ambrose WP, Affleck RL, Goodwin PM, Keller RA, Martin JC (1995) Exp Tech Phys 41:1–12
147. Dunn RC, Holton GR, Mets L, Xie XS (1994) J Phys Chem 98:3094–3098
148. Betzig E, Trautmann JK (1992) Science 257:189–195

149. Trautmann JK, Macklin JJ, Bruns LE, Betzig E (1994) Nature 369:40–42
150. Eigen M, Rigler R (1994) Proc Natl Acad Sci USA 91:5740–5747
151. Nie S, Chiu DT, Zare RN (1995) Anal Chem 67:2849–2857
152. Mertz J, Xu C, Webb W (1995) Opt Lett 20:2532–2534
153. Brown EB, Shear JB, Adams SR, Tsien RY, Webb WW (1999) Biophys J 76:489–499
154. Schwille P, Haupts U, Maiti S, Webb WW (1999) Biophys J 77:2251–2265
155. Schmidt T, Schultz GJ, Baumgartner W, Gruber HJ, Schindler H (1996) Proc Natl Acad Sci USA 93, 2926–2929
156. Gosh RN, Webb WW (1994) Biophys J 66:1301–1318
157. Vale RD, Funatsu T, Pierce DW, Romberg L, Harada Y, Yanagida T (1996) Nature 380:451–453
158. Lakowicz JR (1999) Principles of fluorescence spectroscopy. Plenum, New York, pp 531–572
159. Wolfbeis OS (1991) Fiber optic chemical sensors and biosensors. CRC, Boca Raton
160. Verkman AS (1990) Am J Physiol 253:C375–C388
161. Thomas JA, Buchsbaum RN, Zimniak A, Racker E (1979) Biochemistry 18:2210–2218
162. Rink TJ, Tsien RY, Pozzan T (1982) J Cell Biol 95:189–196
163. Szmacinski H, Lakowicz JR (1994) In: Lakowicz JR (ed) Topics in fluorescence spectroscopy: probe design and chemical sensing. Plenum, New York, pp 295–334

2 Pulse and Phase Fluorometries: An Objective Comparison

B. Valeur

Keywords: Pulse fluorometry; Phase-modulation fluorometry; Lifetime-based decomposition of spectra; Time-resolved emission anisotropy; Lifetime imaging microscopy

Abbreviations

FLIM	Fluorescence lifetime imaging microscopy
MCA	Multichannel analyzer
SPT	Single-photon timing
TAC	Time-to-amplitude converter
TCSPC	Time-correlated single-photon counting

2.1 Introduction

Two time-resolved techniques, *pulse fluorometry* and *phase-modulation fluorometry*, are commonly employed to recover the lifetimes, or more generally the parameters characterizing the *δ-pulse response* of a fluorescent sample (i.e., the response to an infinitely short pulse of light). Pulse fluorometry uses a short exciting pulse of light and gives the δ-pulse response of the sample, convoluted by the instrument response. Phase-modulation fluorometry uses modulated light at variable frequency and gives the *harmonic response* of the sample, which is the Fourier transform of the δ-pulse response. The first technique works in the *time domain*, and the second one in the *frequency domain*. Pulse fluorometry and phase-modulation fluorometry are theoretically equivalent, but the principles of the instruments are different.

Table 2.1 reports the various techniques that can be used in the time domain and the frequency domain. The most widely used technique in the time domain is the *time-correlated single-photon counting technique*, preferably called the *single-photon timing technique*. The time-gated systems are less popular. Streak cameras offer a very good time resolution (a few picoseconds or less) but the dynamic range is smaller than that of the single-photon timing technique. The instruments that provide the best time resolution (about 100 fs) are based on *fluorescence up-conversion*, but they are very expensive and not commercially available. Because of space limitations, only the single-photon timing technique will be presented.

In the frequency domain, the phase and modulation measurements can be done by using either a cw source (lamp or laser) and an optical modulator, or the harmonic content of a pulsed laser. Both techniques will be described.

Table 2.1. Various techniques in pulse and phase fluorometry

Time domain	Frequency domain
- Time-correlated single-photon counting or single-photon timing - Time gated systems (boxcar, strobe) - Streak cameras - Fluorescence up-conversion	Phase and modulation measurements using: - cw lamp or laser + optical modulator - Harmonic content of a pulsed laser

The aim of this article is to make an objective comparison between pulse and phase fluorometries from the theoretical, methodological, and instrumental points of view.

2.2 General Principles of Time-Resolved Fluorometry [1–4]

The principles of time-resolved fluorometry are illustrated in Fig. 2.1. The δ-pulse response $I(t)$ of the fluorescent sample is, in the simplest case, a single exponential whose time constant is the excited-state lifetime, but more frequently it is a sum of discrete exponentials, or a more complicated function; sometimes, the system is characterized by a distribution of decay times.

For any excitation function $E(t)$, the response $R(t)$ of the sample is the convolution product of this function by the δ-pulse response:

$$R(t) = E(t) \otimes I(t) = \int_{-\infty}^{t} E(t')\, I(t - t')\, \mathrm{d}t' \tag{2.1}$$

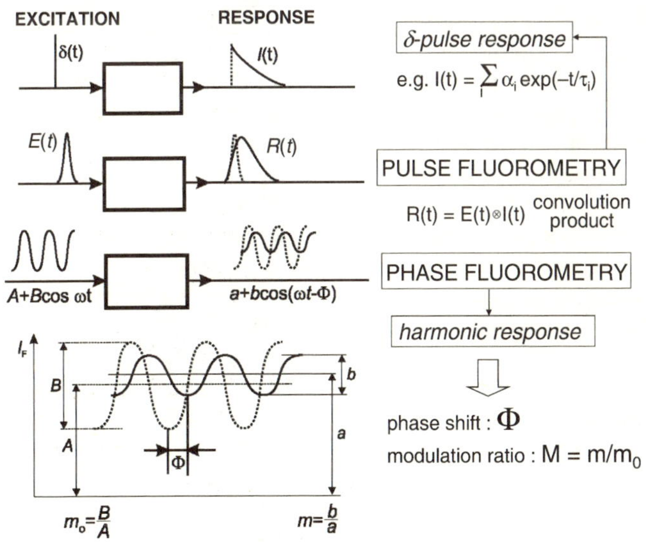

Fig. 2.1. Principles of time-resolved fluorometry

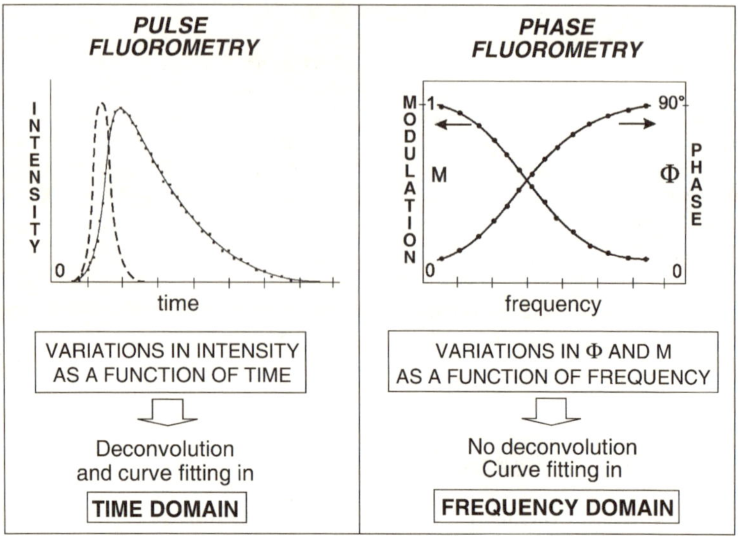

Fig. 2.2. Differences in methodology between pulse and phase-modulation fluorometries. (Adapted from reference 1)

The differences in methodology between pulse and phase fluorometry, as illustrated in Fig. 2.2, will now be discussed.

2.2.1 Pulse Fluorometry

The sample is excited by a short pulse of light and the fluorescence response is recorded as a function of time. If the duration of the pulse is not short with respect to the time constants of the fluorescence decay, the fluorescence response is the convolution product given by Eq. (2.1): the fluorescence intensity increases, goes through a maximum, and becomes identical to the *true* δ-pulse response $i(t)$ as soon as the intensity of the light pulse is negligible (Fig. 2.2). In this case, data analysis for the determination of the parameters characterizing the δ-pulse response requires a deconvolution of the fluorescence response.

2.2.2 Phase-Modulation Fluorometry

The sample is excited by sinusoidally modulated light at high frequency. The fluorescence response, which is the convolution product (Eq. 2.1) of the δ-pulse response by the excitation function, is sinusoidally modulated at the same frequency but delayed in phase and partially demodulated with respect to the excitation. The phase shift Φ and the modulation ratio M (equal to m/m_0), that is the ratio of the modulation depth m (AC/DC ratio) of the fluorescence and the modulation depth of the excitation m_0 (see Fig. 2.1), characterize the harmonic response of the system. These parameters are measured as a function of the modulation frequency. No deconvolution is necessary because the data are directly analyzed in the frequency domain (Fig. 2.2).

2.2.3 Relation Between Harmonic Response and δ-Pulse Response

When the excitation function is sinusoidal, i.e., of the following form

$$E(t) = E_0 \left[1 + m_0 \exp\left(\mathrm{j}\omega t\right)\right] \tag{2.2}$$

where ω is the angular frequency ($=2\pi f$), the response of the system can be calculated using Eq. (2.1). It can then be shown that

$$M \exp\left(-\mathrm{j}\Phi\right) = \int_0^\infty i(t) \exp\left(-\mathrm{j}\omega t\right) \mathrm{d}t \tag{2.3}$$

where $i(t)$ is the normalized δ-pulse response according to

$$\int_0^\infty i(t)\, \mathrm{d}t = 1 \tag{2.4}$$

This important expression shows that *the harmonic response expressed as* $M \exp(-\mathrm{j}\Phi)$ *is the Fourier transform of the δ-pulse response.*

It is convenient to introduce the sine and cosine transforms P and Q of the δ-pulse response:

$$P = \int_0^\infty i(t) \sin\left(\omega t\right) \mathrm{d}t \tag{2.5}$$

$$Q = \int_0^\infty i(t) \cos\left(\omega t\right) \mathrm{d}t \tag{2.6}$$

If the δ-pulse response is not normalized according to Eq. (2.4), then Eqs. (2.5) and (2.6) should be replaced by

$$P = \frac{\int_0^\infty I(t) \sin\left(\omega t\right) \mathrm{d}t}{\int_0^\infty I(t)\, \mathrm{d}t} \tag{2.7}$$

$$Q = \frac{\int_0^\infty I(t) \cos\left(\omega t\right) \mathrm{d}t}{\int_0^\infty I(t)\, \mathrm{d}t} \tag{2.8}$$

Since Eq. (2.3) can be rewritten as $M\cos\Phi - jM\sin\Phi = Q - jP$, it is easy to show that the phase shift and the modulation ratio are given by

$$\Phi = \tan^{-1}\left(\frac{P}{Q}\right) \tag{2.9}$$

$$M = [P^2 + Q^2]^{1/2} \tag{2.10}$$

2.2.4
General Relations for Single Exponential and Multiexponential Decays

For a single exponential decay, the δ-pulse response is

$$I(t) = \alpha \exp(-t/\tau) \tag{2.11}$$

where τ is the decay time and α is the preexponential factor or amplitude. The phase shift and relative modulation are related to the decay time by

$$\tan\Phi = \omega\tau \tag{2.12}$$

$$M = \frac{1}{(1 + \omega^2\tau^2)^{1/2}} \tag{2.13}$$

For a multiexponential decay with n components, the δ-pulse response is

$$I(t) = \sum_{i=1}^{n} \alpha_i \exp(-t/\tau_i) \tag{2.14}$$

Note that the fractional intensity of component i, i.e., the fractional contribution of component i to the total steady-state intensity, is

$$f_i = \frac{\int_0^\infty I_i(t)\,\mathrm{d}t}{\int_0^\infty I(t)\,\mathrm{d}t} = \frac{\alpha_i\tau_i}{\sum_{i=1}^{n}\alpha_i\tau_i} \tag{2.15}$$

with, of course, $\sum_{i=1}^{n} f_i = 1$.

Using Eqs. (2.7) and (2.8), the sine and cosine Fourier transforms, P and Q, are given by

$$P = \frac{\omega \sum_{i=1}^{n} \frac{\alpha_i\tau_i^2}{1+\omega^2\tau_i^2}}{\sum_{i=1}^{n}\alpha_i\tau_i} = \omega\sum_{i=1}^{n}\frac{f_i\tau_i}{1+\omega^2\tau_i^2} \tag{2.16}$$

$$Q = \frac{\sum_{i=1}^{n} \frac{\alpha_i\tau_i}{1+\omega^2\tau_i^2}}{\sum_{i=1}^{n}\alpha_i\tau_i} = \sum_{i=1}^{n}\frac{f_i}{1+\omega^2\tau_i^2} \tag{2.17}$$

2.3 Pulse Fluorometers [1–4]

Most of the pulse fluorometers are based on the *time-correlated single-photon counting* (TCSPC) method, better called *single-photon timing* (SPT). The basic principle relies on the fact that the probability of detecting a single photon at time *t* after an exciting pulse is proportional to the fluorescence intensity at that time. After timing and recording the single photons following a large number of exciting pulses, the fluorescence intensity decay curve is reconstructed.

Figure 2.3 shows a schematic diagram of a conventional single-photon counting instrument. The excitation source can be either a flash lamp or a pulsed laser. An electrical pulse associated with the optical pulse is generated (e.g., by a photodiode or the electronics associated with the excitation source) and routed – through a discriminator – to the start input of the time-to-amplitude converter (TAC). Meanwhile, the sample is excited by the optical pulse and emits fluorescence. The optics is tuned (e.g., by means of a neutral density filter) so that the photomultiplier detects no more than one photon for each exciting pulse. The corresponding electrical pulse is routed – through a discriminator – to the stop input of the TAC. The latter generates an output pulse whose amplitude is directly proportional to the delay time between the start and the stop pulses. The height analysis of this pulse is achieved by an analog-to-digital converter and a multichannel analyzer (MCA), which in-

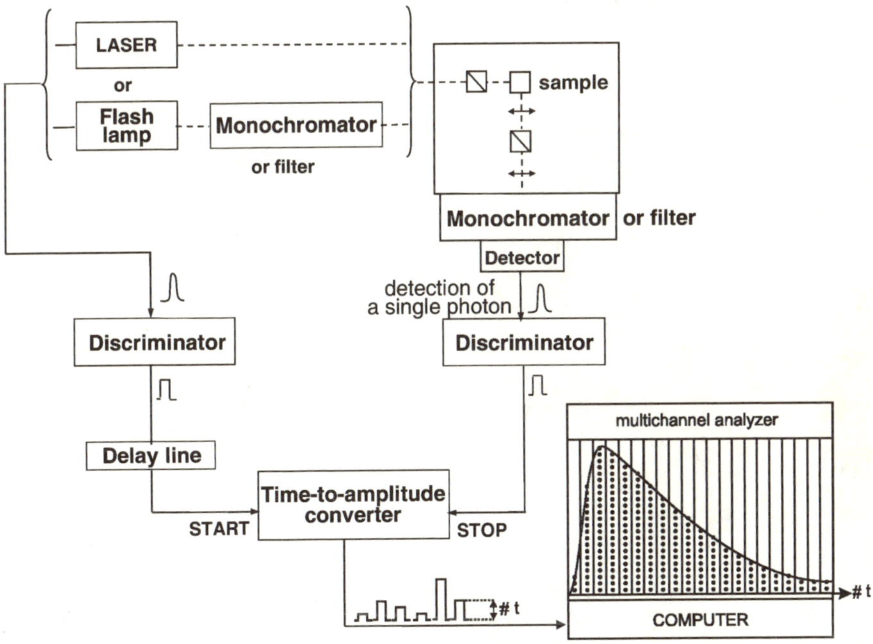

Fig. 2.3. Schematic diagram of a single-photon timing fluorometer

creases by one the contents of the memory channel corresponding to the digital value of the pulse. After a large number of excitation and detection events, the histogram of pulse heights represents the fluorescence decay curve. Obviously, the larger the number of events, the better the accuracy of the decay curve. The required accuracy depends on the complexity of the δ-pulse response of the system; for instance, a high accuracy is of course necessary for recovering a distribution of decay times.

When deconvolution is required, the time profile of the exciting pulse is recorded under the same conditions by replacing the sample by a scattering solution (solution of glycogen or suspension of colloidal silica "Ludox").

It is important to note that the number of fluorescence pulses must be kept much smaller than the number of exciting pulses (<0.01–0.05 stops per pulse), so that the probability of detecting two fluorescence pulses per exciting pulse is negligible. Otherwise, the TAC will take into account only the first fluorescence pulse and the counting statistics will be distorted: the decay will appear shorter than it is in reality. This effect is called the "pile-up effect".

A typical test experiment is shown in Fig. 2.4.

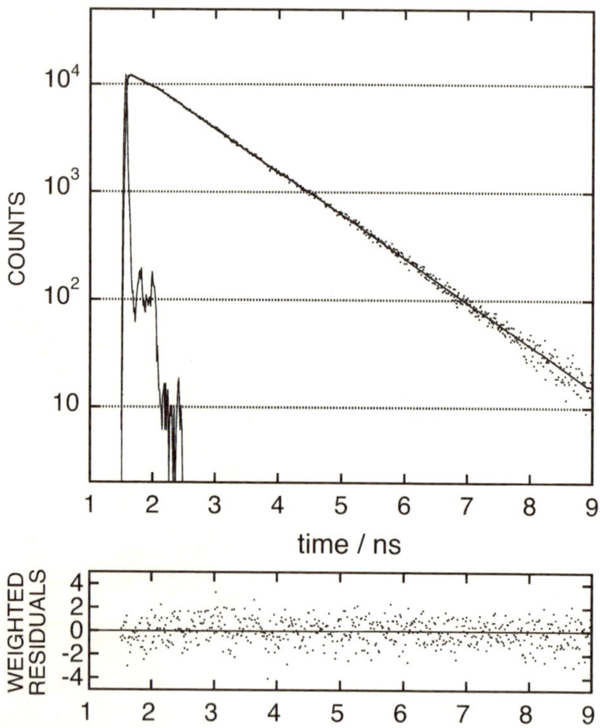

Fig. 2.4. Data obtained by the single-photon timing technique using a Ti:Sa laser and a MCP photomultiplier. Sample: solution of POPOP in cyclohexane (undegassed). Excitation wavelength: 327 nm. Reference: scattering solution (Ludox). Channel width: 12.2 ps. Result: τ=1.057±0.003 ns; χ_r^2=1.006

The SPT technique offers numerous advantages:

- High sensitivity.
- Outstanding dynamic range and linearity: three or four decades are common and five is possible.
- Well-defined statistics (Poisson distribution) allowing proper weighting of each point in data analysis.

The excitation source is of major importance. Flash lamps running in air, or filled with N_2, H_2 or D_2, are not expensive but the excitation wavelengths are restricted to the 200–400 nm range. They deliver nanosecond pulses, so that decay times of a few hundreds of picoseconds can be measured. Furthermore, the repetition rate is not high (10^4–10^5 Hz) and since the number of fluorescence pulses per exciting pulse must be kept less than 5%, the collection period may be quite long depending on the required accuracy (a few tens of minutes to several hours). For long collection periods, lamp drift may be a serious problem.

Lasers as excitation sources are of course much more efficient and versatile, at the penalty of high cost. Mode-locked dye lasers or Ti:sapphire lasers can generate pulses over broad wavelength ranges. The pulse widths are in the picosecond range or less with a high repetition rate. This rate must be limited to a few megahertz in order to let the fluorescence of long lifetime samples vanish before a new exciting pulse is generated.

The time resolution of the instrument is governed not only by the pulse width but also by the electronics and the detector. Microchannel plate photomultipliers are to be preferred to standard photomultipliers, but they are much more expensive. They exhibit faster time responses (10- to 20-fold faster). With mode-locked lasers and microchannel plate photomultipliers, the instrument response in terms of pulse width is about 30–40 ps so that decay times as short as 10 ps can be measured.

2.4 Phase-Modulation Fluorometers [1–3, 5–7]

Before describing the instruments, it is worth making two preliminary remarks:

1. The optimum frequency for decay time measurements using either the phase shift or the modulation ratio is, according to Eqs. (2.9) and (2.10), such that $\omega\tau$ is close to 1, i.e., $f \approx 1/(2\pi\tau)$. Therefore, for decay times of 10 ps, 1 ns, and 100 ns the optimum frequencies are about 16 GHz, 160 MHz, and 1.6 MHz, respectively.
2. In the case of a single exponential decay, Eqs. (2.9) and (2.10) provide two independent ways of measuring the decay time:
   - By phase measurements:

$$\tau_\Phi = \frac{1}{\omega} \tan \Phi \qquad (2.18)$$

   - By modulation measurements:

$$\tau_M = \frac{1}{\omega}\left(\frac{1}{M^2} - 1\right)^{1/2} \qquad (2.19)$$

The values measured in these two ways should of course be *identical* and *independent of the modulation frequency*. This provides two criteria to check whether an instrument is correctly tuned by using a lifetime standard whose fluorescence decay is known to be a single exponential. Note that the measurement of a decay time is fast (a fraction of a second) for a single exponential decay since a single frequency suffices. Note also that a significant difference between the values obtained by Eqs. (2.18) and (2.19) is compelling evidence of the nonexponentiality of the fluorescence decay.

2.4.1 Phase Fluorometers Using a Continuous Light Source and an Optical Modulator [5]

The light source can be a xenon lamp associated with a monochromator or a cw laser. The optical modulator (usually a Pockels cell) works better with a cw laser whose cost is not very high as compared to pulsed lasers.

Figure 2.5 shows a schematic diagram of a multifrequency phase-modulation fluorometer. A beam splitter reflects a few percent of the incident light toward a reference photomultiplier (via, or not, a cuvette containing a reference scat-

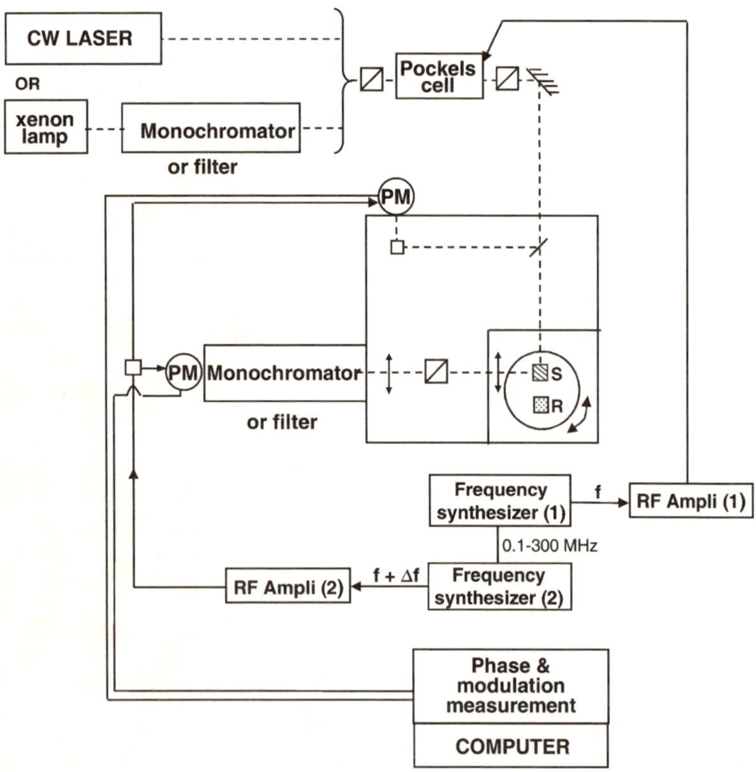

Fig. 2.5. Schematic diagram of a multifrequency phase-modulation fluorometer (*S*: sample; *R*: reference)

tering solution). The fluorescent sample and reference solution (containing either a scatterer or a reference fluorescent compound) are placed in a rotating turret. The emitted fluorescence or scattered light is detected by a photomultiplier through a monochromator or an optical filter. The Pockels cell is driven by a frequency synthesizer and the photomultiplier response is modulated by varying the voltage at the second dynode by means of another frequency synthesizer locked in phase with the first one. The two synthesizers provide modulated signals that differ in frequency by a few tens of hertz in order to achieve cross-correlation (heterodyne detection). This procedure offers excellent accuracy because the phase and modulation information contained in the signal is transposed to the low frequency domain where phase shifts and modulation depths can be measured with a much better accuracy than in the high frequency domain.

Practically, the phase delay φ_R and the modulation ratio m_R of the light emitted by the scattering solution are measured with respect to the signal detected by the reference photomultiplier. Then, after rotation of the turret, the phase

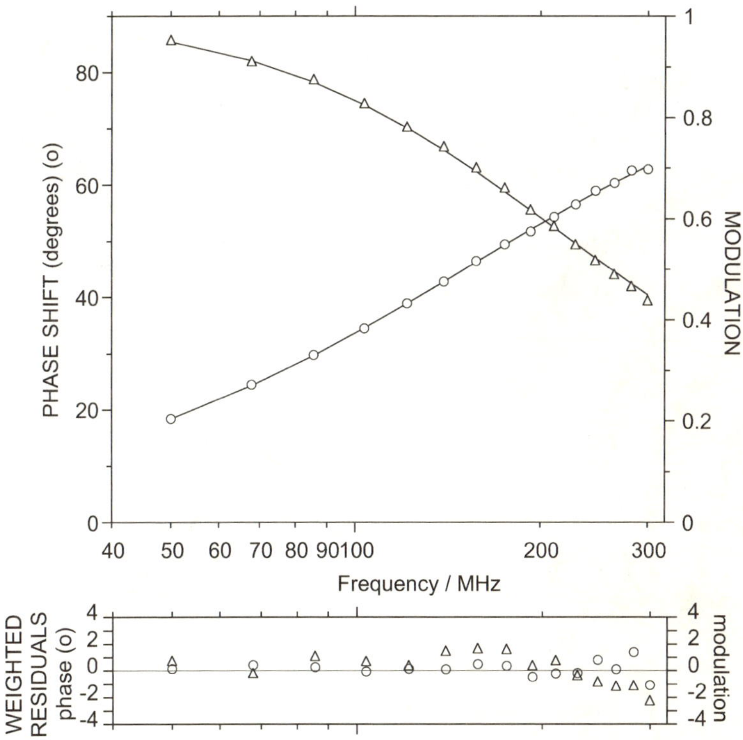

Fig. 2.6. Data obtained by the phase-modulation technique with a Fluorolog tau-3 instrument (Jobin Yvon-Spex) operating with a xenon lamp and a Pockels cell. Sample: solution of POPOP in cyclohexane (undegassed). Excitation wavelength: 340 nm. Reference: scattering solution (Ludox). Number of frequencies: 15 frequencies from 50 to 300 MHz. Result: τ=1.06±0.01 ns; χ_r^2=0.924. (Adapted from reference 1)

delay φ_F and the modulation ratio m_F for the sample fluorescence are measured with respect to the signal detected by the reference photomultiplier. The absolute phase shift and modulation ratio of the sample are then $\Phi=\varphi_F-\varphi_R$ and $M=m_F/m_R$, respectively.

Figure 2.6 shows a typical test experiment using POPOP whose fluorescence decay is a single exponential. Note that, since the decay is a single exponential, a single appropriate modulation frequency suffices for the lifetime determination. The broad set of frequencies permits control of the proper tuning of the instrument.

2.4.2
Phase Fluorometers Using the Harmonic Content of a Pulsed Laser [6, 7]

The type of laser source that can be used is exactly the same as for single-photon timing pulse fluorometry (see above). Such a laser system that delivers pulses in the picosecond range with a repetition rate of a few megahertz can be considered as an intrinsically modulated source. The harmonic content of the pulse train – which depends on the width of the pulses (as illustrated in Fig. 2.7) – extends to several gigahertz.

For high-frequency measurements, the usual photomultipliers are too slow and microchannel plate photomultipliers are required. However, internal cross-correlation is not possible with the latter and an external mixing circuit must be used. The time resolution of a phase fluorometer using the harmonic content of a pulsed laser and a microchannel plate photomultiplier is comparable to that of a single-photon timing instrument using the same kind of laser and detector.

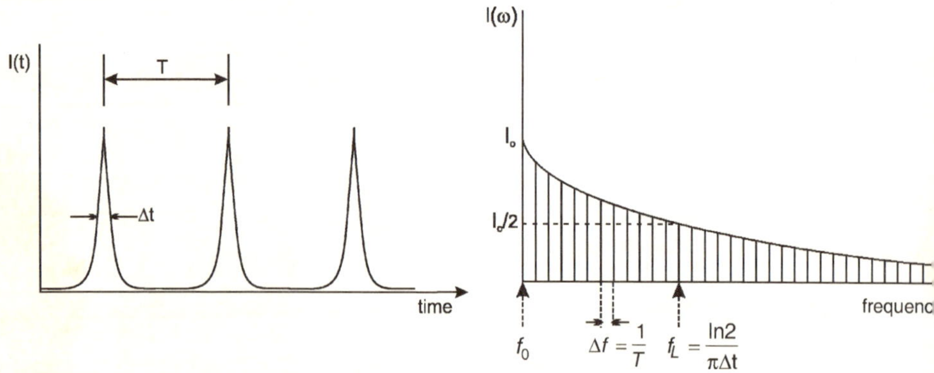

Fig. 2.7. Harmonic content of a pulse train. Example of numerical values: Δf=4 MHz; Δt=4 ps; f_L=44 GHz

2.5 Data Analysis [1, 2, 4, 8]

In both pulse and phase fluorometries, the most widely used method of data analysis is based on a nonlinear least-squares method. The basic principle of this method is to minimize a quantity which expresses the mismatch between data and fitted function. This quantity is the reduced chi-square, χ_r^2, defined as the weighted sum of the squares of the deviations of the experimental response $R(t_i)$ from the calculated ones $R_c(t_i)$:

$$\chi_r^2 = \frac{1}{\nu} \sum_{i=1}^{N} \left[\frac{R(t_i) - R_c(t_i)}{\sigma_i\,(i)} \right]^2 \tag{2.20}$$

where N is the total number of data points and $\sigma(i)$ is the standard deviation of the ith data point, i.e., the uncertainty expected from statistical considerations (noise). ν is the *number of degrees of freedom* ($\nu=N-p$, where p is the number of fitted parameters). The value of χ_r^2 should be close to 1 for a good fit.

In addition to the value of χ_r^2, it is useful to display graphical tests. The most important of them is the plot of the *weighted residuals* defined as

$$W(t_i) = \frac{R(t_i) - R_c(t_i)}{\sigma_i\,(i)} \tag{2.21}$$

with $\sigma(i)=[R(t_i)]^{1/2}$ for single-photon counting data. If the fit is good, the weighted residuals are randomly distributed around zero.

When the number of data points is large (i.e., in the single-photon timing technique, or in phase fluorometry when using a large number of modulation frequencies), the autocorrelation function of the residuals defined as

$$C(j) = \frac{\dfrac{1}{N-j} \displaystyle\sum_{i=1}^{N-j} W(t_i)\, W(t_{i+j})}{\dfrac{1}{N} \displaystyle\sum_{i=1}^{N} [W(t_i)]^2} \tag{2.22}$$

is also a useful graphical test of the quality of the fit. $C(j)$ expresses the correlation between the residual in channel j and channel $i+j$. A low-frequency periodicity is a symptom of radiofrequency interferences.

In *single-photon timing experiments*, the statistics obey the Poisson distribution and the expected deviation $\sigma(i)$ is approximated to $[R(t_i)]^{1/2}$ so that Eq. (2.20) becomes

$$\chi_r^2 = \frac{1}{\nu} \sum_{i=1}^{N} \frac{[R(t_i) - R_c(t_i)]^2}{R(t_i)} \tag{2.23}$$

In practice, initial guesses of the fitting parameters (e.g., preexponential factors and decay times in the case of a multiexponential decay) are used to calculate the decay curve; the latter is reconvoluted with the instrument response for comparison with the experimental curve. Then, a minimization algorithm (e.g., Marquardt method) is employed to search the parameters giving the best fit. At each step of the iteration procedure, the calculated decay is reconvoluted with the instrument response. Several pieces of software are commercially available.

In *phase fluorometry*, no deconvolution is required: curve fitting is indeed performed in the frequency domain, i.e., directly using the variations of the phase shift Φ and the modulation ratio M as functions of the modulation frequency. Phase data and modulation data can be analyzed separately, or simultaneously. In the latter case the reduced chi-square is given by

$$\chi_r^2 = \frac{1}{\nu}\left[\sum_{i=1}^{N}\left[\frac{\Phi(\omega_i) - \Phi_c(\omega_i)}{\sigma_\Phi(\omega_i)}\right]^2 + \sum_{i=1}^{N}\left[\frac{M(\omega_i) - M_c(\omega_i)}{\sigma_M(\omega_i)}\right]^2\right] \tag{2.24}$$

where N is the total number of frequencies. In this case, the number of data points is twice the number of frequencies, so that the number of degrees of freedom is $\nu=2N-p$.

Data analysis in phase fluorometry requires knowledge of the sine and cosine Fourier transforms of the δ-pulse response. This is of course not a problem for the most common case of multiexponential decays (see above), but in some cases, the Fourier transforms may not have analytical expressions, and numerical calculations of the relevant integrals are then necessary.

Global analysis of data consists in simultaneous analysis of several curves – recorded for instance at different wavelengths – by constraining the decay times to be the same for all sets, but allowing different values of the preexponential factors. Such a method that improves the accuracy of the recorded parameters can be used in both pulse and phase fluorometries.

The recovery of lifetime distributions deserves particular attention. Methods without a priori assumption of the shape of the distribution, such as the maximum entropy method [9], are to be preferred when no physical model is available.

2.6 Specific Applications

Pulse and phase fluorometries are now to be compared for specific applications: time-resolved spectra, time-resolved emission anisotropy, lifetime-based decomposition of spectra, and lifetime imaging microscopy.

2.6.1 Time-Resolved Spectra [1, 4, 10]

The evolution of fluorescence spectra during the lifetime of the excited state provides in some cases interesting information. Such an evolution occurs for instance when a fluorescent compound is excited and then evolves toward a new configuration whose fluorescent decay is different. A typical case is the solvent

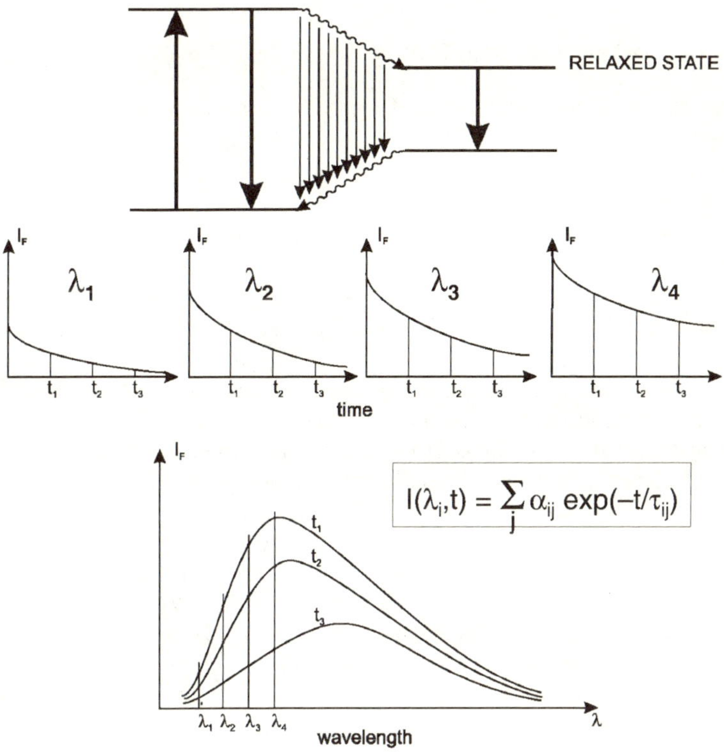

Fig. 2.8. Principles of determination of time-resolved fluorescence spectra

relaxation around an excited state compound whose dipole moment is higher in the excited state than in the ground state: the relaxation results in a gradual red shift of the fluorescence spectrum, and information on the polarity of the microenvironment around a fluorophore is thus obtained (e.g., in biological macromolecules). Figure 2.8 shows how the fluorescence spectra at various times are recovered from the fluorescence decays at several wavelengths.

In *phase fluorometry*, the phase (and modulation) data are recorded at a given wavelength and analyzed in terms of a multiexponential decay (without a priori assumption on the shape of the decay). Then, the fitting parameters are used to calculate the fluorescence intensities at various times t_1, t_2, t_3, … The procedure is repeated for each observation wavelength λ_1, λ_2, λ_3, … It is then easy to reconstruct the spectra at various times.

In *pulse fluorometry*, the same methodology can be applied, but one can also take advantage of the fact that the amplitudes of the output pulses of the time-to-amplitude converter are proportional to the times of arrival of the fluorescence photons on the photomultiplier. Selection of a given height of pulse, i.e., of a given time of arrival, is electronically possible (by means of a single-channel analyzer) and allows one to record the fluorescence spectra at a given time t after the excitation pulse. This is repeated for various times. The method described above for phase fluorometry can also be used in pulse fluorometry.

2.6.2 Time-Resolved Emission Anisotropy [1, 4, 10]

Following an infinitely short pulse of light, the time dependence of emission anisotropy is defined as

$$r(t) = \frac{I_{\parallel}(t) - I_{\perp}(t)}{I_{\parallel}(t) + 2I_{\perp}(t)} = \frac{I_{\parallel}(t) - I_{\perp}(t)}{I(t)} \tag{2.25}$$

where $I_{\parallel}(t)$ and $I_{\perp}(t)$ are the polarized components parallel and perpendicular to the direction of polarization of the incident light, and $I(t)$ is the total fluorescence intensity.

In *pulse fluorometry*, the polarized components are recorded as a function of time. The emission anisotropy can be calculated by means of Eq. (2.25), but only if the decay times are much larger than the width of the excitation pulse. Otherwise, Eq. (2.25) cannot be used because the responses $R_{\parallel}(t)$ $(=E(t)\otimes I_{\parallel}(t))$ and $R_{\perp}(t)$ $(=E(t)\otimes I_{\perp}(t))$ must be deconvoluted (see Fig. 2.9).

In *phase fluorometry*, one measures the differential polarized phase angle $\Delta(\omega)=\Phi_{\parallel}-\Phi_{\perp}$ between these two polarized components and the polarized modulation ratio $\Lambda(\omega)=m_{\parallel}/m_{\perp}$. The basic equations (see Fig. 2.9) involve the sine and cosine Fourier transforms of the δ-pulse response of $I_{\parallel}(t)$ and $I_{\perp}(t)$.

In the case of complex expressions of the δ-pulse response of emission anisotropy, pulse fluorometry appears to be more straightforward than phase fluorometry.

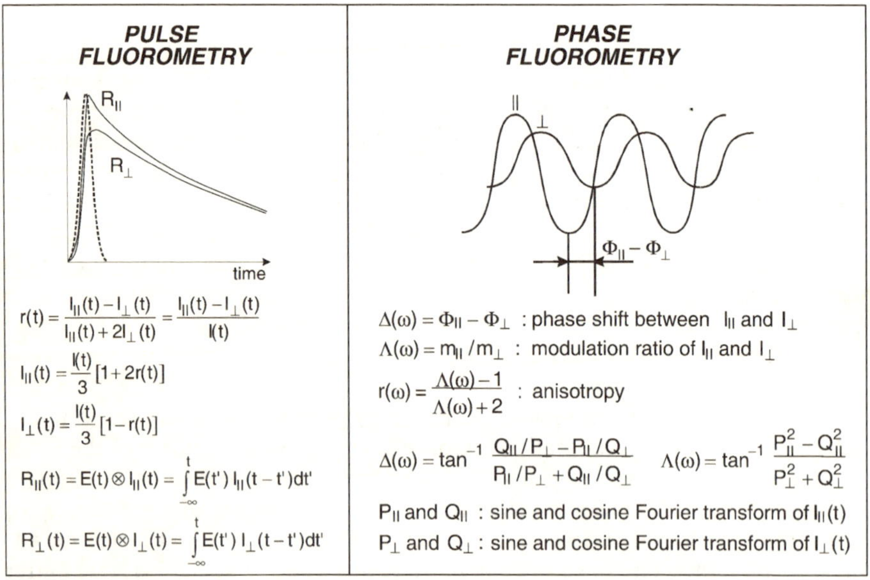

Fig. 2.9. Principles of time-resolved emission anisotropy measurements in time and frequency domains

2.6.3 Lifetime-Based Decomposition of Spectra [11]

A fluorescence spectrum may result from overlapping spectra of several fluorescent species (or several forms of a fluorescent species). If each of them is characterized by a single lifetime, it is possible to decompose the overall spectrum into its components.

Let us consider for instance a spectrum consisting of three components whose lifetimes, τ_1, τ_2, and τ_3, have been determined separately. Decomposition of the fluorescence spectrum is possible in pulse fluorometry by analyzing the decay with a three-exponential function at each wavelength

$$I_\lambda(t) = \alpha_{1\lambda} \exp(-t/\tau_1) + \alpha_{2\lambda} \exp(-t/\tau_2)\alpha_{3\lambda} \exp(-t/\tau_3) \tag{2.26}$$

and by calculating the fractional intensities, $f_{1\lambda}$, $f_{2\lambda}$, and $f_{3\lambda}$, as follows:

$$f_{i\lambda} = \frac{\alpha_{i\lambda}\,\tau_i}{\sum_{i=1}^{3} \alpha_{i\lambda}\,\tau_i} \tag{2.27}$$

The procedure in phase-modulation fluorometry is more straightforward. The sine and cosine Fourier transforms of the δ-pulse response are, according to Eqs. (2.16) and (2.17), given by

$$P_\lambda = \frac{f_{1\lambda}\,\omega\tau_1}{1+\omega^2\tau_1^2} + \frac{f_{2\lambda}\,\omega\tau_2}{1+\omega^2\tau_2^2} + \frac{f_{3\lambda}\,\omega\tau_3}{1+\omega^2\tau_3^2} \tag{2.28}$$

$$Q_\lambda = \frac{f_{1\lambda}}{1+\omega^2\tau_1^2} + \frac{f_{2\lambda}}{1+\omega^2\tau_2^2} + \frac{f_{3\lambda}}{1+\omega^2\tau_3^2} \tag{2.29}$$

with

$$1 = f_{1\lambda} + f_{2\lambda} + f_{3\lambda} \tag{2.30}$$

Decomposition *in real time* is possible by measuring Φ_λ and M_λ as a function of wavelength at a single frequency and by calculating $P_\lambda = M_\lambda \sin \Phi_\lambda$ and $Q_\lambda = M_\lambda \cos \Phi_\lambda$. $f_{1\lambda}, f_{2\lambda}, f_{3\lambda}$ are then solutions of the system of Eqs. (2.28), (2.29), and (2.30).

2.6.4 Fluorescence Lifetime Imaging Microscopy (FLIM) [12–15]

The excited-state lifetime of a fluorophore does not depend on its concentration and is sensitive to its microenvironment; this parameter is thus of great interest for mapping in fluorescence microscopy. When several fluorophores are used simultaneously, distinction between them is often easier on the basis of lifetime than on the basis of intensity. The contrast in the fluorescence images can thus be greatly improved. FLIM has been developed using either time-domain or frequency-domain methods [12–14]. FLIM is an outstanding tool for the study of single cells with the possibility of coupling multiparameter imaging of cellular structures with spectral information. Moreover, recently it was demonstrated that FLIM is an excellent tool to monitor Förster resonance energy transfer in cells [15].

In *wide-field fluorescence microscopy*, the sample is uniformly illuminated. Time resolution is obtained by using a pulsed or modulated exciting light, and a gated or modulated image intensifier, as shown in Fig. 2.10. An image intensifier uses a multichannel plate and a CCD camera placed behind a phosphor screen. Gated detection or gain modulation can be achieved by driving the grid (located behind the photocathode) by a gate pulse or a sine wave signal.

In the *time domain*, only photons emitted from the sample that arrive at the photocathode of the photomultiplier during the gate time will be detected. The fluorescence intensity as a function of time can be constructed by moving the time window after each pulse. For a single exponential decay, two delays are sufficient (provided that deconvolution is not necessary).

In the *frequency domain*, a frequency synthesizer is used for driving the modulator and to modulate the gain of the image intensifier via the voltage of the grid. The use of a variable phase shifter is convenient, as shown in Fig. 2.10. The

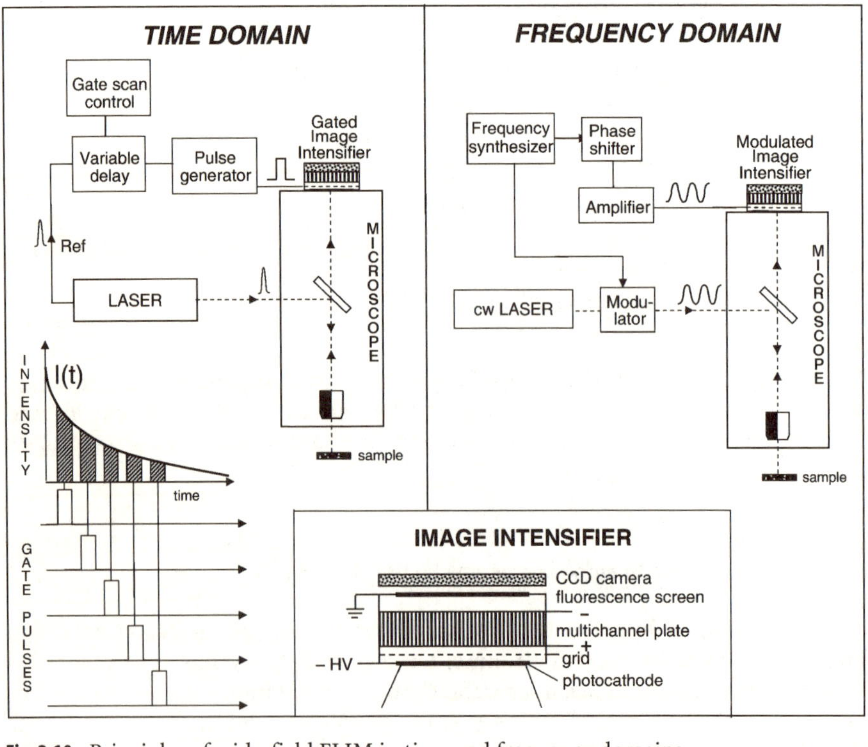

Fig. 2.10. Principles of wide-field FLIM in time and frequency domains

phase and modulation are calculated from a series of images taken at different phase delays between the excitation light and the intensifier. Alternatively, heterodyne detection can be used (see Sect. 2.4.1). In the case of a single exponential decay, a single frequency is sufficient. For multiexponential decays, several images have to be acquired at different frequencies.

Confocal FLIM is possible by using laser scanning microscopes either in the time domain or frequency domain. The time domain method based on single-photon timing requires lasers with high repetition rates to acquire an image in a reasonable time, because each pixel requires many photon events to generate a decay curve. The frequency domain method using a cw laser coupled with an electro-optical modulator is less expensive. The harmonic content of a high repetition pulsed laser can also be used. Such lasers permit two-photon excitation.

2.7 Concluding Remarks

Comparison between the two techniques can be presented from three points of view: theoretical, instrumental, and methodological [1]:

1. Pulse and phase fluorometries are *theoretically equivalent*: they provide the same kind of information since *the harmonic response is the Fourier transform of the δ-pulse response*.
2. From the *instrumental point of view*, the latest generations of instruments use both pulsed lasers and microchannel plate detectors. Only the electronics is different. Since the *time resolution* is mainly limited by the time response of the detector, this parameter is the same for both techniques. Moreover, the optical module is identical so that the total cost of the instruments is similar.
3. The *methodologies* are quite different because they are relevant to the time domain and the frequency domain. The advantages and drawbacks are the following:
   - Pulse fluorometry permits *visualization of the fluorescence decay*, whereas visual inspection of the variations of the phase shift versus frequency does not allow the brain to visualize the inverse Fourier transform!
   - Pulse fluorometry has an outstanding *sensitivity*: experiments with a very low level of light (e.g., owing to low quantum yields or strong quenching) simply require longer acquisition times (but attention has to be paid to the possible drift of the excitation source), whereas in phase fluorometry, the fluorescence intensity must be high enough to get an analog signal whose zero crossing (for phase measurements) and amplitude (for modulation measurements) can be measured with sufficient accuracy.
   - No *deconvolution* is necessary in phase fluorometry, while this operation is often necessary in pulse fluorometry and requires great care in recording the instrument response, especially for very short decay times.
   - The *well-defined statistics* in single-photon counting is an advantage for data analysis. In phase fluorometry, the evaluation of the standard deviation of phase shift and modulation ratio may not be easy.

- *Time-resolved emission anisotropy* measurements are more straightforward in pulse fluorometry.
- *Time-resolved spectra* are more easily recorded in pulse fluorometry.
- *Lifetime-based decomposition of spectra* into components is simpler in phase fluorometry.
- The *time of data collection* depends on the complexity of the δ-pulse response. For a single exponential decay, phase fluorometry is more rapid. For complex δ-pulse responses, the time of data collection is about the same for the two techniques: in pulse fluorometry, a large number of photon events are necessary, and in phase fluorometry, a large number of frequencies are to be selected. It should be emphasized that the short acquisition time for phase shift and modulation ratio measurements at a given frequency is a distinct advantage in several situations, especially for lifetime-imaging spectroscopy.

In conclusion, pulse and phase fluorometries have their own advantages and drawbacks. They appear to be equivalent or complementary but by no means competitive.

References

1. Valeur B (2002) Molecular fluorescence. Principles and applications. Wiley-VCH, Weinheim
2. Demas JN (1983) Excited-state lifetime measurements. Academic, New York
3. Lakowicz JR (ed) (1991) Topics in fluorescence spectroscopy, vol 1. Techniques. Plenum, New York
4. O'Connor DV, Phillips D (1984) Time-correlated single photon counting. Academic, London
5. Jameson DM, Gratton E, Hall RD (1984) The measurement and analysis of heterogeneous emissions by multifrequency phase and modulation fluorometry. Appl Spectrosc Rev 20:55–106
6. Alcala JR, Gratton E, Jameson DM (1985) A multifrequency phase fluorometer using the harmonic content of a mode-locked laser. Anal Instrum 14:225–250
7. Lakowicz JR, Laczko G, Gryczynski I (1986) 2 GHz frequency-domain fluorometer. Rev Sci Instrum 57:2499–2506
8. Eaton DF (1990) Recommended methods for fluorescence decay analysis. Pure Appl Chem 62:1631–1648
9. Brochon JC (1994) Maximum entropy method of data analysis in time-resolved spectroscopy. Meth Enzymol 240:262–311
10. Lakowicz JR, Gratton E, Cherek H, Maliwal BP, Laczko G (1984) Determination of time-resolved emission spectra and anisotropies of a fluorophore–protein complex using frequency-domain phase-modulation fluorometry. J Biol Chem 259:10967–10972
11. Gratton E, Jameson DM (1985) New approach to phase and modulation resolved spectra. Anal Chem 57:1694–1697
12. Wang XF, Periasamy A, Herman B, Coleman DM (1992) Fluorescence lifetime imaging microscopy (FLIM) – Instrumentation and Applications. Crit Rev Anal Chem 23:369–395
13. Lakowicz JR, Szmacinski H, Nowaczyk K, Berndt KW, Johnson M (1992) Fluorescence lifetime imaging. Anal Biochem 202:316–330
14. Draaijer A, Sanders R, Gerritsen HC (1995) Fluorescence lifetime imaging: a new tool in confocal microscopy. In: Pawley J (ed) Handbook of biological confocal microscopy, 2nd edn. Plenum, New York, pp 491–505
15. Wouters FS, Bastiaens PIH (1999) Fluorescence lifetime imaging of receptor tyrosine kinase activity in cells. Curr Biol 9:1127–1130

3 Non-Exponential Fluorescence of Electronically Coupled Donors Contains Distance Information

S. Kalinin, M. Isaksson and L. B.-Å. Johansson*

Keywords: Donor-donor energy migration (DDEM); Homotransfer; Non-exponential fluorescence; Fluorescence lifetimes; Distance information

Abbreviations

BODIPY	Derivatives of *N*-4,4-difluoro-4-bora-3a,4a-diaza-*s*-indacene
DDEM	Donor-donor energy migration
PAI-2	Plasminogen activator inhibitor of type 2
PDDEM	Partial donor-donor energy migration
TCSPC	Time-correlated single photon counting

3.1 Introduction

Besides traditional energy-transfer experiments [1, 2], donor–donor energy migration (DDEM) studies [3, 4] were recently used in studies of proteins [4–7] and lipid membranes [8, 9]. According to a rule rather than an exception, the fluorescence relaxation of donor and acceptor groups in the absence of electronic dipole-dipole coupling is not a single-exponential function. For years the origin of this fact has been and still is a subject of discussion. In practice the fluorescence decay is most often fitted to a sum of exponential functions [2], while models that assume distributions [10, 11] are also used. The former approach would mean that chromophoric groups experience a few distinct physicochemical environments, each representing a unique fluorescence lifetime. This is reasonable for proteins of well-defined structure in which the bound chromophoric group occupies a few preferential orientations in the binding site. However, the explanation seems less reasonable for loosely attached groups that undergo local rotational motions in the binding site, or for which the binding site is flexible. Usually the local rotational correlation times and fluorescence lifetimes take place on similar timescales, making a distribution of lifetimes physically more tractable.

Unfortunately, models assuming discrete or continuous distributions of lifetimes can be statistically very well fitted to most realistic experimental data, implying that one cannot definitely distinguish the nature of lifetime distributions. In order to distinguish between the models, one would need stable experimental conditions providing extremely good statistics with the number of counts in the order of 10^6 or higher [12]. In a recent study we showed that, irrespective of the model used to describe the fluorescence

* Corresponding author.

decay, the extracted rate of energy migration as well as the related distance remain invariant [13]. The latter study implies that the measured fluorescence decay contains information about the mutual distance between the interacting donor molecules for which the individual donors exhibit a non-exponential fluorescence. In practice one observes that the average fluorescence lifetime decreases for the coupled system as compared to that of the non-interacting donors.

3.2 Theory

Equations that will be presented here are based on the PDDEM model [14, 15]. Here we consider a special case of identical donors (D). The non-exponential fluorescence relaxation of both donors can be approximated by a sum of discrete exponentials

$$F(t) = \sum_i a_i \exp(-t/\tau_i) \tag{3.1}$$

The rate of the energy migration ω depends on the distance R between the centres of mass of D_α and D_β according to [16]

$$\omega = \frac{3\langle\kappa^2\rangle}{2\tau}\left(\frac{R_0}{R}\right)^6 \tag{3.2}$$

In Eq. (3.2), τ and $\langle\kappa^2\rangle$ represent the fluorescence lifetime of the donor and the square of the angular part of the dipole-dipole interaction, respectively. The Förster radius R_0 is defined by

$$R_0 = \left(\frac{9000 \ln 10 (2/3) \Phi J}{128 \pi^5 n^4 N_A}\right)^{1/6} \tag{3.3}$$

In Eq. (3.3), N_A, n, Φ and J denote the Avogadro constant, the refractive index, the quantum yield of the donor and the overlap integral, respectively. We assumed that the energy migration rate is the same for the whole ensemble of donors. This approximation is not the only one possible [2], but it leads to the exact expression for a non-exponential decay caused by a dynamic quenching process ($\Phi \propto \tau$ in Eqs. (3.2 and 3.3)) [14, 17].

For a subpopulation of D-D pairs for which the lifetimes of donors are τ_i and τ_j, the fluorescence relaxation is given by [14]

$$F_{ij}(t) = [(-2\lambda_2 - \tau_i^{-1} - \tau_j^{-1})\exp(\lambda_1 t) + (2\lambda_1 + \tau_i^{-1} + \tau_j^{-1})\exp(\lambda_2 t)]/(\lambda_1 - \lambda_2) \tag{3.4}$$

where

$$\lambda_{1,2} = \frac{1}{2}[-1/\tau_i - 1/\tau_j - 2\omega \pm \sqrt{(1/\tau_i - 1/\tau_j)^2 + 4\omega^2}] \tag{3.5}$$

For the coupled system we assume that the photophysics is

$$F_{DD}(t) = \sum_{i,j} a_i a_j F_{ij}(t,\tau_i,\tau_j) \tag{3.6}$$

Here $F_{ij}(t,\tau_i,\tau_j)$ is calculated from Eqs. (3.4) and (3.5). For the protein systems in the present study the fluorescence decays were typically fitted to the sum of two or three exponential functions.

3.3 Methods

The synthetic time-correlated single photon counting (TCSPC) data sets were generated as described by Chowdhury et al. [18]. The decay curves contained 25,000 counts in the peak maximum. The excitation pulse was taken to be Gaussian with the full width at half maximum height being 0.1 τ_0. For simplicity, the radiative lifetime τ_0 was chosen the same for all fluorescent species in the synthetic "experiments". The time resolution was 0.005 τ_0/channel.

The distances (R) were extracted by fitting Eqs. (3.1) to (3.6) to the fluorescence decay of the coupled $D_\alpha \leftrightarrow D_\beta$ system by using a common deconvolution procedure [19]. Three data sets corresponding to the individual fluorescence relaxations and the coupled system were analysed globally [20] for the $D_\alpha \leftrightarrow D_\beta$ system. The synthetic data sets were analysed in the same way as the experimental data. The quality of the global fitting was judged by standard χ^2 and Durbin-Watson parameters, as well as by the weighted residuals and autocorrelation function graphs. All calculations were performed on a 733 MHz PC. The most time-consuming global analyses of PDDEM data required about 1–2 min.

Single photon counting measurements were performed on a PRA 3000 system (Photophysical Research Ass. Inc., Ontario, Canada). The excitation source was a Nanoled-01 pulsed diode, operating at 800 kHz (IBH, Glasgow, Scotland). The excitation and emission wavelengths were selected with interference filters centred at 500 and 550 nm, respectively. Lifetime data were collected with an emission polariser set to the "magic angle" (54.7°) relative to the excitation polariser. At least 25,000 counts were collected in the maximum peak. The resolution was 50 ps/channel and the fluorescence decay was collected over 1,024 channels.

3.4 Results and Discussion

3.4.1 Synthetic Data

Even fluorophores that are exceptionally insensitive to their microenvironment, such as BODIPY [21], often exhibit complex fluorescence relaxation. Typically the fluorescence decay of BODIPY in proteins is a biexponential decay with lifetimes

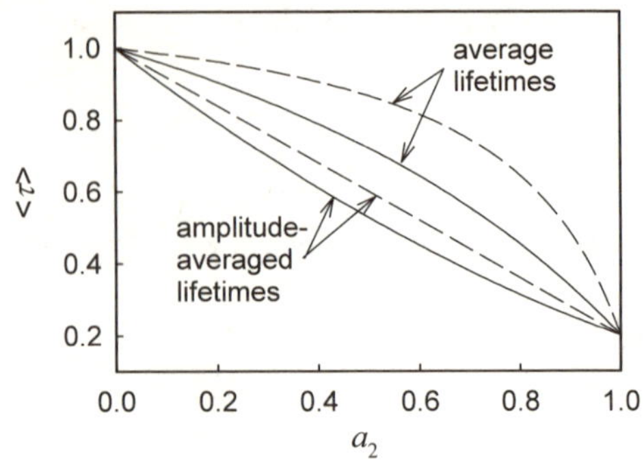

Fig. 3.1. Average and amplitude-averaged lifetimes of coupled donor–donor systems (*solid lines*) plotted vs. the fraction of the shorter lifetime a_2. The corresponding average lifetimes in the absence of energy transfer are shown as *dashed lines*. The fluorescence relaxation of both donors was assumed to be biexponential, τ_1=1, τ_2=0.2, a_1=1−a_2, and the energy migration rate ω was taken to be 1 (=1/τ_1)

of about 1–3 and 6 ns (90%). To mimic TCSPC experiments, synthetic data sets were generated in the absence and presence of electronic coupling within donor pairs, for which each donor exhibits a biexponential fluorescence decay. For different contributions of two lifetimes the average and amplitude-averaged lifetimes were calculated (see Fig. 3.1). The results show that both kinds of lifetime averages decrease in the presence of energy migration. The analysis of TCSPC data also clearly reveals the influence of energy migration, as is exemplified in Fig. 3.2.

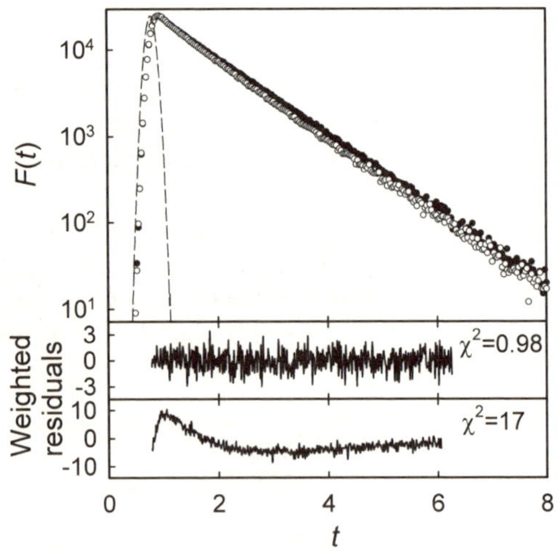

Fig. 3.2. Synthetically generated fluorescence decay curve of donor–donor system in the absence (●) and presence (○) of energy migration. The parameters of the modelled fluorescence relaxation were a_1=0.8, a_2=0.2, τ_1=1, τ_2=0.2, ω=1. The *upper* weighted residuals plot is a result of fitting a biexponential function $F(t)$ to the donor's decay. The *lower* plot shows the weighted residuals obtained when fitting the same function $F(t)$ to the fluorescence decay of the coupled system

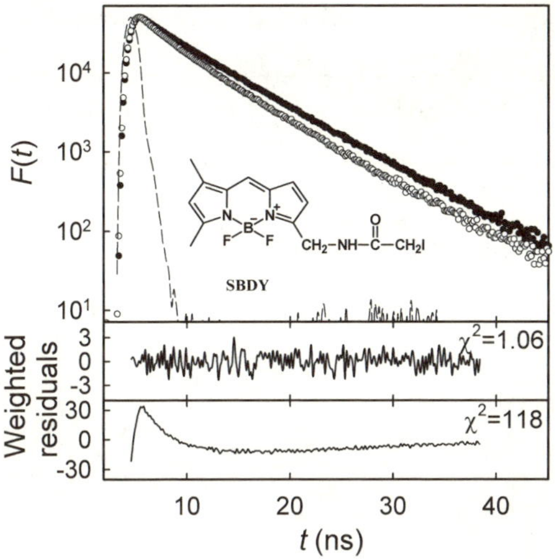

Fig. 3.3. Experimental data showing the fluorescence decay of SBDY-labelled 171Cys (●) and 79Cys/171Cys (○) PAI-2 mutants. The chemical structure of *N*-(4,4-difluoro-5,7-dimethyl-4-bora-3a,4a-diaza-*s*-indacene-3-yl)methyl iodoacetamide (SBDY) is inserted. The weighted residuals for SBDY-171Cys and SBDY-79Cys/171Cys mutants are also shown in the *panel* below the decay curves. The residual plot shown at the *bottom* (yielding χ^2=118) is the result of fitting the SBDY-79Cys/171Cys fluorescence decay to the arithmetic average of the 79Cys and 171Cys decays

3.4.2 Experimental Data

A sulphhydryl-specific BODIPY (see Fig. 3.3) was covalently linked to three different Cys residues of mutant forms of PAI-2. The fluorescence decays were measured for BODIPY attached to the 79Cys, 171Cys and 347Cys residues of PAI-2. TCSPC measurements of the decays showed that the photophysics can be described by a sum of two exponential functions. Lifetime experiments were also performed on the double mutants 79Cys/171Cys and 79Cys/347Cys, each labelled with two BODIPY groups. The average and amplitude-averaged lifetimes calculated differ slightly between the BODIPY groups located at the 79Cys, 171Cys and 347Cys residues. The corresponding averages calculated from experiments on the BODIPY-labelled double mutants are significantly shorter than the arithmetically averaged lifetime of the corresponding singly labelled mutants. This is compatible with non-reversible donor–donor energy migration. We also applied a previously developed model [14] (Eqs. (3.1) to (3.6)) to estimate the distances *R* between BODIPY groups within PAI-2 mutants. The analyses yielded rates of energy migration compatible with those obtained from independent experiments [22] (Table 3.1).

Table 3.1. Fluorescence lifetimes of BODIPY fluorophores covalently linked to Cys residues of mutant forms of PAI-2

Experiments on PAI-2 mutants	a_1	τ_1 (ns)	a_2	τ_2 (ns)	ω (ns^{-1})	$\langle\bar{\tau}\rangle^{a}$ (ns)	$\bar{\tau}^{b}$ (ns)
79Cys	0.69	5.94	0.25	3.22		5.49	5.22
171Cys	0.76	5.85	0.23	1.77		5.51	4.92
347Cys	0.85	6.25	0.05	2.92		6.15	6.05
79Cys/171Cys					0.21[c]	4.78	3.93
79Cys/347Cys					0.54[c]	5.57	5.27
Simulated systems							
D	0.9	1	0.1	0.2		0.98	0.95
D–D					1	0.92	0.89
D[d]	0.8	1	0.2	0.2		0.96	0.84
D–D[d]					1	0.91	0.79
D	0.6	1	0.4	0.2		0.91	0.68
D–D					1	0.79	0.61

[a] Average lifetimes.
[b] Amplitude-averaged lifetimes.
[c] Calculated from the distances measured before [22].
[d] Also shown in Fig. 3.2.

3.5 Conclusions

Synthetically generated and experimental data sets were studied for chemically identical fluorescent groups that exhibit non-exponential fluorescence relaxation. In the presence of Förster coupling between adjacent fluorescent moieties of this kind the average lifetimes decrease as compared to the average lifetimes of the non-coupled systems. As a consequence one can expect that average lifetimes of chemically identical fluorophores, such as tryptophans in proteins, are shorter than the average formed by the individual contributions. It means that the fluorescence relaxation of the coupled system of fluorophores contains information about the distances between the interacting groups.

Acknowledgements
The Swedish Research Council and the Kempe Foundations financially supported this work.

References

1. Van der Meer BW, Coker G III, Chen S-YS (1994) Resonance energy transfer: theory and data. VCH, Weinheim
2. Lakowicz JR (1999) Principles of fluorescence spectroscopy. Kluwer/Plenum, New York
3. Karolin J, Johansson LB-Å (1997) Trends Phys Chem 6:171
4. Bergström F, Hägglöf P, Karolin J, Ny T, Johansson LB-Å (1999) Proc Natl Acad Sci 96:12477
5. Karolin J, Fa M, Wilczynska M, Ny T, Johansson LB-Å (1998) Biophys J 74:11
6. Fa M, Bergström F, Hägglöf P, Wilczynska M, Johansson LB-Å, Ny T (2000) Structure 8:397

7. Runnels LW, Scarlata SF (1995) Biophys J 69:1569
8. Mikhalyov I, Bogen S-T, Johansson LB-Å (2001) Spectrochim Acta A 57:1839
9. Bogen S-T, Karolin J, Molotkovsky JG, Johansson LB-Å (1998) J Chem Soc Faraday Trans 94:2435
10. Livesey AK, Brochon JC (1987) Biophys J 52:693
11. Lakowicz JR, Cherek H, Gryczynski I, Joshi N, Johnson ML (1987) Biophys Chem 28:35
12. James DR, Ware WR (1985) Chem Phys Lett 120:455
13. Kalinin S, Johansson LB-Å (2004) J Phys Chem B 108:3092–3097
14. Kalinin SV, Molotkovsky JG, Johansson LB-Å (2002) Spectrochim Acta A 58:1087
15. Kalinin S, Molotkovsky JG, Johansson LB-Å (2003) J Phys Chem B 107:3318
16. Förster T (1948) Ann Phys 2:55
17. Albaugh S, Steiner RF (1989) J Phys Chem 93: 8013
18. Chowdhury FN, Kolber ZS, Barkley MD (1991) Rev Sci Instrum 62:47
19. O'Connor DV, Phillips D (1984) Time-correlated single photon counting. Academic, London
20. Knutson JR, Beechem JM, Brand L (1983) Chem Phys Lett 102:501
21. Bergström F, Mikhalyov I, Hägglöf P, Wortmann R, Ny T, Johansson LB-Å (2002) J Am Chem Soc 124:196
22. Isaksson M, Kalinin S, Lobov S, Wang S, Ny T, Johansson LB-Å (2004) Phys Chem Chem Phys 6:3001–3008

4 Fluorescence Nanotomography: Recent Progress, Constraints and Opportunities

O. J. Rolinski and D. J. S. Birch

Keywords: Fluorescence nanotomography; FRET; fluorescence decay; Förster-type models; Nafion 117; phospholipid bilayers

Abbreviations

BODIPY	Derivatives of 4-bora-3a,4a-diaza-*s*-indacene
DMPC	Dimyristoylphosphatidylcholine
DPPC	Dipalmitoylphosphatidylcholine
FN	Fluorescence nanotomography
FRET	Fluorescence resonance energy transfer
PBS	Phosphate-buffered saline
TCSPC	Time-correlated single photon counting

4.1 Introduction

The ubiquitous challenge, which today presents itself, to understanding molecular functions is that of being able to track molecular structure, dynamics and reactions. Given the molecular diversity which exists, the scale of the task is daunting. The very nature of molecular fluorescence presents us with a generic opportunity to track molecular events down to the single-molecule level and in complex media such as tissue, cells, DNA and proteins, which underpin the very fundamentals of life, as well as synthetic media such as amorphous solids, colloids, polymers and glasses, which permeate much of modern technology.

If we consider the stage we are at in using molecular fluorescence to monitor molecular events, then what we have at present is a series of snapshots, often obtained under ideal conditions, which represent only part of the big picture. So, for example, we can label proteins with fluorophores at discrete sites using site-directed mutagenesis and monitor folding/unfolding, but we are still clearly only scratching a very small part of an extremely large multidimensional surface.

Fluorescence microscopy in all its forms clearly represents the first step to learning about the big picture, but how do we obtain the angstrom resolution that is needed? Conventional far-field microscopy has a $\lambda/2$ diffraction limit and in the near field ~50 nm is the present day limit – a long way from the angstrom resolution we seek.

Fluorescence resonance energy transfer (FRET) stands out as one technique with the potential to unravel molecular events with angstrom resolution and for this reason it has latterly found a rebirth in molecular biology. When combined with pico- to nanosecond time frames from fluorescence decays, the exciting opportunity is now emerging that it might be possible to make a fluorescence nanomovie which tells the story of nanosecond events with nanometre resolution.

FRET is often used to detect the proximity between donors and acceptors by means of the quenching of fluorescence which results. However, this usually involves some guesswork as to the distribution of donors and acceptors, which pertains to the interpretation – a guessing game frequently yielding multiple possible answers and not the unique solution we seek. Hence recently we have developed numerical methods of treating fluorescence decay data and revealed near-angstrom distances without assuming the structural form in the first place. Here we present some of our latest developments towards the making of a fluorescence nanomovie using FRET and demonstrate the resultant nanotomography on some model systems.

4.2 Fluorescence Resonance Energy Transfer

Excitation energy transfer occurs from an initially excited donor molecule D to the acceptor molecule A of the same or different kind (see Fig. 4.1) and is a non-radiative process. The D–A distance-dependent FRET rate $k_{DA}(r)$ can be calculated from time-dependent perturbation theory. If the dipole–dipole interaction between them is assumed, then

$$k_{DA}(r) = \frac{1}{\tau_D}\left(\frac{R_0}{r}\right)^6 \tag{4.1}$$

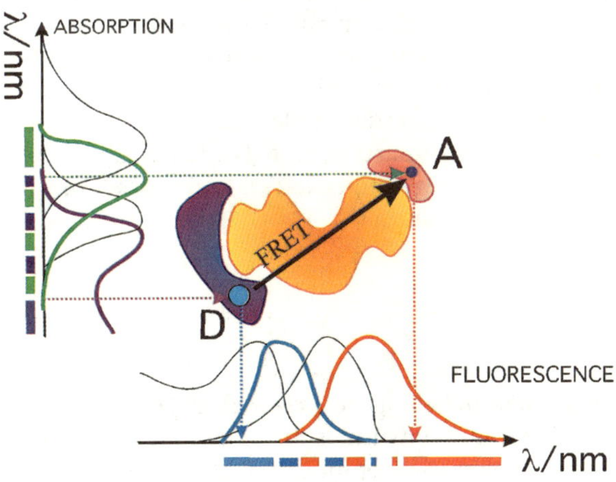

Fig. 4.1. The principle of the time-resolved FRET experiment. The donor decay function depends on the FRET rate and thus carries information on the mutual donor-acceptor distribution

where τ_D is the unquenched donor fluorescence lifetime and R_0 is the critical transfer distance, defined as

$$R_0^6 = \frac{9000(\ln 10)\kappa^2 \Phi}{128\pi^5 n^4 N_A} \int_0^\infty \frac{F_D(\tilde{\nu})\varepsilon_A(\tilde{\nu})}{\tilde{\nu}^4} \tag{4.2}$$

Φ is the donor fluorescence quantum yield in the absence of acceptor and n is the refractive index of the solvent. $F_D(\tilde{\nu})$ is the normalised fluorescence spectrum of the donor and $\varepsilon_A(\tilde{\nu})$ is the molar absorption coefficient of the acceptor. κ^2 in Eq. (4.2) is the orientational factor which, in general, varies from 0 to 4. For fast and freely rotating donor and acceptor molecules this term may be isotropically averaged over all possible orientations. In this dynamic case, κ^2 is a constant equal to 2/3 and we make such an assumption here. The so-called overlap integral appearing in Eq. (4.2) provides the opportunity for high sensitivity for imaging and sensing (e.g. large overlap with the analyte of interest gives a high FRET rate) as well as high selectivity (e.g. poor overlap with the other analytes present).

4.3 FRET Sensors

The presence of FRET affects the donor fluorescence. This gives an opportunity for developing FRET-based fluorescence images for molecular separations on the angstrom scale. In the simplest approach, the D–A proximity can be assessed from the ratio of fluorescence intensities of the donor and acceptor. Although simple in application, this method carries all the drawbacks of the steady-state experiment (sensitivity for scatter and background fluorescence, leak or bleaching of the fluorophores, etc.) and provides limited structural information, as the donor time response function, rich in structural information, is averaged over time in a steady-state experiment. The lifetime approaches are much more information effective and are based on the kinetics of the system being significantly dependent on the donor and acceptor distribution, as well as on the extent of diffusion in the system if the diffusional displacement of the donor or acceptor molecule during the donor lifetime is comparable to the critical transfer distance R_0. In a basic two-particle model diffusion is neglected, and a low donor concentration (no energy migration) is considered. In such a case the donor fluorescence decay function is

$$I_D(t) = \exp\left[-t/\tau_D - \int_0^\infty \rho_{DA}(r)(1-\exp[-(R_0/r)^6(t/\tau_D)]dr\right] \tag{4.3}$$

This equation belongs to a class of ill-posed Fredholm integral equations of the first kind and constitutes an underdetermined inverse problem, in which an unknown donor–acceptor distribution function $\rho_{DA}(r)$ is to be determined from the experimentally available $I_D(t)$. The solution to this problem requires adding some a priori information to the integral equation, which contains additional constraints on the possible solutions, to make the problem mathematically solvable. The a priori information used in numerous existing theories concerns an assumed analytical expression for $\rho_{DA}(r)$.

Traditionally (the Förster-type approach), $\rho_{DA}(r)$ is guessed from the expected donor and acceptor distributions in the investigated medium. Then, by using the D and A spectral parameters, the decay $I_D(t)$ is calculated on the basis of Eq. (4.3). $I_D(t)$ is related to the experimental decay function $F(t)$ by the convolution integral

$$F(t) = \int_0^t L(t')I_D(t-t')dt' \tag{4.4}$$

where $L(t)$ is an excitation pulse profile. Both $F(t)$ and $L(t)$ functions are available in a time-correlated single photon counting (TCSPC) experiment [1]. $I_D(t)$ is convoluted with the excitation pulse profile $L(t)$ and then compared with the experimental decay $F(t)$ in terms of the χ^2 goodness-of-fit criteria. The best-fit parameters of $\rho_{DA}(r)$ are found by minimising the χ^2 function. For example, the Förster-type result for a random n-dimensional distribution of acceptors around the donor ($\rho_{DA}(r)\sim r^{n-1}$) is

$$I_D(t) = \exp\left[-\frac{t}{\tau_D} - 2\gamma\left(\frac{t}{\tau_D}\right)^{n/6}\right] \tag{4.5}$$

where γ is proportional to the bulk concentration of acceptors. Similar approaches to fractal, Gaussian or cylindrical distributions have been used in many sensing applications.

The main drawback of approaches of this type is in assuming specific $\rho_{DA}(r)$ and then determining parameters of this distribution. Although after introducing an analytical formula the problem becomes well defined, thus solvable mathematically, we lose the generality of our search for $\rho_{DA}(r)$. Indeed, this approach is acceptable for the systems of known distributions, when we want to determine specific parameters, like concentration of the acceptor. However, when applied to complex and unknown distributions, this approach may lead to misinterpretation, as the recovered parameters of the assumed $\rho_{DA}(r)$ may have no meaning in the real system. For example, the fluorescence decays of the donor in a system with the Gaussian D–A distribution, but analysed using the Förster model (Eq. (4.5)), show increasing dimensionality n with increasing acceptor concentration.

In numerous sensing applications in our lab, we have used FRET for developing transition metal ion sensors based on the dye→metal ion and dye→ ligand:metal ion FRET [2–6] in a porous polymer matrix. The experimental data obeyed the Förster-type equations for the random distributions of acceptors in bulk solutions, but some deviations were observed for the donor–acceptor systems placed in the polymer. However, the Förster γ-type parameters (Eq. (4.5)) recovered in data analysis were still useful for sensing. The problem of the Förster-type models not being relevant to the experimental FRET systems became more severe in work on glucose sensing [7, 8]. In this sensor, glucose recognition occurs by competitive binding of glucose and acceptor-labelled polysugar to the donor-labelled protein. Thus, the change in glucose level does not affect the bulk concentration of acceptor, but changes its distribution in the system, contrary to Förster-type theories. Indeed, these models poorly fitted the experimental data. Although γ vs [glucose] calibration curves can be used for sensing purposes [8], it was clear that a new approach, dealing with the complex distributions in biological media, is required. Moreover, once such distributions can be determined then there is clearly new scope for imaging molecular events with extremely high (<1 nm) resolution.

The other drawback of the Förster-type approaches is redetermining of the parameters like τ_D or R_0, which could be measured in separate experiments (τ_D in the lifetime measurement of the donor and R_0 from the spectral overlap integral). Therefore, our strategic goal was to develop alternative experimental methods for determining the full structural information achievable by FRET and then use them to study the morphology and dynamics of nanostructures. We achieved this in a fluorescence nanotomography (FN) approach by careful and more adequate choice of the a priori information required to solve the inverse problem (Eq. (4.3)), without sacrificing the generality of $\rho_{DA}(r)$.

4.4 Fluorescence Nanotomography Theory

In FN, the master equation (Eq. (4.3)) is solved by including different additional information on the system. Instead of using the assumed analytical formula for $\rho_{DA}(r)$, Eq. (4.3) is supported by generic properties of the $\rho_{DA}(r)$ as a distribution function, information validated by the previous experiments, sample preparation and known parameters of the donor and acceptor. Consequently, the only assumptions regarding molecular distributions we make in FN include some limitations to the kinetics of FRET introduced during derivation of Eq. (4.3) [9]. These limitations result in the requirements of the low donor and low to average acceptor concentrations, which can be easily satisfied in most real systems.

In the first stage of the analysis, reconvolution is performed to reveal the donor impulse response function $I_D(t)$. A well-researched multiexponential representation and relevant reconvolution procedure can be used in a first attempt. However, a higher precision in determining the experimental $I_D(t)$ is required than in traditional lifetime data analysis, as instead of determining a few parameters of the assumed distribution, we aim to determine a whole $\rho_{DA}(r)$. Therefore, a more advanced approach of fluorescence lifetime distribution, e.g. one based on the maximum entropy method, seems to be most appropriate.

A few alternative theories of FN developed so far differ mainly in the choice of mathematical expression for $\rho_{DA}(r)$. It has to be general enough to accommodate all possible donor–acceptor distributions in the investigated medium. For example, in our recent approach [10], the infinite series of orthonormal Laguerre polynomials was used and an assumption was made that FRET was limited to donor–acceptor transitions only. The first version of the FN approach [9] included values of R_0 and τ_D, measured in separate experiments, as a priori information and enabled determination of $\rho_{DA}(R_0)$. This method applied to the glucose sensor data [11] provided essential information on the evolution of the D–A distribution on the appearance of glucose and enabled monitoring of changes in sol–gel pore sizes while shrinking [12]. Further development of the method [11] enabled determination of the $\rho_{DA}(r)$ in the region up to ~$2R_0$, offering much richer structural information. The new glucose studies [13] based on this approach revealed morphological details on the protein used for sensing. Also, we revisited the dye/metal-ion systems in polymers [14] and in phospholipid bilayers and found that our data did not fit the Förster models, as the actual donor–acceptor distributions significantly differed from random.

In this chapter we present a more recent, alternative version of FN, based on separation of variables, and discuss the constraints and new opportunities it may lead to.

4.4.1 An Inverse Problem

A simple rearrangement of Eq. 4.3 allows the FRET inverse problem to be expressed in the form of the Fredholm integral equation of the first kind

$$F_{DA}(t) = \int_0^\infty \rho_{DA}(r)K(r,t)dr \tag{4.6}$$

where

$$F_{DA}(t) = -t/\tau_D - \ln I_D(t) \tag{4.7}$$

is an experimentally available dependence, as it is determined by the donor response function $I_D(t)$. The time and distance kernel function

$$K(r,t) = 1 - \exp\left[-\left(\frac{R_0}{r}\right)^6\left(\frac{t}{\tau_D}\right)\right] \tag{4.8}$$

reflects a character of the FRET influence on the donor excited state kinetics. Figure 4.2 shows how this function depends on the donor–acceptor distance r for a number of moments of time after excitation.

As can be seen from Eq. (4.6) and Fig. 4.2, the $F_{DA}(t)$ at a given t is an integral over distance r of the product of the sought-for function, $\rho_{DA}(r)$, and the evolving in time function $K(r,t)$. Therefore, the $F_{DA}(t)$ contains information on $\rho_{DA}(r)$, but limited by the nature of FRET. Moreover, the efficiency of FRET, $E(r)=1/(1+(R_0/r)^6)$, also shown in Fig. 4.2, indicates that the distribution function can be experimentally determined only for $\sim R_0/2<r<\sim 3R_0/2$. Indeed, for the acceptors at $r<\sim R_0/2$ the FRET efficiency $E\cong 1$, which means that FRET occurs immediately ($k_{FRET}\gg 1/\tau_D$) after excitation and will be detected in TCSPC experiments of a typical sub-nanosecond resolution, just as a reduction in fluorescence

Fig. 4.2. The $K(r,t)$ and efficiency of FRET $E(r)=R_0^6/(R_0^6+r^6)$ as functions of the donor–acceptor distance. At the flat parts of this dependence, at small r and when r approaches $2R_0$, FRET is too fast and too weak, respectively, to provide essential structural information

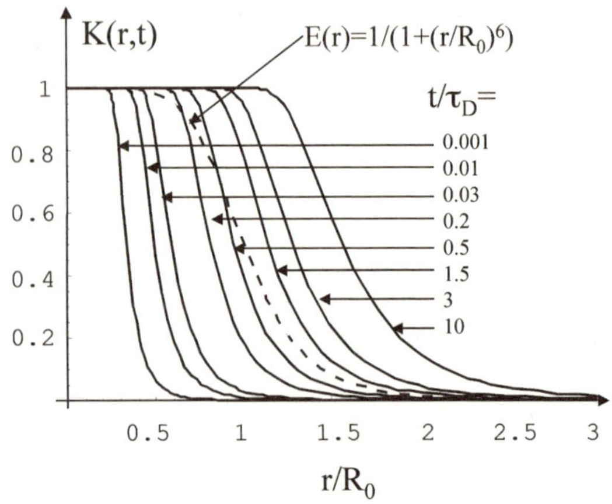

intensity. For the acceptors at $r>\sim 3R_0/2$ the efficiency of FRET is $E\cong 0$, thus $I_D(t)$ is only slightly affected by FRET and $\rho_{DA}(r)$ cannot be accurately determined at these distances.

To conclude, if any approach to lifetime data analysis recovers $\rho_{DA}(r)$ at $r<\sim R_0/2$ or $r>\sim 3R_0/2$, this is obtained by extrapolation of the result received for the $(\sim R_0/2, \sim 3R_0/2)$ region, or a relevant assumption regarding $\rho_{DA}(r)$ had been made in this approach. No genuine experimental evidence at the too short or too large D–A distances is available. In other words, a FRET pair characterised by a specific R_0 offers a spatial window $(R_0/2, 3R_0/2)$ where $\rho_{DA}(r)$ can be determined, and using a picosecond resolution in the experiment or improving the signal-to-noise ratio can broaden this window only slightly.

4.4.2 Separation of Variables Approach

Separation of variables in a kernel function Eq. (4.8) implies that there exist functions of time $f_n(t)$ and functions of distance $\phi_n(r)$, such that

$$K(r,t) = \sum_{n=0}^{\infty} f_n(t)\varphi_n(r) \tag{4.9}$$

The advantage of such an approach becomes clear when we replace the kernel function in Eq. (4.6) by Eq. (4.9), giving

$$F_{DA}(t) = \sum_{n=0}^{\infty} a_n f_n(t) \tag{4.10}$$

where

$$a_n = \int_0^{\infty} \rho_{DA}(r)\varphi_n(r)dr \tag{4.11}$$

Indeed, the parameters a_n can be determined from fitting the sum of components $a_n f_n(t)$, Eq. (4.10), to the experimental function $F_{DA}(t)$ and then can be used to get the function $\rho_{DA}(r)$. At this stage we have to specify $f_n(t)$ and $\varphi_n(r)$ functions. There are two options: to define $f_n(t)$ and then, using Eq. (4.9), calculate the resulting $\varphi_n(r)$, or vice versa, define $\varphi_n(r)$ and calculate $f_n(t)$. In this paper we express the $\varphi_n(r)$ functions by a series of orthonormal Legendre polynomials $P_n(x)$ and then calculate the relevant $f_n(t)$.

The novelty of this version of the analysis over our previous treatment [9, 10] is in assuming the "effective" $\rho_{DA}(r)$ of the form

$$\rho_{DA}(r) = \begin{cases} \rho(r), & r \leq \sigma R_0 \\ 0 \quad , & r > \sigma R_0 \end{cases} \tag{4.12}$$

where $\sigma \cong 2$. The assumption that $\rho_{DA}(r)$ vanishes at distances larger than σR_0 originates from the fact that FRET practically does not occur at donor–acceptor distances larger than σR_0; thus, it is not possible to determine $r_{DA}(r)$ at these distances in FRET experiments. On the other hand, focusing on the region of the donor–acceptor distances which have an impact on FRET, namely $0<r<\sigma R_0$, enables more precise determination of the $\rho_{DA}(r)$.

Inserting Eq. (4.12) into Eq. (4.6) gives

$$F_{DA}(t) = \int_0^{\sigma R_0} dr \rho(r) K_{DA}(r,t) dr \tag{4.13}$$

to express both the $\varphi_n(r(x))$ functions and the distribution function (see Fig. 4.3), namely

$$\rho(x) = \sum_{m=0}^{M} \rho_m P_m(x) \tag{4.14}$$

where the length of the series M has to be large enough to represent $\rho(x)$ properly, and

$$\rho_n = \frac{2n+1}{2} \int_{-1}^{+1} \rho(x) P_n(x) dx \tag{4.15}$$

leads to

$$F_{DA}(t) = \sum_{n=0}^{\infty} \rho_n G_n(t) \tag{4.16}$$

where

$$G_n(t) = \frac{\sigma R_0}{2} \int_{-1}^{+1} \left(1 - \exp\left[-\left(\frac{2}{\sigma(x+1)} \right)^6 \frac{t}{\tau_D} \right] \right) P_n(x) dx \tag{4.17}$$

is a series of standard functions, parametrically dependent on the experimental constants σ, R_0 and τ_D. Finally, inserting Eq. (4.16) into Eq. (4.7) gives a donor decay function

$$I_D(t) = \exp\left[-\frac{t}{\tau_D} - \sum_{n=0}^{\infty} \rho_m G_n(t) \right] \tag{4.18}$$

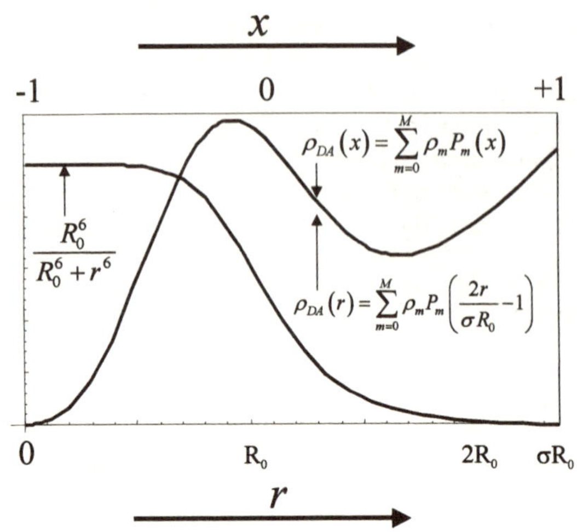

Fig. 4.3. A hypothetical distribution $\rho_{DA}(r)$ and the efficiency of FRET $E(r)$ as functions of the donor–acceptor distance using variables r and x

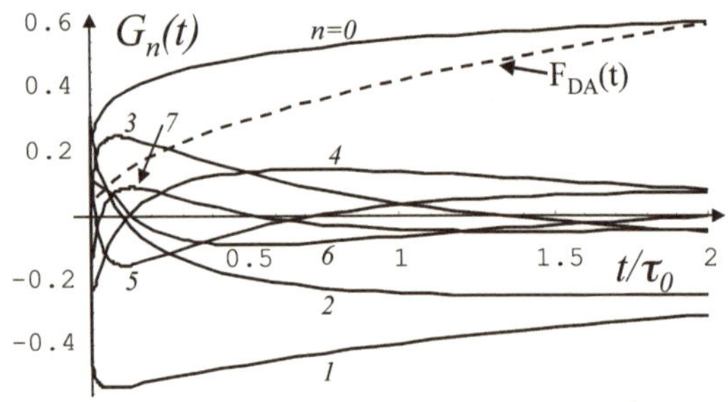

Fig. 4.4. $G_n(t)$ functions for σ=1.5 and R_0=1 Å. The $F_{DA}(t)$ function characteristic for the random distribution $\rho_{DA}(r) \sim r^2$ is also shown

Thus, the parameters ρ_n, introduced in Eq. (4.15) to express the distribution function, also constitute the representation of the experimental decay function $F_{DA}(t)$. This finding implies the method of recovery of the ρ_n parameters from the fluorescence decay. As the $G_n(t)$ are known functions dependent only on the previously determined donor lifetime τ_D, a critical transfer distance R_0 and a fixed σ value, the ρ_n parameters can be obtained by fitting the linear combination of the $G_n(t)$ functions to $F_{DA}(t)$, while $F_{DA}(t)$ is calculated from the experimental $I_D(t)$ according to Eq. (4.7) (Fig. 4.4).

4.4.3
Numerical Simulations

The derivation presented in the previous section sets up a method to recover the distribution function $\rho_{DA}(r)$ from the real fluorescence decay $I_D(t)$. The procedure includes the following steps:

1. Reconvolution of the raw data obtained from the fluorescence decay measurement.

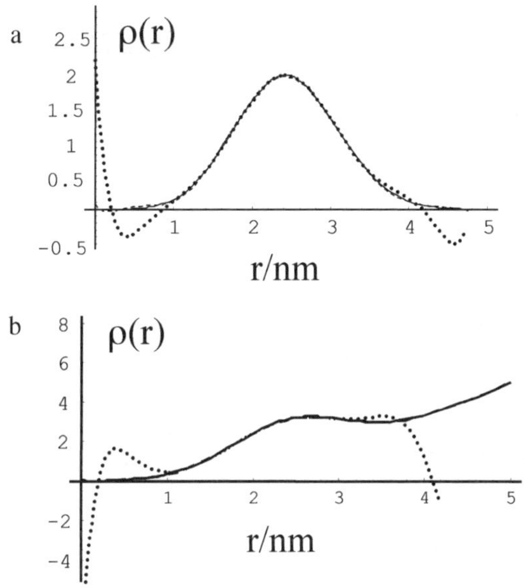

Fig. 4.5a,b. The assumed (*solid lines*) and recovered (*dotted lines*) Gaussian (**a**) and Gaussian+random (**b**) distributions. The critical transfer distance R_0 was 2.5 nm and τ_D was 2 ns

2. Calculating the $F_{DA}(t)$ function according to Eq. (4.7).
3. Fitting the parameters ρ_n according to Eq. (4.16).
4. Plotting $\rho_{DA}(r)$ using Eq. (4.14).

In the simulations below we assume that both the measurements and deconvolution were performed without systematic error. In order to assess what structural information can be recovered from the FRET-based fluorescence decay measurements, we focus on the accuracy of the recovery of the $\rho_{DA}(r)$ function from the $I_D(t)$. Therefore, in all cases described here, $I_D(t)$ was calculated for the given $\rho_{DA}(r)$, according to Eq. (4.3).

Figure 4.5 shows our findings obtained for the different types of distribution functions, namely the assumed and recovered distributions for the Gaussian (a) and Gaussian+random (b) distributions. In both cases the critical transfer distance R_0 was 2.5 nm and τ_D was 2 ns. It was found that the recovered distribution was in good agreement with the actual distribution for $0.4R_0<r<1.6R_0$ only. This limitation is not caused by the approach applied, but by the nature of FRET, as discussed in the previous section. In our previous theory on FN [10], the distribution functions recovered from the simulated fluorescence decays demonstrated a good agreement with the original distributions (for which fluorescence decays were simulated) at the small distances $r \ll R_0$, contrary to the results obtained here. However, in that case the distribution function was expressed as a series of Laguerre polynomials, and it was assumed to be zero at $0<r<a$, where a was a minimum donor–acceptor distance. Thus, the zero value of the recovered $\rho_{DA}(r)$ for $r \ll R_0$ was forced by mathematical formalism, rather than actually determined from the time-resolved experiment. In our opinion, the current approach is more appropriate, as we are interested in the recovery of the real $\rho_{DA}(r)$, not in modelling it, so any limitations regarding the shape of the distribution

function should be avoided. It must be clear, however, what is the minimum donor–acceptor distance r_{min} for which $\rho_{DA}(r)$ can be determined. Our simulations show that $r_{min} \cong 0.4R_0$.

4.5 Experimental

The applicability of the proposed FN method to real systems has been confirmed in a number of experiments. Although the approach is still a subject of some computational improvements, the principle of FN has proved to be effective in determining molecular distributions in a number of materials.

Time-correlated single photon counting (TCSPC) with sub-nanosecond resolution [1] was used to measure the fluorescence decay functions in the systems investigated. A number of pulsed excitation sources were used, according to the donor absorption bands. They included a Hamamatsu 670 nm diode laser and IBH NanoLEDs for 490 and 590 nm. The laser source was characterised by a half width of the instrumental response function of ~300 ps and a corresponding figure of ~1 ns for the LEDs. The excitation pulses were vertically polarised and the polariser at the magic angle was used in the detection channel in order to avoid orientational effects in the recorded decays. The IBH reconvolution software for multiexponential analysis was applied for the recovery of the $F_{DA}(t)$ functions.

4.5.1 Bulk Solutions

When donor and acceptor molecules are dissolved in a bulk solution, their mutual distribution is usually random. Therefore, the fluorescence decay of the donor is expected to be consistent with Eq. 4.3 (with $n=3$) and, consequently, the FN approach can be tested to reveal this random distribution, namely $\rho_{DA}(r) \sim r^2$. The donor was a near-infrared dye, DTDCI, which was quenched by hydrated copper ions $Cu \cdot 5H_2O^{2+}$ (acceptor). Propylene glycol was used as a solvent, in which the spectral overlap integral of the dye resulted in $R_0=19.5$ Å. Our previous research on this D–A pair [4] demonstrated a perfect consistency of the lifetime data with Eq. (4.3). The Förster behaviour of this system was also confirmed by FN. The expected parabolic ($\sim r^2$) distribution function obtained from FN is shown in Fig. 4.6. An inset presenting a $\rho_{DA}(r)$ vs r^2 dependence shows near-straight lines, confirming random distribution.

4.5.2 Porous Polymer Nafion 117

We used FN to investigate the structure of a perfluorosulphonated porous polymer Nafion 117, due to its previous use as a medium for the fluorescence-based sensors for a number of analytes [2–5, 14], including Cu^+, Cu^{2+} and Co^{2+} and a synthetic model of a tissue. The DTDCI-Cu^{2+} pair was chosen for this research because it exhibits a random distribution in a bulk solvent; thus, any change

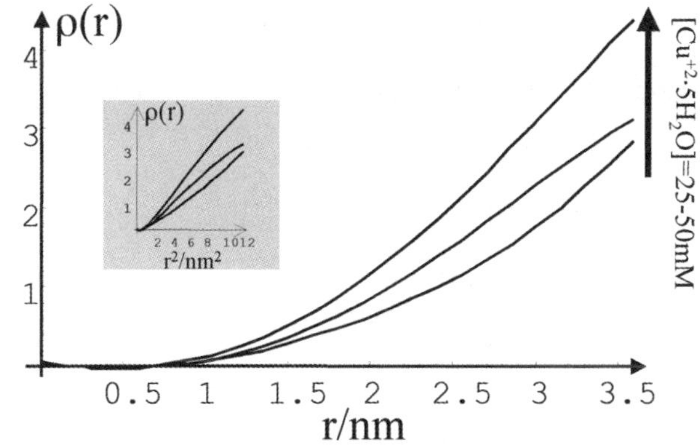

Fig. 4.6. The DTDCI-copper distribution functions in propylene glycol for increasing concentrations of copper ions

observed in a polymer can be attributed to the polymer structure. Also, the relatively low R_0 in Nafion, 19.2 Å [4], enables the use of high concentrations of copper up to 0.05 M, and is relevant to the structure of Nafion, which is believed to contain water pools of diameter ~40 Å and inter-connecting channels of diameter ~10 Å [15].

Figure 4.7 shows the distribution function determined from the fluorescence decays of DTDCI in Nafion, quenched by hydrated copper ions of different concentrations. Comparing this result with the one obtained in bulk solution, Fig. 4.6, gives an indication of the limited access of acceptor molecules to the donors, as the distribution function increases slower than according to $\rho(r) \sim r^2$. This finding explains our previous observations, based on the Förster analysis of a number of Nafion FRET measurements [4], that the critical transfer distance found according to Eq. (4.3) is less than the theoretical value predicted by the overlap integral (in the case of DTDCI and copper, 14.2 Å as compared to 19.2 Å). At present we interpret this deviation from random distribution as being due to the donors trapped in the Nafion channels, while metal ions are freely distributed within

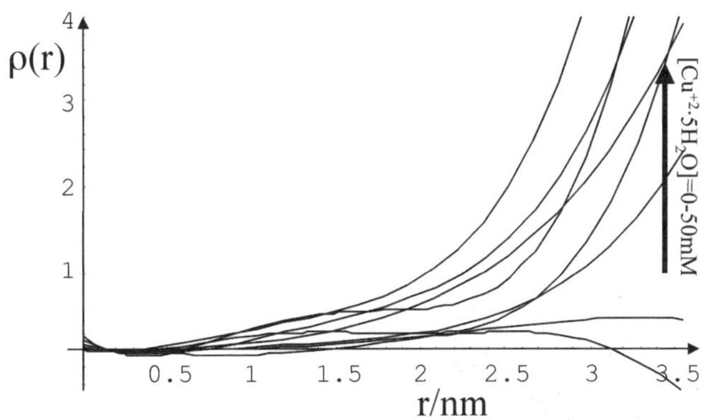

Fig. 4.7. The DTDCI-copper distribution functions in Nafion 117 polymer for increasing concentrations of copper ions

water pools, leading to an offset in their mutual distribution function. A similar result for the Nafion structure was obtained in our later studies using the Rhodamine B-cobalt FRET pair [14].

4.5.3
Phospholipid Bilayers

Phospholipid bilayers have been studied intensively using FRET for about 30 years because of their importance as models of cellular membranes. However, a Förster-type model of the decay functions used in these studies [16], derived for random distributions of acceptors, provided only limited information on these highly non-homogeneous structures. Using FN, we detected the donor–acceptor distribution in a phospholipid bilayer (DPPC) below and above the transition temperature. The BODIPY (Molecular Probes) donor and $Co \cdot 6H_2O^{2+}$ acceptor pair was used. The spectral overlap integral gives R_0=28.9 Å. The BODIPY molecule is thought to lie within the hydrocarbon region of the bilayer, while the hydrated cobalt ions, like many of the hydrated transition metal ions, are expected to be bound to the anionic lipid headgroups or remain free in solution.

Small DPPC unilamellar vesicles were prepared using methanol injection at 50 °C, which is above the phase transition temperature (T_g=41 °C). The lipid-to-probe ratio in all cases was 200:1. The concentration of the lipid in the buffer was 6×10^{-4} M; the buffer solution was PBS. The BODIPY fluorescence decay measurements were taken for different bulk concentrations of cobalt (0, 10, 50 and 80 mM) at two temperatures of the sample, 25 and 50 °C.

The resulting BODIPY-cobalt distribution functions are presented in Fig. 4.8. They show three main features: an ~15 Å offset of the distribution, maximum at about 30 Å at 25 °C, which increases to 35 Å at 50 °C, and parabolic (random) distribution at larger distances. The existence of an offset confirms different locations of the donors and acceptors, with the donor located within a bilayer. The

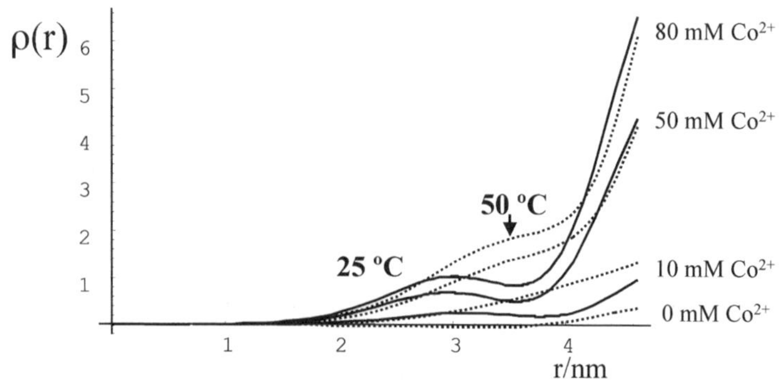

Fig. 4.8. BODIPY-cobalt distribution functions in phospholipid bilayer for 0, 10, 50 and 80 mM of Co^{2+} at temperatures below (25 °C, *solid lines*) and above (50 °C, *dashed lines*) the transition temperature

minimum D–A separation is about 15 Å, while the most probable distance is 30 Å (at 25 °C). The last number is probably related to the thickness of the bilayer, and the peak at 3 nm for $T<T_g$ suggests local clustering of $Co{\cdot}6H_2O^{2+}$ ions at this location. A parabolic dependence at larger distances corresponds to these cobalt ions, which remain free in the solution. Increasing the temperature above the transition temperature, T_g, modifies the distribution function. As the system becomes more random, the distribution becomes closer to the parabolic dependence of a random distribution. Further FN studies of DPPC and DMPC bilayers are currently in progress in our laboratory, but clearly Fig. 4.8 indicates the rich seam of new information which is potentially available using FN. Mother nature is unlikely to abide by the dictates of simple geometries, and FN would seem to offer an exciting opportunity for studying complex structures.

4.6 Conclusions

Time-resolved FRET experiments enable the recovery of more information on molecular distributions than is perhaps commonly appreciated. We demonstrated here that instead of determining only the parameters of the assumed $\rho_{DA}(r)$, the same lifetime data, when analysed in terms of fluorescence nanotomography, can produce real $\rho_{DA}(r)$ for $0.4R_0<r<1.6R_0$. By choosing the donor–acceptor pairs of the specific R_0, it becomes possible to detect or monitor changes in the real distribution function in the selected region of distances. This new opportunity of FRET-based sensing and imaging would provide a step change in fluorescence methodology, perhaps eventually leading to nanomovies of molecular events.

Acknowledgements
We wish to thank EPSRC and The Wellcome Trust for financial support. The work of Chinnaraj Mathivanan in providing fluorescence decay measurements in phospholipids is gratefully acknowledged.

References

1. Birch DJS, Imhof RE (1991) In: Lakowicz JR (ed) Topics in fluorescence spectroscopy, vol 1. Techniques. Plenum, New York, pp 1–95
2. Birch DJS, Rolinski OJ, Hatrick DA (1996) Fluorescence lifetime sensor of copper ions in water. Rev Sci Instrum 67:2732–2737
3. Rolinski OJ, Hatrick DA, Volkmer A, Birch DJS (1997) The photophysics of xanthene dyes in Nafion polymer membrane: application to copper sensing. J Fluoresc 7:207–209
4. Rolinski OJ, Birch DJS, Holmes AS (1998) Metal ion quenching kinetics of DTDCI in viscous solution and Nafion membranes: a model system for near-infrared fluorescence sensing. J Biomed Opt 3:346–356
5. Rolinski OJ, Birch DJS (1999) A fluorescence lifetime sensor for Cu(I) ions. Meas Sci Technol 10:127–136
6. Birch DJS, Rolinski OJ (2001) Fluorescence resonance energy transfer sensors. Res Chem Intermed 27:425–446
7. Rolinski OJ, Birch DJS, McCartney LJ, Pickup JC (2000) A time-resolved near-infrared fluorescence assay for glucose: opportunities for transdermal sensing. J Photochem Photobiol B 54:26

8. McCartney LJ, Pickup JC, Rolinski OJ, Birch DJS (2001) Near-infrared fluorescence lifetime assay for serum glucose based on allophycocyanin-labelled concanavalin A. Anal Biochem 292:216–221
9. Rolinski OJ, Birch DJS (2000) Determination of acceptor distribution from fluorescence resonance energy transfer. Theory and simulation. J Chem Phys 112:8923–8933
10. Rolinski OJ, Birch DJS (2002) Structural sensing using fluorescence nanotomography. J Chem Phys 116:10411–10418
11. Rolinski OJ, Birch DJS, McCartney LJ, Pickup JC (2001) Molecular distribution sensing in fluorescence resonance energy transfer based affinity assay for glucose. Spectrochim Acta A 57:2245–2254
12. Birch DJS, Geddes CD, Karolin J, Leishman R, Rolinski OJ (2002) Fluorescence nanometrology in sol-gels. In: Kraayenhof R, Visser AJWG, Gerritsen HC (eds) Fluorescence spectroscopy, imaging and probes. Springer Ser Fluoresc Appl 2:69–85
13. Rolinski OJ, Birch DJS, McCartney LJ, Pickup JC (2001) Fluorescence nanotomography using resonance energy transfer: demonstration with a protein-sugar complex. Phys Med Biol 46:N221-N226
14. O'Hagan WJ, McKenna M, Sherrington DC, Rolinski OJ, Birch DJS (2002) MHz LED source for nanosecond fluorescence sensing. Meas Sci Technol 13:84–91
15. Lee PC, Meisel D (1980) Luminescence quenching in the cluster network of perfluorosulphonate membrane. J Am Chem Soc 102:5477–5481
16. Holmes AS, Suhling K, Birch DJS (1993) Fluorescence quenching by metal ions in lipid bilayers. Biophys Chem 48:193–204

5 Solvent Relaxation as a Tool for Probing Micro-Polarity and -Fluidity

J. Sýkora, R. Hutterer and M. Hof

Keywords: Solvent relaxation; Time-correlated single photon counting; Time-resolved emission spectra

Abbreviations

Laurdan	6-Lauroyl-2-dimethylaminonaphthalene
n-AS	*n*-(9-Anthroyloxy)stearic acids
Patman	6-Hexadecanoyl-2-(((2-(trimethylammonium)ethyl)methyl)amino)naphthalene chloride
PC	Phosphatidylcholine
Prodan	6-Propionyl-2-dimethylaminonaphthalene
SR	Solvent relaxation
TCSPC	Time-correlated single photon counting
TRES	Time-resolved emission spectra

5.1 Introduction

Time-resolved fluorescence spectroscopy is one of the major tools for the characterization of microenvironments in supramolecular assemblies. While fluorescence techniques like quenching, energy transfer, lifetime (distributions), excimer formation, and determination of fluorescence anisotropy have already been applied in those studies for more than two decades, the solvent relaxation (SR) method found its application just recently. Despite several advantages of the SR method in probing micro-polarities and -"viscosities" using time-correlated single photon counting (TCSPC) equipment, the reader might not be familiar with this technique. Moreover, recent developments in pulsed picosecond laser diodes with high (MHz) repetition rates might lead to a more intensive application of this approach. Thus, we explain in this chapter the basic principles of the SR method and give some examples of its application.

5.2 Basic Principles of the SR Method

A short introduction into the matter of SR has already been given in Chap. 1. SR refers to the dynamic process of solvent reorganization in response to an abrupt change in charge distribution of a dye via electronic excitation. The temporal response can be monitored through the observation of the dynamic Stokes shift

$\nu(t)$ of the dye's emission maximum frequency. The complete time-dependent Stokes shift $\Delta\nu$ ($\Delta\nu=\nu(0)-\nu(\infty)$) increases with increasing solvent polarity. Linear proportionality between $\Delta\nu$ and a dielectric measure of the solvent polarity has been verified experimentally [1]. Provided that this proportionality has been characterized for a certain probe, the determination of $\Delta\nu$ may yield quantitative information on the polarity of the probed microenvironment, giving the first major information provided by the SR technique.

The second piece of information obtainable from the measurement of time-dependent spectral shifts is based on the fact that the SR kinetics is determined by the mobility of the dye environment. At ambient temperatures, a typical relaxation process $C(t)$ ($C(t)=(\nu(t)-\nu(\infty))/\Delta\nu$) in an isotropic polar solvent starts with a fast inertial motion on the 0.05 to 0.5 ps timescale, followed by rotational and translational diffusion occurring on the pico- to sub-nanosecond timescale [1]. An (integral) average SR time can be defined according to

$$\langle\tau_r\rangle \equiv \int_0^\infty C(t)\mathrm{d}t \tag{5.1}$$

which has been determined to be about 0.3 ps in pure water [2]. If the dye is located in a viscous medium, like in solvents at low temperatures, or is associated with supramolecular assemblies, like phospholipid membranes, micelles, or polymers, a substantial part of the SR is monitored on the nanosecond timescale, and is thus observable with TCSPC equipment. In vitrified solutions the solvation dynamics is much slower than nanosecond fluorescence decay times and thus the fluorescence might originate from the Franck–Condon state.

Although there have been several attempts to simplify the characterization of the SR process, the determination of the normalized spectral response function $C(t)$ is certainly the most general and most precise way to characterize the time course of the solvent response. The $C(t)$ function is usually determined by "spectral reconstruction". The primary data consist of a set of emission decays recorded at a series of wavelengths spanning the steady-state emission spectrum. The time-resolved emission spectra (TRES hereafter) are obtained by relative normalization of the fitted decays to the steady-state emission spectrum. Log-normal fitting of TRES yields various characteristic parameters, such as full width at half maximum height (FWHM) and emission maxima ($\nu(t)$) of TRES [1]. In order to get a complete picture on any SR process, precise knowledge about the emission spectrum of the Franck–Condon state is essential. The group of Maroncelli [1] showed that the frequency of the t=0 peak emission, $\nu(t_0)$, of a chromophore in the system of interest can be calculated quite accurately when the absorption and fluorescence spectra in a nonpolar reference solvent, as well as the absorption spectrum in the system of interest, are known [1]. The combination of the latter method, which is exclusively based on steady-state data and TCSPC, allows the characterization of SR processes in a large range of supramolecular assemblies.

5.3 Applications of the SR Technique by Using Time-Correlated Single Photon Counting

5.3.1 SR in Phospholipid Bilayers

A major requirement for valid application or physical interpretation of SR studies in bilayers is knowledge about the location of the chromophore used. It has been demonstrated [3, 5–7] that $\Delta\nu$ and τ_r are strongly dependent on the location of the chromophore within the bilayer. The SR kinetics of dyes like 6,8-difluoro-4-heptadecyl-7-hydroxycoumarin, which are probing the external interface of the bilayer, occurs on a rather short timescale. About 50% of the relaxation probed by 6,8-difluoro-4-heptadecyl-7-hydroxycoumarin in phosphatidylcholine (PC) small unilamellar vesicles at room temperature is faster than 50 ps. The second part of the SR, on the other hand, occurs on the nanosecond timescale, characterized by two relaxation times (τ_1=0.5 ns; τ_2=2.6 ns). This finding can be understood when considering the following facts. Firstly, the biologically relevant fluid phase of the bilayer is characterized by the intrinsic presence of dynamic fluctuations. Thus, the appropriate description for the positions of atoms in lipids is that of broad statistical distribution functions, and the membrane surface should not be considered as "flat". Secondly, the location of the ensemble of chromophores is characterized by a distribution along the z-axis of the membrane. Thus, at a certain time, some of the chromophores might face an environment characterized by a large amount of "bulk" water molecules and thus monitor picosecond or even sub-picosecond relaxation kinetics. On the other hand, a significant part of the chromophores might be located several angstroms closer to the hydrophobic part of the membrane and thus, as explained later in this article, probe SR kinetics on the nanosecond timescale. This rather complex relaxation kinetics is certainly limiting the validity of the SR approach for the characterization of physiologically relevant changes in micro-polarity and -viscosity within the external interface (Fig. 5.1).

SR in the headgroup region probed for example by 6-propionyl-2-dimethylaminonaphthalene (Prodan), 6-lauroyl-2-dimethylaminonaphthalene (Laurdan), or 6-hexadecanoyl-2-(((2-(trimethylammonium)ethyl)methyl)amino)naphthalene chloride (Patman) in PC bilayers is purely a nanosecond process and in all three cases almost monoexponential [3]. The resulting SR times are 1.0, 1.7, and 1.7 ns for Prodan, Laurdan, and Patman, respectively. The observation that an "ultrafast" component is missing indicates that there is no bulk water present in the environment of the chromophore of those dyes. The direct influence of the chemical composition of the headgroup on the SR characteristics indicates that the chromophores are surrounded by hydrated functional groups of the phospholipid molecules.

SR within the hydrophobic backbone region has been monitored by a set of n-(9-anthroyloxy)stearic acids (n-AS) [3, 5]. Again, SR is occurring on the nanosecond timescale. For example, the relaxation time assigned to the SR for 9-AS is about 4.0 ns [3]. In summary, we observe a slowing down of SR when starting

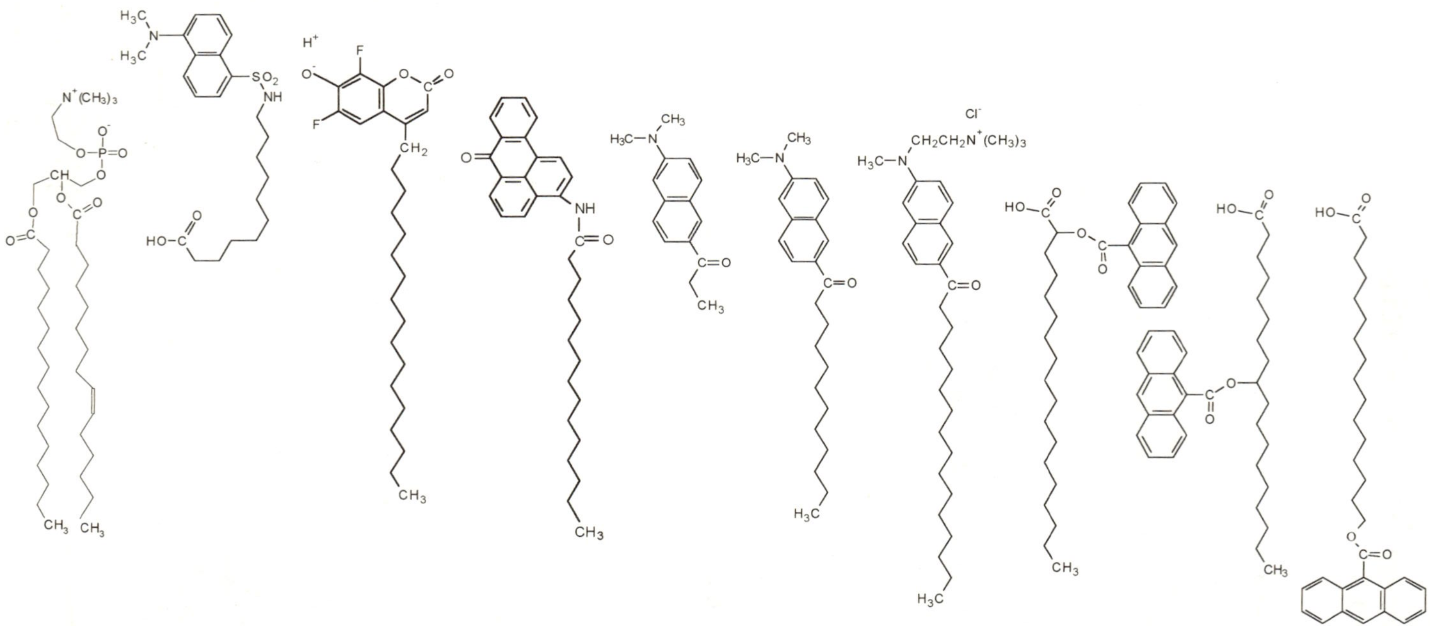

Fig. 5.1. Distribution of dyes used relative to the phosphatidylcholine molecule along the *z*-axis in the phospholipid bilayer. From the *right*: phosphatidylcholine, 11-((5-dimethylaminonaphthalene-1-sulfonyl)amino)undecanoic acid (Dauda) [3], 6,8-difluoro-4-heptadecyl-7-hydroxycoumarin [3], *N*-palmitoyl-3-aminobenzanthrone (ABA-C_{15}) [16], 6-propionyl-2-dimethylaminonaphthalene (Prodan) [3, 4, 7, 8], 6-lauroyl-2-dimethylaminonaphthalene (Laurdan) [3], 6-hexadecanoyl-2-(((2-(trimethylammonium)ethyl)methyl)amino)naphthalene chloride (Patman) [3, 4, 7–9], 2-(9-anthroyloxy)stearic acid (2-AS) [3, 5], 9-(9-anthroyloxy)stearic acid (9-AS) [3, 5], (16-(9-anthroyloxy)palmitoic acid (16-AP) [5]

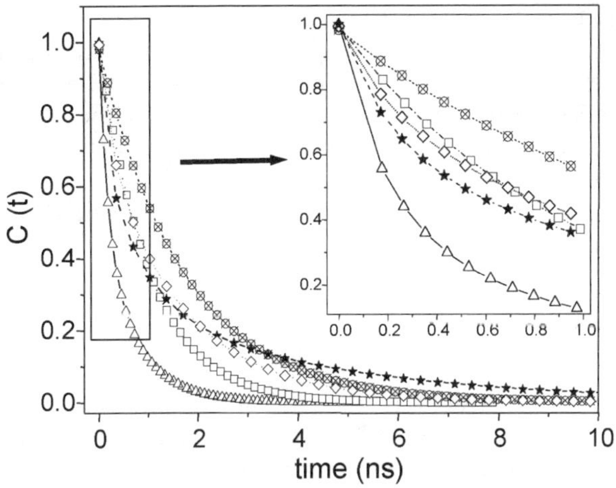

Fig. 5.2. Correlation functions $C(t)$ for C_{17}DiFU (△), Prodan (□), Patman (⊗), 2-AS (◇), and 9-AS (*) in PC small unilamellar vesicles at ambient temperature. Shown are the (multi)-exponential fits to the experimental data, using $\nu(0)$ obtained by the time-zero spectrum estimation [1]

from bulk water (τ_r in the sub-picosecond domain), passing the external interface (τ_r in the sub-nanosecond domain) and the headgroup region (τ_r in the nanosecond domain), and finally reaching the hydrophobic backbone (τ_r is equal to several nanoseconds) of the bilayer. A comparison of $\Delta\nu$ values for different membrane labels is only valid if the labels contain the identical chromophore. Comparison of three "Prodan"-like dyes [3] and five n-AS dyes [3, 5] demonstrates a decreasing $\Delta\nu$ with deeper location, which gives evidence for a polarity gradient within the bilayer in the liquid-crystalline phase. The SR technique has been shown to detect micro-"fluidity" (τ_r) and -polarity ($\Delta\nu$) changes in the bilayer due to temperature [4, 6, 7], ethanol addition [8], membrane curvature [4], and lipid composition variations [5–8] as well as due to the binding of calcium ions [5, 7], membrane-active peptides [9], and blood coagulation proteins [7]. A defined dye location provides the possibility to investigate selectively the protein interaction with different domains of the bilayer. It has been demonstrated that this technique is more sensitive to the binding of peripheral membrane proteins than standard fluorescence techniques, like anisotropy or excimer formation studies [7]. Some representative correlation functions $C(t)$ for C_{17}DiFU, Prodan, Patman, 2-AS, and 9-AS in PC small unilamellar vesicles at ambient temperature are given in Fig. 5.2.

5.3.2 SR in Reverse Micelles

Reverse micelles are often used to study the physical nature of confined water molecules. The Aerosol-OT/water/hydrocarbon system is certainly the most studied reverse micellar system. The SR kinetics of a large variety of probes has been characterized [10] and occurs apparently on the nanosecond timescale. Since the probes used are located within the water pool formed by the surfactant aggregates, the SR characteristics are reasonably independent of the probe. On the other hand, the SR kinetics depends on the size of the water pool, which might suggest a practical application of the SR method. However, the motivation for SR studies in reverse micelles so far is to elucidate the origin of slow relaxation kinetics in constrained water.

5.3.3
SR in Polymers

Recently, polyethylene glycol-based block polymers forming swellable macromolecular networks have been used as microreactors. The reaction rates in these microreactors are strongly affected by the local polarity and viscosity of the reaction centers. Substitution of the reactive centers by the SR probe dimethylnaphthylsulfonamide allowed the characterization of micro-polarities and -"fluidities" of reaction centers in the presence of different liquid phases [11]. It should be mentioned that there is a series of SR studies in hydrogels or sol–gel matrices. Despite the high bulk viscosity, the SR kinetics in those systems has been shown to occur on a timescale faster than 50 ps [12]. Low micro-"viscosity" together with high bulk viscosity in similar systems has also been found by recent fluorescence anisotropy studies [11].

5.3.4
SR in Ionic Liquid

In recent years the use of room-temperature ionic liquids as a replacement for conventional solvents for carrying out chemical reactions has been discussed. Their practical use, however, is still rather limited in view of the fact that many of their physical properties are unknown. In a recent paper the SR behavior of coumarin 153 was used to gain polarity and fluidity information on the room-temperature ionic liquid 1-butyl-3-methylimidazolium tetrafluoroborate. While the polarity of this system has been found to be similar to that of 2-propanol, the SR kinetics appears to be biphasic, occurring on the nanosecond timescale (0.3 and 4.0 ns, respectively) [13].

5.3.5
SR in DNA

Fluorescent DNA intercalators are distributed rather randomly throughout the (spatially inhomogeneous) DNA molecule and usually do not exhibit large Stokes shifts. For both reasons these dyes have only limited suitability for the defined probing of micro-polarity and -"fluidity" in DNA molecules. A specifically and covalently attached dye, however, may yield defined information on the probed microenvironment. A step in this direction is represented by a study on coumarin 480 covalently attached to DNA oligonucleotides [14]. The authors observed logarithmic relaxation for over three decades (40 ps–40 ns), indicating a complex relaxation among a large number of conformational substates.

5.3.6
SR in Proteins

There are several SR dyes which have been shown to bind rather nonspecifically to proteins. SR studies on such systems give rather unspecific information. A representative example of specific information gained by the SR technique is a study of the single tryptophan residue of the enzyme protein subtilisin Carlsberg [15]. Since the tryptophan residue is significantly exposed to water, the obtained SR behavior is rather fast with two relaxation times of about 1 and 40 ps. The discussed tryptophan residue is certainly an extreme example of an exposed tryptophan. Thus, studies on less exposed tryptophan residues might yield slower SR kinetics, which will be quantitatively detectable by the TCSPC approach.

Acknowledgement
The authors thank the Ministry of Education, Youth and Sports of the Czech Republic (via LN 00A032) for financial support.

References

1. Horng ML, Gardecki JA, Papazyan A, Maroncelli M (1995) Subpicosecond measurements of polar solvation dynamics: coumarin 153 revisited. J Phys Chem 99:17311–17337
2. Jimenez R, Fleming GR, Kumar PV, Maroncelli M (1994) Femtosecond solvation dynamics of water. Nature 369:471–473
3. Sýkora J, Kapusta P, Fidler V, Hof M (2002) On what timescale does SR in phospholipid bilayers happen? Langmuir 18:571–574
4. Hof M, Hutterer R, Perez N, Ruf H, Schneider FW (1994) Influence of vesicle curvature on fluorescence relaxation kinetics of fluorophores. Biophys Chem 52:165–72
5. Hutterer R, Schneider FW, Lanig H, Hof M (1997) SR behaviour of *n*-anthroyloxy fatty acids in PC vesicles and paraffin oil: a TRES study. Biochim Biophys Acta 1323:195–207
6. Hutterer R, Schneider FW, Hof M (1997) Time-resolved emission spectra and anisotropy profiles for symmetric diacyl- and diether phosphatidylcholines. J Fluoresc 7:27–33
7. Hutterer R, Schneider FW, Hermens WT, Wagenvoord R, Hof M (1998) Binding of prothrombin and its fragment 1 to phospholipid membranes studied by the SR technique. Biochim Biophys Acta 1414:155–164
8. Hutterer R, Hof M (2002) Probing ethanol-induced phospholipid phase transitions by the polarity-sensitive fluorescence probes Prodan and Patman. Z Phys Chem 216:333–346
9. Sheynis T, Sýkora J, Benda A, Kolusheva S, Hof M, Jelinek R (2003) Bilayer localization of membrane-active peptides studied in biomimetic vesicles by visible and fluorescence spectroscopies. Eur J Biochem 270:4478–4487
10. Bhattacharyya K, Bagchi B (2000) Slow dynamics in constrained water in complex geometries. J Phys Chem A 104:10603–10613
11. Egelhaaf HJ, Lehr B, Hof M, Hafner A, Fritz H, Schneider FW, Bayer E, Oelkrug D (2000) Solvation and SR in swellable copolymers as studied by time-resolved fluorescence spectroscopy. J Fluoresc 10:383–392
12. Birch DJS, Geddes CD (2000) Sol-gel particle growth studied using fluorescence anisotropy: an alternative to scattering techniques. Phys Rev E 62:2977–2980
13. Karmakar R, Samanta A (2002) Solvation dynamics of coumarin 153 in a room-temperature ionic liquid. J Phys Chem A 106:4447–4452

14. Brauns EB, Madaras ML, Coleman RS, Murphy CJ, Berg MA (2002) Complex local dynamics in DNA on the picosecond and nanosecond timescales. Phys Rev Lett 88:158101–158104
15. Kumar Pal S, Peon J, Zewail AH (2002) Biological water at the protein surface: dynamical solvation probed directly with femtosecond resolution. Proc Natl Acad Sci USA 99:1763–1768
16. Sýkora J, Mudogo V, Hutterer R, Nepraš M, Vaněrka J, Kapusta P, Fidler V, Hof M (2002) ABA-C15: a new dye for probing solvent relaxation in phospholipid bilayers. Langmuir 18:9276–9282

6 Total Internal Reflection Fluorescence Microscopy: Applications in Biophysics

N. L. Thompson and J. K. Pero

Keywords: Evanescent; Surface binding equilibria; Surface binding kinetics; Cell–substrate contact regions; Single molecule fluorescence; Fluorescence recovery after photobleaching; Fluorescence polarization; Fluorescence correlation spectroscopy; Fluorescence resonance energy transfer

Abbreviations

AFM	Atomic force microscopy
CCD	Charge coupled device
FRET	Fluorescence resonance energy transfer
P-TIRFM	Polarized total internal reflection fluorescence microscopy
TIRAF	Total internal reflection aqueous fluorescence
TIR-FCS	Total internal reflection fluorescence correlation spectroscopy
TIRFM	Total internal reflection fluorescence microscopy
TIR-FPPR	Total internal reflection fluorescence pattern photobleaching recovery
TIR-FRAP	Total internal reflection fluorescence recovery after photobleaching
VA-TIRF	Variable angle total internal reflection fluorescence

6.1 Introduction

6.1.1 Overview

Evanescent illumination may be combined with a number of techniques in fluorescence microscopy and employed in a wide variety of biophysical measurements. Progress in this field has been reviewed recently in several contexts [1–12]. This review is focused on areas in which various types of total internal reflection fluorescence microscopy (TIRFM) have been developed and applied to questions of interest in biophysics. Several important areas in which total internal reflection fluorescence has experienced recent development are not discussed in detail, including but not limited to optical biosensors [13], protein-substrate adsorption [14], and polymer-surface interactions [15].

6.1.2 Optical Principles

TIRFM is based on the surface-associated evanescent electromagnetic field that is created when light is internally reflected at a planar interface between two transparent materials with different refractive indices. Consider a plane wave traveling in the x–z plane in a higher refractive index medium which impinges on an interface with a lower refractive index medium, where the interface is defined as the x–y plane (Fig. 6.1). If the plane wave is incident on the interface at an angle (defined from the normal to the interface) greater than the critical angle (see below), the wave completely reflects into the higher refractive index medium. However, the reflection also generates a surface-associated "evanescent" field in the lower refractive index medium that propagates parallel to the interface and penetrates into the lower refractive index medium with a distance on the order of the incident light wavelength.

For an ideal plane wave, the evanescent intensity as a function of the distance from the interface is $I(z)=I_0\exp(-z/d)$, where I_0 is the intensity at the interface and d is the characteristic distance of evanescent penetration into the lower refractive index medium. The distance d equals $\lambda_0/[4\pi(n_1^2\sin^2\alpha-n_2^2)^{1/2}]$, where λ_0 is the vacuum wavelength of the incident light, and n_1 and n_2 are the higher and lower refractive indices, respectively. Parameter α is the incidence angle and the critical angle is given by $\alpha_c=\sin^{-1}(n_2/n_1)$. Typical values for these parameters are $\lambda_0\approx 500$ nm (visible light), $n_1=1.5$ (glass or fused silica), $n_2=1.33$ (water), $\alpha=75°$, $\alpha_c=65°$, and $d=69$ nm. The evanescent intensity at the interface, I_0, depends on the two refractive indices and the incidence angle. For typical conditions, I_0 is on the order of the intensity of the incident light [16, 17].[1]

The polarization of the electric field is somewhat unusual for evanescent waves created by total internal reflection as compared to freely propagating plane waves [17]. When the incident light is polarized perpendicular to the incidence plane (s-polarized; $\psi=90°$ in Fig. 6.1), the evanescent field is also polarized perpendicular to the incidence plane (along the y-axis). However, an atypical evanescent polarization is predicted when the incident light is polarized in the incidence plane (p-polarized; $\psi=0°$ in Fig. 6.1). Here, the evanescent electric field, although it lies completely in the incidence plane, contains components both perpendicular to the direction of evanescent propagation (along the z-axis) and parallel to the direction of propagation (along the x-axis). This latter electric field component defines the evanescent wave as being nontransverse. These as well as other properties of evanescent fields have been discussed previously in extensive detail [18–20].

The primary advantage of using evanescent illumination is that fluorescent molecules that are on or near the interface and in the medium with lower refractive index are selectively excited by the evanescent illumination. This feature has been used extensively during the past 25 years to examine a variety

[1] In fluorescence spectroscopy, one is usually interested in the intensity of the incident light as a proportionality factor for the excited fluorescence. This intensity is proportional to $|\mathbf{E}|^2$ where $\mathbf{E}$ is the electric field associated with the light. Conventionally, the intensity of an electromagnetic field is described by the Poynting vector, $\mathbf{S}\propto\mathbf{E}\times\mathbf{B}$ where $\mathbf{E}$ is the electric field and $\mathbf{B}$ is the magnetic field. For freely propagating, transverse light, $|\mathbf{S}|$ is proportional to $|\mathbf{E}|^2$. However, for evanescent light, this relationship does not apply. Therefore, the evanescent intensity $I(z)$ in the text, as described, is defined as proportional to $|\mathbf{E}|^2$ rather than $|\mathbf{S}|$ [17].

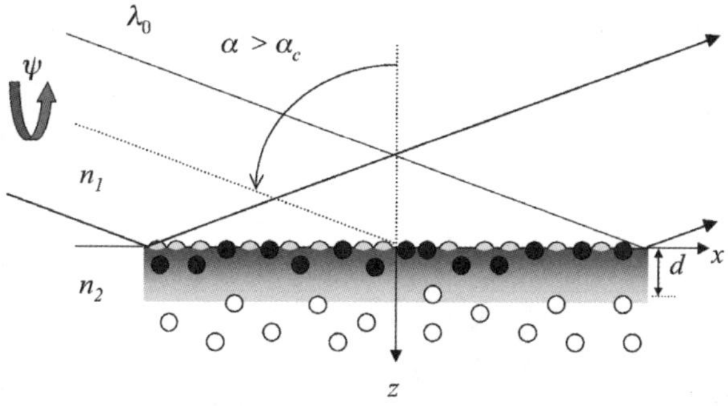

Fig. 6.1. Total Internal Reflection Fluorescence. A laser beam with vacuum wavelength λ_0 traveling from a high refractive index (n_1) medium into a lower refractive index (n_2) medium is totally internally reflected at a planar interface when the incidence angle α is greater than the critical angle α_c. The incidence plane is defined as the x–z plane and the interfacial plane as the x–y plane. Internal reflection creates a thin layer of light in the lower refractive index medium, called the "evanescent field", which decays exponentially in intensity along the z-axis with a characteristic depth *d*. The depth *d* depends on α, n_1, and n_2 (see text). The polarization of the incident light is defined by angle ψ measured from the incidence plane. When ψ=0°, the light is p-polarized; when ψ=90°, the light is s-polarized. Only molecules very close or bound to the surface are illuminated and subsequently fluoresce

of surface-associated phenomena. The unusual polarization characteristics of the evanescent electric field have also been useful for studies of orientation distributions and rotational mobilities at surfaces.

6.1.3 Apparatus

The past two decades have seen a rapid surge in the development and application of evanescent-related spectroscopies. This review is restricted primarily to the use of evanescent illumination in fluorescence microscopy (TIRFM). Experimental arrangements based on converted spectrofluorometers [21] or fiber optics [13] are not extensively discussed. Most, but not all, microscopy measurements employ one of two alignments organized around inverted optical microscopes [6, 7]. In the traditional scheme (Fig. 6.2a), a laser beam is brought over the microscope stage and internally reflected through a prism attached to the top of a sample chamber [22]. An alternative geometry (Fig. 6.2b) is to internally reflect a laser beam through a high-aperture objective [23, 24]. Two primary advantages of through-prism optics are the relative simplicity and reproducibility of optical alignment and a lower background signal arising from the fact that the incident beam does not enter the interior of the microscope. In addition, prior theoretical work has demonstrated that when a Gaussian-shaped laser beam is internally reflected through a prism, the characteristics of the evanescent wave are very similar to those of an internally reflected plane wave unless the beam is tightly focused [25]. Thus, quantitative data can in theory be more accurately interpreted. Advantages of through-objective optics include sample top accessibility,

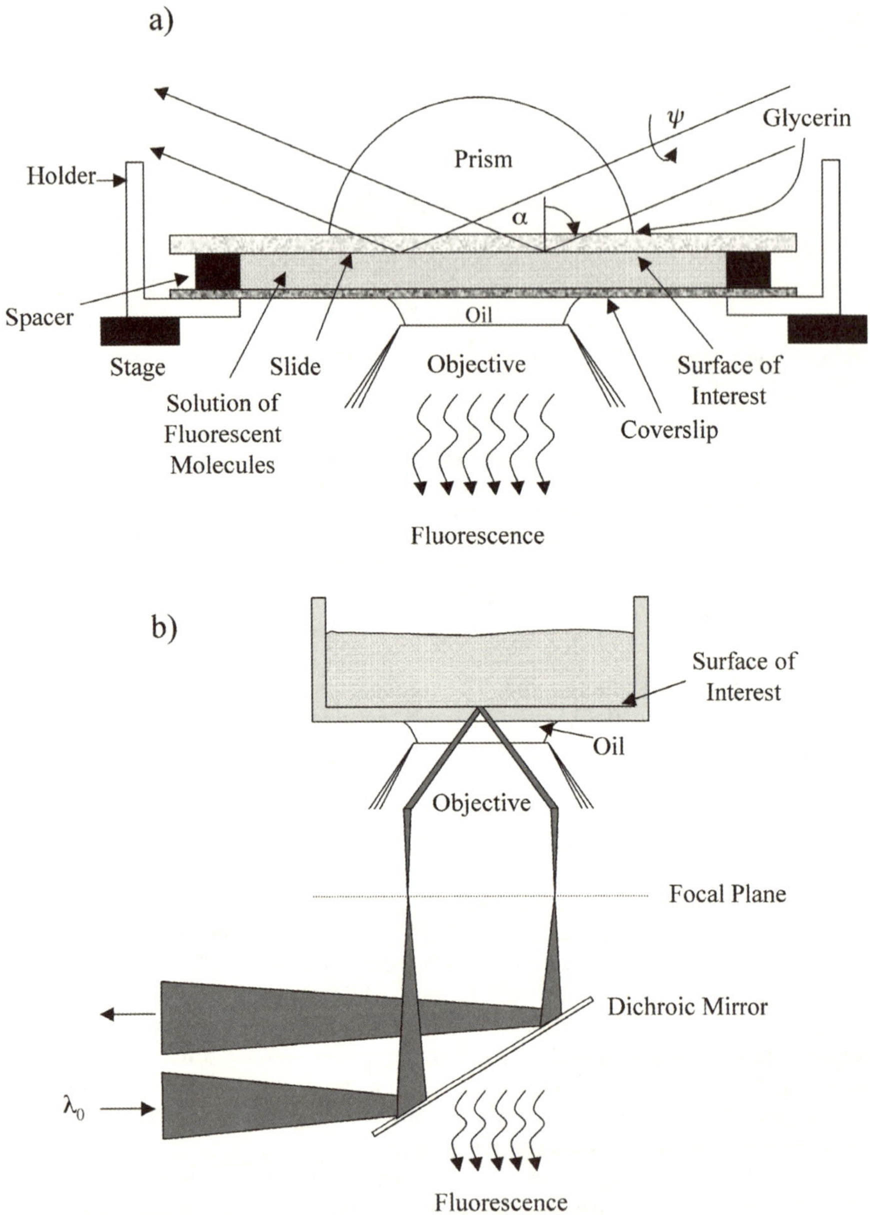

Fig. 6.2a,b. TIRFM Configurations. The two most commonly used methods for creating evanescent fields in fluorescence microscopy are to internally reflect a laser beam through **a** a prism or **b** a high numerical aperture objective. Both schemes use inverted optical microscopes

better image quality due to proper index matching of the objective with the coverslip and sample plane, and the fact that several systems are now commercially available. In both arrangements, the evanescently excited fluorescence is collected by the microscope objective and detected by a photomultiplier, a silicon avalanche photodiode, or an imaging device such as a CCD camera.

6.1.4 Sample Types

One type of sample to which TIRFM has been extensively applied is substrate-supported planar phospholipid membranes [26]. The fact that these model membranes can be deposited on transparent planar substrates makes them particularly amenable to evanescent illumination. Langmuir–Blodgett methods can be used to construct phospholipid monolayers on planar hydrophobic surfaces [27] or phospholipid bilayers on planar hydrophilic surfaces [28]. An alternative technique in which small, preformed phospholipid vesicles adsorb and fuse at planar hydrophilic surfaces has also been widely employed [29]. Two primary advantages of this method are its simplicity (relative to the Langmuir–Blodgett technique) and the fact that it can be used to incorporate functional membrane proteins into planar membranes [30]. A second type of sample on which TIRFM has been widely used is adherent cells, i.e., the use of the evanescent field to selectively excite cell–substrate contact regions [31]. TIRFM has also been used to examine other types of modified surfaces including, for example, thin polymer films [15] and surfaces containing adsorbed biological molecules such as membrane receptors [32] or DNA [33].

6.2 Combination of TIRFM with Other Methods

6.2.1 Fluorescence Recovery after Photobleaching

One of the first applications in fluorescence microscopy of using evanescent illumination was its combination with fluorescence recovery after photobleaching (TIR-FRAP) [34]. In this method, fluorescent molecules are in equilibrium between solution and binding sites on an adjacent, evanescently illuminated surface. Surface-bound molecules are bleached with a brief, intense pulse; as time proceeds, bleached molecules desorb and are replaced with unbleached molecules from solution or nonilluminated, surrounding surface areas (Fig. 6.3).

For a simple bimolecular reaction occurring at the surface, the rate and shape of the fluorescence recovery depend on four characteristic rates: the intrinsic dissociation rate, a rate related to surface rebinding, the rate for diffusion in solution through the illuminated area, and the rate for diffusion on the surface through the illuminated area [35, 36]. For large areas, the latter two rates do not contribute to the recovery curve. Furthermore, the slower of the two remaining rates dominates the fluorescence recovery. For high solution concentrations, low surface site densities, and rapid solution diffusion coefficients, the propensity for

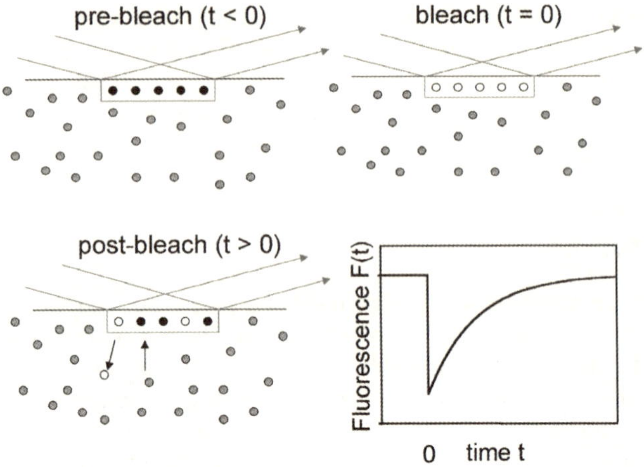

Fig. 6.3 TIR-FRAP. In TIR-FRAP, fluorescent molecules are in equilibrium between solution and sites on the evanescently illuminated surface. A brief, intense pulse of the laser photobleaches surface-bound molecules. Subsequently, the surface-associated fluorescence increases in time as bleached molecules desorb and are replaced by unbleached molecules from solution or the surrounding, nonilluminated area. The rate and shape of the fluorescence recovery curve contain information about the surface association and dissociation rate constants, as well as the diffusion coefficients in solution and on the surface. *Black circles* represent evanescently excited, unbleached molecules; *grey circles* represent unbleached molecules which are not evanescently excited; *white circles* represent bleached molecules

rebinding is reduced and it is possible to directly measure the intrinsic dissociation rate from TIR-FRAP data [37, 38]. For lower solution concentrations, higher surface site densities, and lower solution diffusion coefficients, TIR-FRAP data can be used to verify the existence of surface rebinding and quantitatively characterize this phenomenon of potential biological importance [39].

TIR-FRAP has now been used to examine surface kinetics in a variety of biochemical systems including antibodies at supported, haptenated planar membranes [37, 39], antibodies at supported membranes containing Fc receptors [38, 40], soluble Fc receptors at antibody-coated planar membranes [41], bovine prothrombin and its fragment 1 at planar membranes containing negatively charged phospholipids [29, 42], α-actinin at phospholipid Langmuir–Blodgett films [43], and protein 4.1 and spectrin at planar phospholipid membranes [44].

TIR-FRAP has also been used to monitor reversible kinetics of both extracellular and intracellular proteins with the plasma membranes of intact cells. In an early study, this method allowed measurement of the dissociation kinetic rate of fluorescein-labeled EGF with its specific receptors on the outer surface of intact but mildly fixed A431 carcinoma cells [45]. This early work provides a fairly comprehensive discussion of different experimental factors that need to be considered, as well as methods for doing so, when using TIR-FRAP on intact cells. In a subsequent investigation, TIR-FRAP was used to monitor the kinetics of protein 4.1 with cytoskeletal components presented by adhering and hemolyzing red blood cells at surfaces [46, 47]. These initial studies were extended in a very elegant and thorough investigation of the reversible association kinetics of microinjected, rhodamine-labeled G-actin and phalloidin with the cytoplasmic face of muscle-like cells [48]. Of particular interest in this work is the extension to the use of an imaging detector, which provided a spatial map of the membrane based on the kinetic dissociation rates as well as analyses in which the kinetic rates could be correlated with other factors such as the local ligand concentration. This area is likely to see increased development in the near future as it becomes more urgent to monitor the kinetics of biochemical reactions in intact, live cells. Two other applications of TIR-FRAP to intracellular dynamics include studies in which the method was used to monitor the diffusion coefficient of small fluorescent molecules close to the basal membrane of adherent MDCK epithelial cells [49, 50] and to monitor the exchange between membrane-bound and intracellular integrins at focal contacts of endothelial cells [51].

6.2.2
Evanescent Interference Patterns

When two laser beams collide at the point of internal reflection, a periodic pattern is generated in the evanescent intensity (Fig. 6.4). The nature of the interference pattern has been theoretically described in depth [52]. In the special case that the two beams have equal incidence angles and intensities, and are both either s-polarized or p-polarized, the intensity oscillation is along the y-axis (see Figs. 6.1 and 6.4) and has the form $I(y)=1+V\cos(2\pi y/P)$. The spatial period, P, equals $\lambda_0/[2n_1 \sin(\alpha) \sin(\varphi)]$ where 2φ is the collision angle. Parameter V, the visibility, is related to the pattern contrast and ranges from one to zero. The visibility depends on the polarization of the two incident and interfering beams, the incidence and collision angles, and the two refractive indices. When $n_1=1.467$ (fused silica), $n_2=1.334$ (water), $\alpha=75°$, $\lambda_0=500$ nm, and φ ranges from 2 to 45 to 88°, P ranges from 5.06 to 0.250 to 0.177 µm, V ranges from 0.998 to 0 to 0.998 for s-polarized light, and V ranges from 1.00 to 0.898 to 0.796 for p-polarized light.

One of the initial applications of interfering evanescent waves was as an experimental control to check for scattered evanescent light when examining the contact regions between rat basophil leukemia cells and antibody-coated planar membranes [53]. Subsequently, this optical arrangement was used to monitor lipid diffusion during the fusion of phospholipid vesicles with supported planar membranes [54, 55]. Interfering evanescent waves have been combined with fluorescence photobleaching recovery (TIR-FPPR) to determine the translational diffusion coefficients of molecules transiently and weakly bound to surfaces in a variety of biochemical and nonbiochemical systems [56–64]. More recent applications include the use of interfering evanescent fields to track small colloidal beads bound to actin filaments moving over surfaces coated with myosin [65], to enhance lateral resolution in imaging fluorescence microscopy [66], and to create lateral excitation profiles particularly suitable for examining rotational mobilities of molecules bound or close to an interface [67].

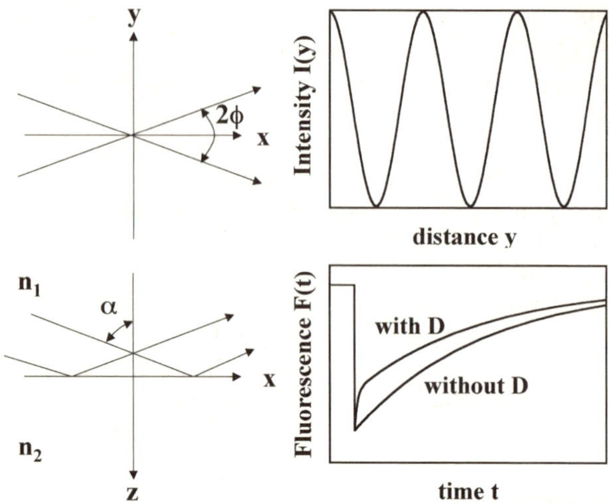

Fig. 6.4. Evanescent Interference Patterns. Two internally reflecting laser beams collide with angle 2φ in the interfacial plane. The evanescent intensity decays exponentially with distance z from the interface and also oscillates sinusoidally along the direction perpendicular to the incidence plane. The *lower right panel* illustrates theoretically predicted forms of TIR-FRAP recovery curves for systems in which a fluorescent ligand reversibly binds to surface sites [58, 161]. If the ligand also diffuses along the surface during its residency time, the data contain information about the diffusion coefficient. This coefficient can be obtained by measuring recovery curves for different collision angles φ, which produce interference patterns with different spatial periods P

6.2.3 Fluorescence Correlation Spectroscopy

Total internal reflection can also be combined with fluorescence correlation spectroscopy (TIR-FCS). In this method, a small sample volume is defined by the depth of an evanescent field created by an internally reflected laser beam and a confocal pinhole (Fig. 6.5). The fluorescence fluctuations from the sample volume are monitored and autocorrelated, and the shape of the autocorrelation function yields information about the rates of the processes causing the fluctuations. By fitting experimental autocorrelation data to theoretically predicted expressions appropriate for the system being studied, properties such as kinetic rate constants, surface-adjacent diffusion coefficients, and the average number of particles within the detection volume can be determined.

Although fluorescence correlation spectroscopy is a fairly mature technique [68–70, 10], its combination with evanescent illumination has thus far been limited to only a handful of experimental applications. TIR-FCS was initially demonstrated as a viable method by examining the nonspecific binding of tetramethylrhodamine-labeled immunoglobulin and insulin to serum albumin-coated fused silica [22]. More recently, TIR-FCS has been used to characterize the reversible adsorption kinetics of rhodamine 6G to C18-modified silica surfaces [71, 72], to examine local diffusion coefficients and concentrations of fluores-

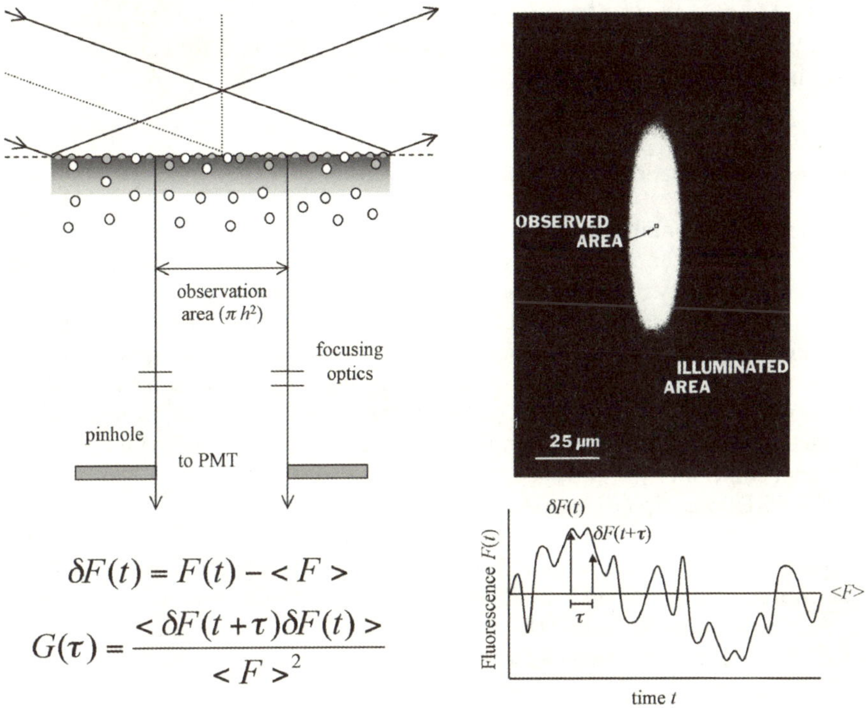

Fig. 6.5. TIR-FCS. A small sample volume is defined by the depth of the evanescent intensity in combination with a circular aperture placed at an intermediate image plane of the microscope that defines an area of radius h in the sample plane. The fluorescence measured from the small sample volume adjacent to the surface fluctuates with time as individual fluorescent ligands diffuse into the volume, bind to surface-associated receptors, dissociate, and diffuse out of the volume. These fluorescence fluctuations, $\delta F(t)$, are autocorrelated as $G(\tau)$. The autocorrelation function contains information about the surface association and dissociation rate constants, the diffusion coefficients in solution and on the surface, the density of surface-bound molecules, and the concentration of molecules in solution

cently labeled, monoclonal IgG in solution very close to substrate-supported phospholipid bilayers [73], to measure mass transport rates of small fluorescent molecules through thin sol–gel films [74], and to directly measure the dissociation kinetic rate for fluorescently labeled antibodies specifically and reversibly interacting with Fc receptors in substrate-supported planar membranes [75]. A number of fairly comprehensive theories for interpreting TIR-FCS data have also been presented [35, 76–78].

Areas of possible future development include application to a wider range of model membrane systems, measurement of kinetic rates associated with the reversible adsorption of intracellular, fluorescent molecules with the inner leaflet of the plasma membrane of adherent cells, and use of the method to screen for the kinetics of nonfluorescent compounds which compete with a fluorescent reporter for sites on immobilized membrane receptors of pharmacological interest. The possible uses of high-order autocorrelation or cross-correlation with

TIR-FCS have not yet been considered, theoretically or experimentally, but hold potential promise for a variety of applications (e.g., the detection and characterization of oligomerized or co-localized species).

6.2.4 Fluorescence Resonance Energy Transfer

Fluorescence resonance energy transfer (FRET) is a spectroscopic technique that monitors fluorescence quenching of a donor fluorophore by an acceptor molecule. The efficiency of energy transfer depends sharply on the distance between donor and acceptor and, consequently, can be used to examine distance-dependent events at the molecular level such as conformational changes, molecular association, and spatial co-localization. By combining evanescent illumination with energy transfer (TIR-FRET), these types of events can be monitored for surface-associated molecules while they are in chemical equilibrium with an adjacent solution.

TIR-FRET was first demonstrated as a viable method by observing a conformational change in doubly labeled bovine serum albumin upon nonspecific adsorption to a fused silica surface [21]. A similar, subsequent study used TIR-FRET to examine the molecular characteristics of IgG adsorbed to thin nylon films [79]. At the single molecule level, TIR-FRET was initially demonstrated by visualizing FRET from Cy3 to Cy5 attached to different chains in the α-tropomyosin coiled-coil [80]. Recently, a novel method of using a wedge prism to image the emission spectrum with a CCD, thereby possibly increasing throughput, has been developed and demonstrated by observing FRET between single, evanescently excited rhodamine red and Cy5 probes in a double-labeled myosin motor domain [81].

TIR-FRET is also applicable to molecular events in cell–substrate contact regions. This method has been used to demonstrate complex formation between a labeled, immobilized histocompatibility antigen (I-A^d) and a labeled peptide antigen, as induced by the binding of antigen-specific T cells [82]. Two more recent and very elegant studies have also employed TIR-FRET to monitor molecular events in intact cell membranes. In one study, conformational changes of hetero-tetrameric G protein coupled potassium channels were monitored by observing energy transfer between CFP and YFP conjugated channel subunits upon cellular activation [83]. In a separate study, certain molecular steps, including stoichiometries, preceding and accompanying cellular activation were followed by observing energy transfer between differently labeled EGF ligands when they bound to EGF receptors on adherent A431 carcinoma cells [84]. A technically interesting feature of this work is that it suggests that one can, at least in some cases, distinguish between events occurring on basal vs apical plasma membranes by changing the incidence angle of the incoming beam. It seems likely that TIR-FRET will see increased development in the near future, as this method is more widely used for dissecting the molecular details of signal transduction at the plasma membranes of intact cells and molecular events at the single molecule level.

6.2.5
Variable Incidence Angles

Variable angle total internal reflection fluorescence (VA-TIRF) has now been explored rather extensively in a variety of contexts. One context particularly noteworthy is the development of a quantitative method for accurate depth profiling of fluorophore concentrations next to interfaces. In the simplest analysis, the fluorescence measured as a function of the incidence angle α, $F(\alpha)$, is proportional to $I_0(\alpha)$ as well as the integral of the product $C(z)\ \exp[-z/d(\alpha)]$ over the region of $z=0$ to $z=\infty$. Here, $I_0(\alpha)$ is the intensity at the interface [16], $C(z)$ is the fluorophore concentration as a function of the distance z from the interface, and the exponential factor is from the z dependence of the evanescent intensity (Fig. 6.6). Because $F(\alpha)$ has the form of a Laplace transform, inverse transforming the measured fluorescence in theory gives a measure of $C(z)$ [85]. However, in practice, inverse Laplace transforms can yield results that are artifactual.

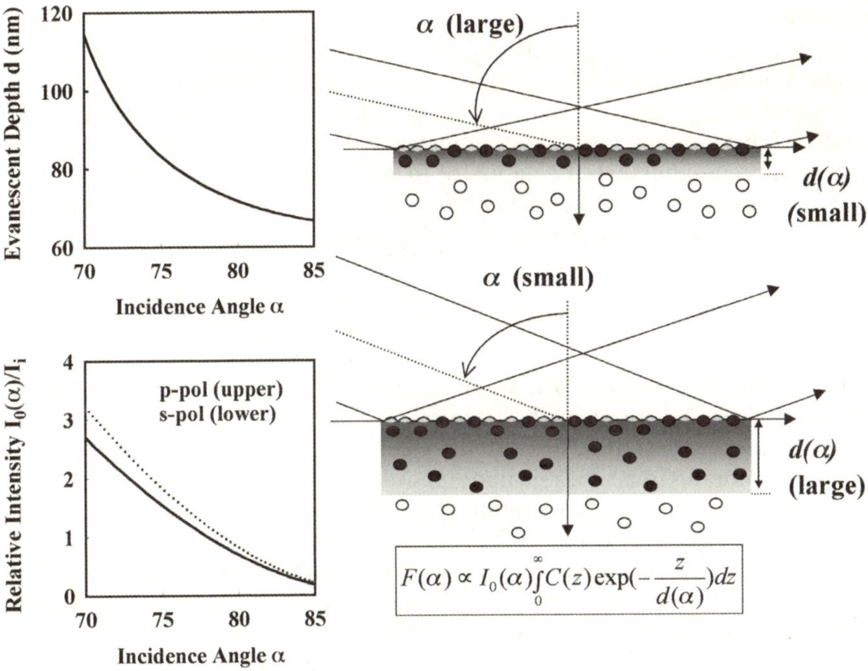

Fig. 6.6. Variable Angle Measurements. When the incidence angle α is decreased from 90° to the critical angle α_c, the evanescent wave depth $d(\alpha)$ increases (*upper left*), probing further into the lower refractive index medium (*upper and center right*). In the simplest case, the measured fluorescence $F(\alpha)$ has the form of a Laplace transform of the concentration of fluorophores as a function of the distance from the interface, $C(z)$, with $z \to [d(\alpha)]^{-1}$ (*lower right*). The fluorescence is also proportional to the evanescent intensity at the interface, $I_0(\alpha)$, which depends on the incident intensity I_i, the incidence angle, the two refractive indices, and the polarization of the incident light (*lower left*). Plots are shown for n_1=1.467 (fused silica) and n_2=1.334 (water), for which α_c=65.4°

Consequently, this method requires careful attention to a number of factors including the degree to which the accuracy of the data influences the accuracy of the inverse transform, the use when possible of reference samples, the possibility that the fluorophores or the medium in which they are embedded alter the local refractive index, the special emission properties of fluorophores next to interfaces (see below), possible deleterious effects arising from scattered evanescent illumination, and instrumental factors such as optical convolution in the x–y plane and image acquisition time [8, 85–93]. Alternative approaches to inverse Laplace transforming are to specify an approximate model for the theoretical form of $C(z)$ [85] or to examine the ratio of fluorescence intensities measured at two different incidence angles [94, 95], although these approaches do not necessarily circumvent all of the potential complicating factors listed above. Higher refractive index substrates are useful in VA-TIRF because they have a lower critical angle for internal reflection, and therefore enable a wider range of incidence angles and evanescent wave depths [89, 90].

VA-TIRF has been used to map out depth-dependent concentration profiles of fluorescein and fluorescein-labeled IgG next to surfaces [85], pyrene in thin polymer films [96], and fluorescein and acridine orange next to surfaces [97]. This method has also been used to map out the topology of cell–substrate contact regions by applying the technology to adherent cells with fluorescently labeled cytoplasms or membranes [89, 94, 98–104]. The goal in these studies has been to determine not only the separation distances between substrates and adjacent cell membranes, but also the nature of focal adhesion contacts. More recently, VA-TIRF has been applied to the problem of tracking secretory vesicle position and motion prior to fusion with the plasma membrane [90, 92, 105].

6.2.6
Inverse Imaging

An interesting variation of TIRFM is called total internal reflection aqueous fluorescence (TIRAF). In this method, the cell of interest is not itself labeled in the membrane or the interior. Instead, the volume surrounding the cell is infused with a membrane-impermeable fluorophore to provide a "negative" image [106, 107]. The basic concept underlying these measurements is that a cell, when close to a surface, will displace the soluble fluorophore, giving rise to a lower evanescently excited fluorescence. In a set of remarkable recent papers, TIRAF was used to monitor separation distances between bacteria and surfaces, with the goal of elucidating the forces governing bacterial adhesion [108] and, in time-resolved imaging studies, to track the separation distance with the goal of understanding the relationship between bacterial motility, basic physical forces, and surface adhesion [109, 110]. TIRAF has also been used successfully to characterize the stationary, short-term and long-term properties of mammalian cell adhesion [111].

6.3 Advanced Topics

6.3.1 High Refractive Index Substrates

The majority of TIRFM measurements have been carried out at the interface of fused silica and water. Three primary reasons for using fused silica are that it is transparent in both the visible and ultraviolet regions, it has a refractive index higher than that of water (n_1=1.467 vs n_2=1.334), and the background fluorescence is much lower than that of glass. In addition, fused silica permits immobilization of biomolecules either by covalent or noncovalent means, the formation of supported planar phospholipid membranes, and (under the proper conditions) cell adhesion and motility. One aspect of the fused silica/water interface is that, for experimentally reasonable incidence angles and visible light, the evanescent penetration depth ranges from 67 to 100 nm; higher refractive index substrates are predicted to generate thinner evanescent wave depths (Table 6.1). Smaller depths are advantageous because they are accompanied by an increase in z-axis resolution. Increased resolution allows TIRFM to be extended to studies measuring weaker equilibrium constants and their associated kinetics [16], to provide better angle-dependent resolution in VA-TIRF studies [89, 90], and to facilitate an increased understanding of the environment extremely close to biosurfaces. Some substrates with refractive indices in the visible region higher than that of fused silica that show promise are TiO_2 and $SrTiO_3$ (n_1=2.5) [112], $LiNbO_3$

Table 6.1. Evanescent Wave Properties for Different Substrate Refractive Indices

Index n_1	Angle α_c (deg)	Angle α (deg)	Depth d (nm)	$I(0)/I_i$ (p-polar)	$I(0)/I_i$ (s-polar)
1,467	65,4	70,4	110	3,10	2,60
		85,0	66,7	0,21	0,18
1,6	56,5	61,5	89,5	4,13	2,99
		85,0	45,6	0,13	0,10
1,8	47,8	52,8	75,7	5,38	3,24
		85,0	33,2	0,10	0,07
2,3	35,5	40,5	59,2	7,78	3,49
		85,0	21,4	0,08	0,05
2,5	32,2	37,2	56,0	8,57	3,55
		85,0	18,9	0,07	0,04

The critical angle, α_c, and the evanescent wave depth, d, were calculated by using equations given in the text. The evanescent intensities at the interface, I(0), relative to the incident intensity, I_i, were calculated by using equations given in reference [16]. The lower refractive index is n_2 = 1,334 (water). Higher substrate refractive indices n_1 have smaller critical angles and allow a larger range of incidence angles α. For the same incidence angle α, d decreases with n_1, $I(0)/I_i$ decreases with both α and n_1, and depends on the incident polarization.

(n_1=2.3) [113], sapphire (n_1=1.8) [90, 114], high-index glass (n_1=1.8) [7], and F2 glass (n_1=1.6) [89]. In many measurements, the ability of different substrates to support immobilized biomolecules, planar membranes, or adherent cells is crucial. For planar membranes, whether or not phospholipid vesicles adsorb and fuse at some of these different surfaces appears to depend critically on whether single crystal substrates or thin films deposited on other materials are employed, as well as on the liposome characteristics and the exact buffer and surface-treatment conditions; the ability of the substrates to support Langmuir–Blodgett films is a related, but different, physiochemical problem [112, 113, 115].

6.3.2 Thin Metal Films and Metallic Nanostructures

Also of significant interest in TIRFM is the use of thin metal films. In certain configurations, metal films can be used to suppress fluorescence arising from very near the interface, and thus to tune z-axis sensitivity away from nonspecifically adsorbed fluorophores and toward those bound to deposited cell fragments [116]. In separate and more recent work, it was found that small, conducting metallic structures deposited on surfaces can dramatically increase the intensity of nearby fluorophores [117–119]. This experimentally demonstrated, very significant metal-induced enhancement of fluorescence is somewhat related to the signal amplification seen in surface-enhanced Raman spectroscopy, in which the naturally low Raman signals are dramatically increased through the use of small metallic structures [120]. Because fluorescence signals even without amplification by small metal structures are already large, the possibilities in this area for fluorescence microscopy, including TIRFM as an intrinsic surface-based method, are quite intriguing. A recent work in which zero-mode waveguides consisting of sub-wavelength holes in a metal film were designed is also quite noteworthy [121]. One particular application in this area is that the new configuration permits studies of single molecule dynamics to be carried out at relatively high (micromolar), as opposed to much lower (nanomolar or picomolar), concentrations.

6.3.3 Fluorescence Emission Near Planar Dielectric Interfaces

The emission properties of fluorescent molecules next to planar dielectric interfaces, modeled as classical oscillating electric dipoles, are dramatically altered from the properties in homogeneous space. A long history of theoretical work has predicted that the nearby interface will cause significant changes in the angular dependence of the emitted fluorescence, the fluorescence lifetime, and the total collected fluorescence given a particular optical geometry [8, 86, 122–127]. These surface-induced changes in fluorescence emission depend on a variety of factors including the emission dipole orientation relative to the interface, the distance from the surface to the dipole, the quantum efficiency of the fluorophore in homogeneous space, and the optical details of fluorescence collection. Careful study of this phenomenon is often required in TIRFM studies. As one example, it has recently been reported that, for certain conditions, when fluorescence is collected through the half-space of lower refractive index, the effective evanescent wave depth is increased because the collected fluorescence increases as a function of z. In contrast, when fluorescence is collected through the half-space of higher refractive index, the effective evanescent wave depth is decreased because the collected fluorescence decreases as a function of z [6]. In the future, it is expected that more thorough consideration of the effects of the nearby interface is likely to stimulate significant new advances in TIRFM. Interfacial effects are expected to be stronger near high-index substrates or thin metal films [128].

6.3.4 Fluorescence Polarization

Evanescent illumination may also be used to obtain information about the orientation distribution of a population of fluorophore transition dipoles at or near an interface (P-TIRFM). In the simplest form of this method, the linear absorption dichroism is measured by examining the dependence of the evanescently excited fluorescence on the polarization of the evanescent field. As the polarization of the incident laser beam is rotated through the plane normal to its direction of propagation, the polarization of the evanescent field rotates (primarily) about the direction of propagation of the evanescent field (Fig. 6.1). The polarization of the evanescent field therefore probes the polar angle (with respect to the normal to the interface) of the dipoles. The use of polarized evanescent excitation is thus particularly applicable to interfacial samples because, in most of these samples, the orientational anisotropy averaged over an optically accessible area is found primarily in the polar rather than the azimuthal angle relative to the normal to the interface. The conceptual and theoretical basis of using P-TIRFM for characterizing transition dipole orientation distributions at surfaces has been extensively developed, including descriptions of measurements not only of the linear absorption dichroism, but also the polarization of the emitted fluorescence [67, 86, 91, 125, 126, 129, 130].

The use of P-TIRFM to examine the orientation distribution of fluorophores at surfaces has a variety of applications. First, when these measurements are made on fluorescent molecules in supported planar membranes, the results may be used to confirm the existence of long-range order in the samples and therefore to provide evidence for membrane integrity. Second, this technique may also be used to monitor changes in fluorophore orientation distributions that occur in response to environmental changes. Third, orientational information should allow more thorough interpretation of energy transfer measurements, which can provide information about conformational changes that occur upon adsorption or other parameters that report molecular structure at interfaces. Finally, orientational measurements give crucial information for interpreting dynamic fluorescence polarization data that can provide information about molecular rotational mobility and/or segmental flexibility at interfaces.

P-TIRFM has been experimentally applied to a variety of fluorescent lipids in supported planar membranes [91, 129, 131–133] as well as cytochrome c at different surfaces [134–138] and plasminogen at modified surfaces[139]. Of particular interest in the future will be the extension of P-TIRFM to the single molecule level [11, 140, 141]. Also of interest is the application of P-TIRFM to intact cells. In a recent and very elegant work, this method was used to vividly image spatial variations in membrane curvature of adherent erythrocytes and macrophages [142]. P-TIRFM has also been exploited in measurements of the dynamics of intracellular secretory vesicles at the plasma membranes of retinal bipolar cells [143].

6.3.5
Fluorescence Lifetimes and Time-Resolved Anisotropies

Like most fluorescence-based investigations, TIRFM studies can be enhanced by measurements of fluorescence lifetimes and/or time-resolved anisotropies. Potential applications of time-resolved fluorescence measurements in biological systems include characterization of rotational mobilities, elucidation of local environmental features, improved accuracy in FRET, and cell membrane imaging. Lifetimes and time-dependent anisotropies of fluorophores close to surfaces have been measured by using evanescent illumination with both time-domain and frequency-domain instruments [144–146]. Many nonbiological applications have used this approach to examine the nature of small fluorescent dyes at various solid/liquid and liquid/liquid interfaces [97, 147–152]. More biological studies have used time-resolved TIRFM to examine the internal flexibility and rotational mobility of proteins adsorbed to surfaces [153–155]. Time-resolved TIRFM has also been used in a set of impressive studies on intact cells. In an early study, the viscosity of the cytoplasm very near the plasma membranes of 3T3 fibroblasts and MDCK cells was determined by monitoring the rotational mobility of small fluorophores introduced into the cell interior [156]. More recently, time-gating has been used to distinguish between fluorescent components with different lifetimes [157]. In one noteworthy example, time-gating significantly reduced background from short-lifetime cellular autofluorescence when used with long-lifetime Tb chelate probes [158, 159]. The combination of TIRFM and fluorescence lifetime imaging holds particular promise for future studies.

6.3.6
Two-Photon Excitation

It has been demonstrated experimentally that evanescent light can excite fluorophores by two-photon absorption [160]. When fluorophores are excited by two-photon absorption, the excitation probability is proportional to the square of the excitation intensity. Thus, the characteristic depth of penetration, d, for two-photon excitation is halved. This halved depth of penetration theoretically gives a twofold better discrimination for surface-adjacent fluorophores as compared to those in solution. However, for the same fluorophore, excited by one-photon or two-photon absorption, the wavelength is approximately doubled in the latter case, giving rise to a depth d which is doubled. Therefore, two-photon absorption is not predicted to significantly enhance surface selectivity. In the simplest analysis, the halving and doubling factors work in opposite directions. Nonetheless, it is possible that other advantages accompanying two-photon excitation, such as reduced cellular autofluorescence, might make this type of spectroscopy useful in evanescent wave studies. One example of a potential application is the use of two-photon excitation with evanescent interference patterns. Because the shape of the excitation intensity in TIR-FPPR is sinusoidal (see above), the square of the intensity is sharper, leading in theory to more accurate extraction of lateral mobility information from TIR-FPPR data [161]. This feature has also been exploited to increase resolution in fluorescence imaging methods using standing evanescent waves [162]. Other applications of two-photon evanescent excitation might also be found in measurements relying on the evanescent polarization [163].

6.4 Other Applications

6.4.1 Single Molecule Imaging

One area of TIRFM that has shown dramatic advances in the past several years is single molecule fluorescence studies [2]. Single molecule measurements are important because bulk spectroscopy provides information about average molecular behavior, but is unable to probe the nuances of individual molecular events. TIRFM has been actively applied in this area for at least two reasons: the background fluorescence can be significantly reduced over some epi-illumination formats, and the method allows one to observe molecules in aqueous, and therefore more biologically relevant, environments [164, 165]. Numerous interesting studies have been published in this field. In many of them, time-lapse fluorescence imaging was used to examine dynamic behavior such as enzyme kinetics (e.g., ATP turnover) or lateral motion (e.g., along biological filaments). A body of initial work demonstrated feasibility and focused on detection details [166, 167], as well as the behavior of individual motor proteins interacting with substrates and regulatory factors, primarily myosin [168–170] and kinesin [171–174].

Single molecule TIRFM has also been used to examine the interaction between the chaperonin GroEL and substrate proteins [175], and the interaction of glucosyltransferase and dextran [176]. Applications to nucleic acid biophysics include a study of the properties of DNA immobilized on surfaces [177], an examination of the interaction of RNA polymerase with DNA [178], and the construction of a dual-beam instrument for watching photoinduced cleavage of individual DNA molecules [179]. Two additional works of interest include the combination of TIRFM with current recordings from single ion channels reconstituted into novel agarose-coated supported bilayers [180] and an observation of single chemical reactions as they produced individual molecules [181]. The connection between single molecule and ensemble measurements for ATP turnover by myosin has also been carefully examined [182]. Some studies have been extended to spectral acquisition; e.g., full emission spectra have been recorded for supramolecular light-harvesting complexes from a photosynthetic bacterium [183, 184] and time-dependent spectral variations arising from a single fluorophore conjugated to a myosin subfragment were observed and attributed to slow protein conformational changes [185]. In a striking work, myosin V was labeled at different positions and tracked as it moved over actin filaments while hydrolyzing ATP. By analyzing fluorescent peaks arising from single myosin molecules to determine their center positions with very high accuracy, a definitive body of data was accumulated supporting the "hand-over-hand" vs "inch-worm" model of myosin mobility. This work used evanescent excitation to localize observation to the small thickness surrounding the immobilized actin plane [186]. The combination of TIRFM and single molecule fluorescence measurements is a growing and fertile area.

6.4.2
Imaging Cell–Substrate Contact Regions

One of the earliest demonstrations of TIRFM as a method for examining processes of importance in biological systems used evanescent illumination to image the regions of contact between cultured cells and substrates to which they were adhered [31]. This work demonstrated that one could selectively probe the cell–substrate contact region as a function of distance from the interface by changing the incidence angle of the internally reflecting light, thereby changing the characteristic depth of evanescent penetration. In two other early works, TIRFM was used to examine the regions of contact between 3T3 fibroblasts and surfaces [94] and between rat basophil leukemia cells and IgE-coated supported planar phospholipid membranes [53].

The role of the structural organization of cell-surface receptors during cellular response to external signals has long been a question of importance in many areas of biology. Because TIRFM allows selective visualization of plasma membrane processes in adherent cells, this method has significant potential for assisting in the dissection of the manner in which oligomerization and/or conformational changes of receptors precedes, accompanies, or follows cellular activation. The possibilities of this strategy have been demonstrated in a variety of works. Time-lapse video recording using evanescent illumination allowed examination of the lateral organization (in particular, clustering) of fluorescently labeled acetylcholine receptors in the membranes of developing rat myotubes during their primary culturing [187]. More recently, single EGF receptors on A431 carcinoma cells during cell stimulation were imaged with TIRFM [84]. TIRFM has also been used to map out the distribution of oligomer sizes for GFP-conjugated cadherin on mouse fibroblasts [188] and for integrins tagged with labeled antibodies on endothelial cells [51]. Structural rearrangements of single, fluorescently labeled ion channels in *Xenopus* oocytes have also been characterized [189], as well as the spatial organization of GFP-dynacortin with respect to the cytoskeleton of *D. discoideum* cells [190].

Intracellular events accompanying cellular activation have also been examined by using TIRFM. In an initial demonstration of this approach, fluorescent calcium indicators were used to compare the relative amplitude and phase of transient changes in intracellular calcium concentrations in the bulk cytosol and close to the plasma membranes of neutrophils stimulated by the chemoattractant peptide *N*-formyl-met-leu-phe [191]. TIRFM has also been used to follow membrane-proximal calcium concentration changes in hybrid neuroblastoma-glioma cells stimulated by uncaging intracellular calcium or IP_3 [192] and in astrocytes stimulated by glutamate [193]. The time dependence of the translocation of GFP-conjugated protein kinase C molecules to the cytoplasmic membrane leaflet, both in glutamate-stimulated astrocytes [193] and in glucose-stimulated pancreatic islet β-cells [194], has also been examined. Other similar works have examined the membrane-proximal behavior of GFP-tagged chemotaxis signal proteins in bacteria [195] and of GFP-tagged pleckstrin homology domains in fibroblasts [196] and myoblasts [197].

Another topic of significant interest in a variety of biological contexts is the manner in which multivalent interactions modulate cellular adhesion to surfaces. In a series of papers, TIRFM was used to monitor the regions of contact between endothelial cells and various types of surfaces coated with fibronectin or RGD peptides while the cells detach under stress induced by flow. These measurements have application to the design of nonthrombogenic synthetic vascular grafts [101, 104, 198, 199].

6.4.3
Exocytosis and Secretion Vesicle Dynamics

One area in which time-lapse TIRFM imaging has been extensively applied in recent years is the interaction of small vesicles with the plasma membranes of intact cells. TIRFM is particularly applicable to this problem because the method focuses specifically on membrane-proximal regions and because the vesicles are larger and brighter than single molecules and therefore readily tracked as they approach and fuse with the cytoplasmic membrane leaflet. A large number of studies in this area have addressed secretory vesicle fusion with the plasma membranes of secretory cells. These studies employed GFP constructs, acridine orange, or fluorescent lipids to label the secretory vesicles and were carried out primarily either in PC12 neuronal cells [200–203], bovine chromaffin cells [204–211], or retinal bipolar cells [143]. In a somewhat related work, TIRFM enabled tracking of GFP-tagged secretory insulin granules in MIN6 β cells and correlation of exocytotic properties with the method of cell triggering [212].

Evanescent illumination has also been used to observe lysosome fusion with the plasma membrane as triggered by calcium in nonsecretory cells, i.e., constitutive exocytosis with the function of membrane repair. In this study, a large variety of GFP-labeled intracellular organelles were examined including the endoplasmic reticulum, the Golgi apparatus, post-Golgi vesicles, early and late endosomes, and lysosomes [213]. In a separate work, TIRFM assisted with visualization of the sorting and trafficking of tubulovesicular transport containers labeled with fluorescent proteins at the apical and basolateral membranes of polarized MDCK and nonpolarized PtK_2 cells [214]. The interaction of post-Golgi carriers labeled with GFP-tagged vesicular stomatitis virus glycoprotein with the plasma membranes of COS-1 [215] and PtK_2 cells [216] has also been characterized by using evanescent illumination. Finally, TIRFM has been used to probe possible coordination between exocytotic and endocytotic events. In this work, which was carried out in PC12 cells, secretory and endocytotic vesicles were labeled, respectively, with synaptobrevin and dynamin I [217].

6.4.4 Emerging Methods

Several areas of current development hold significant promise for future applications. First, because most TIRFM measurements employ a planar interface between a transparent solid and aqueous solution, the possibility exists that optically transparent electrodes (e.g., indium tin oxide) might be used to generate external electric fields across the interface. For example, P-TIRFM measurements have been carried out for cytochrome c [135] and planar membranes [133] deposited on such surfaces. In a similar work, the effects of applied potentials on protein adsorption were examined [218]. Second, the combination of AFM and TIRFM is likely to enable new types of biophysical measurements. In three recent works, dual-purpose AFM/TIRFM instruments were constructed and used to examine the effects of applied forces on cell–substrate contacts [219], to microinject fluorescent particles into adherent cells [220], and to examine radiation pressure effects on dielectric spheres in evanescent fields [114]. Third, microfluidic devices will increase the throughput of TIRFM measurements [221–225]. Finally, evanescent excitation may provide significant advantages in the fields of genomic and proteomic microarrays, particularly in situations where weak interactions are probed [218, 226–230].

6.5 Summary

The use of TIRFM in biophysics has been reviewed. Particular attention has been given to TIRFM samples consisting of covalently or noncovalently adsorbed biomolecules, substrate-supported planar phospholipid model membranes, and adherent cells. A variety of experimental configurations, including the combination of TIRFM with fluorescence recovery after photobleaching, evanescent interference patterns, fluorescence correlation spectroscopy, fluorescence resonance energy transfer, variable-angle measurements, high refractive index substrates, thin metal films or deposited metal nanostructures, fluorescence polarization, fast time-resolved fluorescence spectroscopy, two-photon excitation, AFM measurements, and microfluidics chambers, were described.

Acknowledgements
We thank Alena M. Lieto for assistance with the figures. This work was supported by NSF grant MCB-0130589, NIH grant GM-41402, and ACS-PRF Grant 35376-AC5-7.

References

1. Axelrod D (2003) Meth Enzymol 361:1–33
2. Moerner WE, Fromm DP (2003) Rev Sci Instrum 74:3597–3619
3. Sako Y, Uyemura T (2002) Cell Struct Funct 27:357–365
4. Ross ST (2002) Am Lab 34:22
5. Toomre D, Manstein DJ (2001) Trends Cell Biol 11:298–303
6. Steyer JA, Almers W (2001) Nat Rev Mol Cell Biol 2:268–276
7. Axelrod D (2001) Traffic 2:764–774
8. Oheim M (2001) Lasers Med Sci 16:159–170
9. Krishnan RV, Varma R, Mayor S (2001) J Fluoresc 11:211–226
10. Hovius R, Vallotton P, Wohland T, Vogel H (2000) Trends Pharmacol Sci 21:266–273
11. Forkey JN, Quinlan ME, Goldman YE (2000) Prog Biophys Mol Biol 74:1–35
12. Oheim M, Stühmer W (2000) Eur Biophys J 29:67–89
13. Lee B (2003) Opt Fiber Technol 9:57–79
14. Norde W (1995) Cell Mater 5:97–112
15. Bajpai AK (1997) Prog Polym Sci 22:523–564
16. Thompson NL, Pearce KH, Hsieh HV (1993) Eur Biophys J 22:367–378
17. Axelrod D, Burghardt TP, Thompson NL (1984) Annu Rev Biophys Bioeng 3:247–268
18. Agudin JL, Platzeck AM (1978) J Opt (Paris) 9:101–106
19. Knoll W (1998) Annu Rev Phys Chem 49:569–638
20. Girard C, Joachim C, Gauthier S (2000) Rep Prog Phys 63:893–938
21. Burghardt TP, Axelrod D (1983) Biochemistry 22:979–985
22. Thompson NL, Axelrod D (1983) Biophys J 43:103–114
23. Stout AL, Axelrod D (1989) Appl Opt 28:5237–5242
24. Axelrod D (2001) J Biomed Opt 6:6–13
25. Burghardt TP, Thompson NL (1984) Opt Eng 23:62–67
26. Sackmann E (1986) Science 271:43–48
27. Pisarchick ML, Thompson NL (1990) Biophys J 58:1235–1249
28. Wright LL, Palmer AG, Thompson NL (1988) Biophys J 54:463–470
29. Pearce KH, Hiskey RG, Thompson NL (1992) Biochemistry 31:5983–5995
30. Poglitsch CL, Sumner MT, Thompson NL (1991) Biochemistry 30:6662–6671
31. Axelrod D (1981) J Cell Biol 89:141–145
32. Schmid EL, Tairi AP, Hovius R, Vogel H (1998) Anal Chem 70:1331–1338
33. Beaucage SL (2001) Curr Med Chem 8:1213–1244
34. Burghardt TP, Axelrod D (1981) Biophys J 33:455–467
35. Thompson NL, Burghardt TP, Axelrod D (1981) Biophys J 33:435–454
36. Lagerholm BC, Thompson NL (1998) Biophys J 74:1215–1228
37. Pisarchick ML, Gesty D, Thompson NL (1992) Biophys J 63:215–223
38. Hsieh HV, Thompson NL (1995) Biochemistry 34:12481–12488
39. Lagerholm BC, Starr TE, Volovyk ZN, Thompson NL (2000) Biochemistry 39:2042–2051
40. Sheets ED, Chen L, Thompson NL (1997) Mol Immunol 34:519–526
41. Gesty-Palmer D, Thompson NL (1997) J Mol Recognit 10:63–72
42. Pearce KH, Hof M, Lentz BR, Thompson NL (1993) J Biol Chem 268:22984–22991
43. Fritz M, Zimmerman RM, Bärmann M, Gaub H (1993) Biophys J 65:1878–1885
44. McKiernan AE, MacDonald RI, MacDonald RC, Axelrod D (1997) Biophys J 73:1987–1998
45. Hellen EH, Axelrod D (1991) J Fluoresc 1:113–128
46. Stout AL, Axelrod D (1994) Biophys J 67:1324–1334
47. Stout AL, Axelrod D (1995) Photochem Photobiol 62:239–244
48. Sund SE, Axelrod D (2000) Biophys J 79:1655–1669
49. Swaminathan R, Bicknese S, Periasamy N, Verkman AS (1996) Biophys J 71:1140–1151
50. Kao HP, Verkman AS (1996) Biophys Chem 59:203–210
51. Kawakami K, Tatsumi H, Sokabe M (2001) J Cell Sci 114:3125–3135
52. Abney JR, Scalettar BA, Thompson NL (1992) Biophys J 61:542–552
53. Weis RM, Balakrishnan K, Smith BA, McConnell HM (1982) J Biol Chem 257:6440–6445

54. Kalb E, Tamm LK (1992) Thin Solid Films 210/211:763–765
55. Kalb E, Frey S, Tamm LK (1992) Biochim Biophys Acta 1103:307–316
56. Tilton RD, Gast AP, Robertson CR (1990) Biophys J 58:1321–1326
57. Tilton RD, Robertson CR, Gast AP (1990) J Colloid Interface Sci 137:192–203
58. Huang Z, Pearce KH, Thompson NL (1994) Biophys J 67:1754–1766
59. Gaspers PB, Robertson CR, Gast AP (1994) Langmuir 10:2699–2704
60. Hansen RL, Harris JM (1995) Anal Chem 67:492–498
61. Hansen RL, Harris JM (1996) Anal Chem 68:2879–2884
62. Chan V, Graves DJ, Fortina P, McKenzie SE (1997) Langmuir 13:320–329
63. Wang L, Schönhoff M, Möhwald H (2002) J Phys Chem B 106:9135–9142
64. Miehlich R, Gaub HE (1993) Rev Sci Instrum 64:2632–2638
65. Nakayama H, Yamaga T, Kunioka Y (1998) Biochem Biophys Res Commun 246:261–266
66. Cragg GE, So PTC (2000) Opt Lett 25:46–48
67. Wakelin S, Bagshaw CR (2003) J Microsc-Oxford 209:143–148
68. Haustein E, Schwille P (2003) Methods 29:153–166
69. Thompson NL, Lieto AM, Allen NW (2002) Curr Opin Struct Biol 12:634–641
70. Rigler R, Elson EL (2001) Fluorescence correlation spectroscopy: theory and applications. Springer, Berlin Heidelberg New York
71. Hansen RL, Harris JM (1998) Anal Chem 70:2565–2575
72. Hansen RL, Harris JM (1998) Anal Chem 70:4247–4256
73. Starr TE, Thompson NL (2002) J Phys Chem B 106:2365–2371
74. McCain KS, Harris JM (2003) Anal Chem 75:3616–3624
75. Lieto AM, Cush RR, Thompson NL (2003) Biophys J 85:3294–3302
76. Thompson NL (1982) Biophys J 38:327–329
77. Starr TE, Thompson NL (2001) Biophys J 80:1575–1584
78. Lieto AM, Thompson NL (2004) Biophys J 87:1268–1278
79. Grabbe ES (1993) Langmuir 9:1574–1581
80. Ishii Y, Yoshida T, Funatsu T, Wazawa T, Yanagida T (1999) Chem Phys 247:163–173
81. Suzuki Y, Tani T, Sutoh K, Kamimura S (2002) FEBS Lett 512:235–239
82. Watts TH, Gaub HE, McConnell HM (1986) Nature 320:176–179
83. Riven I, Kalmanzon E, Segev L, Reuveny E (2003) Neuron 38:225–235
84. Sako Y, Minoguchi S, Yanagida T (2000) Nat Cell Biol 2:168–172
85. Reichert WM, Suci PA, Ives JT, Andrade JD (1987) Appl Spectrosc 41:503–508
86. Thompson NL, Burghardt TP (1986) Biophys Chem 25:91–97
87. Burghardt TP (1989) Chem Phys Lipids 50:271–287
88. Gingell D, Heavens OS (1996) J Microsc 182:141–148
89. Ölveczky BP, Periasamy N, Verkman AS (1997) Biophys J 73:2836–2847
90. Rohrbach A (2000) Biophys J 78:2641–2654
91. Tronin A, Blasie JK (2001) Langmuir 17:3696
92. Loerke D, Stühmer W, Oheim M (2002) J Neurosci Methods 199:65–73
93. Stock K, Sailer R, Strauss WSL, Lyttek M, Steiner R, Schneckenburger H (2003) J Microsc 211:19–29
94. Lanni F, Waggoner AS, Taylor DL (1985) J Cell Biol 100:1091–1102
95. Liebmann LW, Robinson JA, Mann KG (1991) Rev Sci Instrum 62:2083–2092
96. Toriumi M, Saito S, Kawaguchi K, Aiki K (1995) Rev Sci Instrum 66:3520–3526
97. Byrne CD, de Mello AJ, Barnes WL (1998) J Phys Chem B 102:10326–10333
98. Reichert WM, Truskey GA (1990) J Cell Sci 96:219–230
99. Heavens OS (1990) J Cell Sci 95:175–176
100. Truskey GA, Burmeister JS, Grapa E, Reichert WM (1992) J Cell Sci 103:491–499
101. Burmeister JS, McKiney VZ, Reichert WM, Truskey GA (1999) J Biomed Mater Res 47: 577–584
102. Heavens OS (1995) J Microsc 180:106–108
103. Burmeister JS, Truskey GA, Reichert WM (1994) J Microsc 173:39–51
104. Olivier LA, Yen J, Reichert WM, Truskey GA (1999) Biotechnol Prog 15:33–42
105. Loerke D, Preitz B, Stühmer W, Oheim M (2000) J Biomed Opt 5:23–30

106. Gingell D, Todd I, Bailey J (1985) J Cell Biol 100:1334–1338
107. Gingell D, Heavens OS, Mellor JS (1987) J Cell Sci 87:677–693
108. Smith LV, Tamm LK, Ford RM (2002) Langmuir 18:5247–5255
109. Vigeant MAS, Wagner M, Tamm LK, Ford RM (2001) Langmuir 17:2235–2242
110. Vigeant MAS, Ford RM, Wagner M, Tamm LK (2002) Appl Environ Microbiol 68: 2794–2801
111. Geggier P, Fuhr G (1999) Appl Phys A Mater Sci Proc 68:505–513
112. Starr TE, Thompson NL (2000) Langmuir 16:10301-10308
113. Ajo-Franklin CM, Kam L, Boxer SG (2001) Proc Natl Acad Sci USA 98:13643–13648
114. Vilfan M, Muševič I, Čopič M (1998) Europhys Lett 43:41–46
115. Reimhult E, Höök F, Kasemo B (2003) Langmuir 19:1681–1691
116. Axelrod D, Fulbright RM, Hellen EH (1986) Adsorption kinetics on biological membranes: measurement by total internal reflection fluorescence. In: Lansing TD (ed) Applications of fluorescence in the biomedical sciences. Liss, New York, p 461
117. Lakowicz JR (2001) Anal Biochem 298:1–24
118. Geddes CD, Lakowicz JR (2003) J Fluoresc 12:121–129
119. Geddes CD, Cao H, Gryczynski I, Gryczynski Z, Fang J, Lakowicz JR (2003) J Phys Chem A 107:3443–3449
120. Tian ZQ, Ren B, Wu DY (2002) J Phys Chem B 106:9463–9483
121. Levene MJ, Korlach J, Turner SW, Foquet M, Craighead HG, Webb WW (2003) Science 299:682–686
122. Carniglia CK, Mandel L, Drexhage KH (1972) J Opt Soc Am 62:479–486
123. Lukosz W, Kunz RE (1977) J Opt Soc Am 67:1607–1614
124. Lukosz W, Kunz RE (1977) J Opt Soc Am 67:1615–1619
125. Burghardt TP, Thompson NL (1984) Biophys J 46:729–737
126. Burghardt TP (1989) Chem Phys Lipids 50:271–287
127. Mertz J (2000) J Opt Soc Am B 17:1906–1913
128. Hellen EH, Axelrod D (1987) J Opt Soc Am B 4:337–350
129. Thompson NL, McConnell HM, Burghardt TP (1984) Biophys J 46:739–747
130. Bos MA, Kleijn JM (1995a) Biophys J 68:2566–2572
131. Timbs MM, Thompson NL (1990) Biophys J 58:413–428
132. Zhai X, Kleijn JM (1997) Biophys J 72:2651–2659
133. Yang J, Kleijn M (1999) Biophys J 76:323–332
134. Fraaije JGEM, Kleijn JM, van der Graaf M, Dijt JC (1990) Biophys J 57:965–975
135. Bos MA, Kleijn JM (1995) Biophys J 68:2573–2579
136. Edmiston PL, Saavedra SS (1998) Biophys J 74:999–1006
137. Edmiston PL, Saavedra SS (1998) J Am Chem Soc 120:1665–1671
138. Tronin A, Edwards AM, Wright WW, Vanderkooi JM, Blasie JK (2002) Biophys J 82: 996–1003
139. Fowers KD, Kopeček J (1997) Colloids Surfs B Biointerfaces 9:315–330
140. Dickson RM, Norris DJ, Moerner WE (1998) Phys Rev Lett 81:5322–5325
141. Vacha M, Kotani M (2003) J Chem Phys 118:5279–5282
142. Sund SE, Swanson JA, Axelrod D (1999) Biophys J 77:2266–2283
143. Zenisek D, Steyer JA, Feldman ME, Almers W (2002) Neuron 35:1085–1097
144. Toriumi M, Masuhara H (1991) Spectrochim Acta Rev 14:353–377
145. Lundgren JS, Bekos EJ, Wang R, Bright FV (1994) Anal Chem 66:2433–2440
146. Phillips D (1994) Analyst 119:543–550
147. Yanagimachi M, Tamai N, Masuhara H (1993) Chem Phys Lett 201:115–119
148. Hamai S, Tamai N, Masuhara H (1995) J Phys Chem 99:4980–4985
149. Bessho K, Uchida T, Yamauchi A, Shioya T, Teramae N (1997) Chem Phys Lett 264:381–386
150. Yao H, Kitagawa F, Kitamura N (2000) Langmuir 16:3454–3461
151. Fujiwara N, Tsukahara S, Watarai H (2001) Langmuir 17:5337–5342
152. Ishizaka S, Kim HB, Kitamura N (2001) Anal Chem 73:2421–2428
153. Wang R, Sun S, Bekos EJ, Bright FV (1995) Anal Chem 67:149–159
154. Lensun L, Smith TA, Gee ML (2002) Langmuir 18:9924–9931

155. Czeslik C, Royer C, Hazlett T, Mantulin W (2003) Biophys J 84:2533–2541
156. Bicknese S, Periasamy N, Shohet SB, Verkman AS (1993) Biophys J 65:1272–1282
157. Schneckenburger H, Stock K, Strauss WSL, Eickholz J, Sailer R (2003) J Microsc 211:30–36
158. Phimphivong S, Kölchens S, Edmiston PL, Saavedra SS (1995) Anal Chim Acta 307: 403–417
159. Phimphivong S, Saavedra SS (1998) Bioconjug Chem 9:350–357
160. Gryczynski I, Gryczynski Z, Lakowicz JR (1997) Anal Biochem 247:69–76
161. Huang Z, Thompson NL (1993) Biophys Chem 47:241–249
162. So PTC, Kwon HS, Dong CY (2001) J Opt Soc Am A 18:2833–2845
163. Fisz JJ (2003) Chem Phys Lett 373:299–307
164. Ambrose WP, Goodwin PM, Nolan JP (1999) Cytometry 36:224–231
165. Conibear PB, Bagshaw CR (2000) J Microsc 200:218–299
166. Xu XH, Yeung ES (1997) Science 275:1106–1109
167. Xu XH, Yeung ES (1998) Science 281:1650–1653
168. Funatsu T, Harada Y, Tokunaga M, Saito K, Yanagida T (1995) Nature 374:555–559
169. Iwane AH, Funatsu T, Harada Y, Tokunanga M, Ohara O, Morimoto S, Yanagida T (1997) FEBS Lett 407:235–238
170. Tokunaga M, Kitamura K, Saito K, Iwane AH, Yanagida T (1997) Biochem Biophys Res Commun 235:47–53
171. Vale RD, Funatsu T, Pierce DW, Romberg L, Harada Y, Yanagida T (1996) Nature 380: 451–453
172. Funatsu T, Harada Y, Higuchi H, Tokunaga M, Saito K, Ishii Y, Vale RD, Yanagida T (1997) Biophys Chem 68:63–72
173. Yamaguchi J, Nemoto N, Sasaki T, Tokumasu A, Mimori-Kiyosue Y, Yagi T, Funatsu T (2001) FEBS Lett 502:79–83
174. Seitz A, Kojima H, Oiwa K, Mandelkow EM, Song YH, Mandelkow E (2002) EMBO J 21:4896–4905
175. Yamasaki R, Hoshino M, Wazawa T, Ishii Y, Yanagida T, Kawata Y, Higurashi T, Sakai K, Nagai J, Goto Y (1999) J Mol Biol 292:965–972
176. Kaseda K, Yokota H, Ishii Y, Yanagida T, Inoue T, Fukui K, Kodama T (2000) J Bacteriol 182: 1162–1166
177. Osborne MA, Barnes CL, Balasubramanian S, Klenerman D (2001) J Phys Chem B 105: 3120–3126
178. Harada Y, Funatsu T, Murakami K, Nonoyama Y, Ishihama A, Yanagida T (1999) Biophys J 76:709–715
179. Lyon WA, Fang MM, Haskins WE, Nie S (1998) Anal Chem 70:1743–1748
180. Ide T, Yanagida T (1999) Biochem Biophys Res Commun 265:595–599
181. Zhang P, Tan W (2000) Chem Eur J 6:1087–1092.
182. Oiwa K, Eccleston JF, Anson M, Kikumoto M, Davis CT, Reid GP, Ferenczi MA, Corrie JET, Yamada A, Nakayama H, Trentham DR (2000) Biophys J 78:3048–3071
183. Saga Y, Wazawa T, Nakada T, Ishii Y, Yanagida T, Tamiaki H (2002) J Phys Chem B 106:1430–1433
184. Saga Y, Wazawa T, Mizoguchi T, Ishii Y, Yanagida T, Tamiaki H (2002) Photochem Photobiol 75:433–436.
185. Wazawa T, Ishii Y, Funatsu T, Yanagida T (2000) Biophys J 78:1561–1569
186. Yildiz, A, Forkey JN, McKinney SA, Ha T, Goldman YE, Selvin PR (2003) Science 300:2061–2065
187. Wang MD, Axelrod D (1994) Dev Dyn 201:29–40
188. Iino R, Koyama I, Kusumi A (2001) Biophys J 80:2667–2677
189. Sonnleitner A, Mannuzzu LM, Terakawa S, Isacoff EY (2002) Proc Natl Acad Sci USA 99:12759–12764
190. Robinson DN, Ocon SS, Rock RS, Spudich JA (2002) J Biol Chem 277:9088–9095
191. Omann GM, Axelrod D (1996) Biophys J 71:2885–2891
192. Suga T, Hirano M, Takayanagi M, Koshimoto H, Watanabe A (1998) Biochem Biophys Res Commun 253:423–430

193. Codazzi F, Teruel MN, Meyer T (2001) Curr Biol 11:1089–1097
194. Pinton P, Tsuboi T, Ainscow EK, Pozzan T, Rizzuto R, Rutter GA (2002) J Biol Chem 277: 37702–37710
195. Khan S, Pierce D, Vale RD (2000) Curr Biol 10:927–930
196. Haugh JM, Codazzi F, Teruel M, Meyer T (2000) J Cell Biol 151:1269–1279
197. Mashanov GI, Tacon D, Knight AE, Peckham M, Molloy JE (2003) Methods 29:142–152
198. Xiao Y, Truskey GA (1996) Biophys J 71:2869–2884
199. Chan BP, Bhat VD, Yegnasubramanian S, Reichert WM, Truskey GA (1999) Biomaterials 20:2395–2403
200. Lang T, Wacker I, Steyer J, Kaether C, Wunderlich I, Soldati T, Gerdes H-H, Almers W (1997) Neuron 18:857–863
201. Han W, Ng YK, Axelrod D, Levitan ES (1999) Proc Natl Acad Sci USA 96:14577–14582
202. Lang T, Wacker I, Wunderlich I, Rohrbach A, Giese G, Soldati T, Almers W (2000) Biophys J 78:2863–2877
203. Scalettar BA, Rosa P, Taverna E, Francolini M, Tsuboi T, Terakawa S, Koizumi S, Roder J, Jeromin A (2002) J Cell Sci 115:2399–2412
204. Steyer JA, Horstmann H, Almers W (1997) Nature 388:474–478
205. Oheim M, Loerke D, Stühmer W, Chow RH (1998) Eur Biophys J 27:83–98
206. Oheim M, Loerke D, Stühmer W, Chow RH (1999) Eur Biophys J 28:91–101
207. Steyer JA, Almers W (1999) Biophys J 76:2262–2271
208. Oheim M, Stühmer W (2000) J Membr Biol 178:163–173
209. Tsuboi T, Kikuta T, Warashina A, Terakawa S (2001) Biochem Biophys Res Commun 282:621–628
210. Johns LM, Levitan ES, Shelden EA, Holz RA, Axelrod D (2001) J Cell Biol 153:177–190
211. Tsuboi T, Kikuta T, Sakurai T, Terakawa S (2002) Biophys J 83:172–183
212. Ohara-Imaizumi M, Nakamichi Y, Tanaka T, Ishida H, Nagamatsu S (2002) J Biol Chem 277:3805–3808
213. Jaiswal JK, Andrews NW, Simon SM (2002) J Cell Biol 159:625–635
214. Keller P, Toomre D, Diaz E, White J, Simons K (2001) Nat Cell Biol 3:140–149
215. Schmoranzer J, Goulian M, Axelrod D, Simon SM (2000) J Cell Biol 149:23–31
216. Toomre D, Steyer JA, Keller P, Almers W, Simons K (2000) J Cell Biol 149:33–40
217. Tsuboi T, Terakawa S, Scalettar BA, Fantus C, Roder J (2002) J Biol Chem 277:15957–15961
218. Brusatori MA, Tie Y, van Tassel PR (2003) Langmuir 19:5089–5097
219. Mathur AB, Truskey GA, Reichert WM (2000) Biophys J 78:1725–1735
220. Nishida S, Funabashi Y, Ikai A (2002) Ultramicroscopy 91:269–274
221. Schult K, Katerkamp A, Trau D, Grawe F, Cammann K, Meusel M (1999) Anal Chem 71:5430–5435
222. Yang T, Jung S, Mao H, Cremer PS (2001) Anal Chem 73:165–169
223. Jakeway SC, de Mello AJ (2001) Analyst 126:1505–1510
224. Hofmann O, Voirin G, Niedermann P, Manz A (2002) Anal Chem 74:5243–5250
225. Yang T, Barysnikova OK, Mao H, Holden MA, Cremer PS (2003) J Am Chem Soc 125: 4779–4784
226. Pawlak M, Schick E, Bopp MA, Schneider MJ, Oroszlan P, Ehrat M (2002) Proteomics 2:383–393
227. Duveneck GL, Abel AP, Bopp MA, Kresbach GM, Ehrat M (2002) Anal Chim Acta 469:49–61
228. Epstein JR, Biran I, Walt DR (2002) Anal Chim Acta 469:3–36
229. Lehr HP, Brandenburg A, Sulz G (2003) Sens Actuators B Chem 92:303–314
230. Lehr HP, Reimann M, Brandenburg A, Sulz G, Klapproth H (2003) Anal Chem 75:2414–2420

7 Single Molecule Spectroscopy: Basics and Applications

J. Enderlein

Keywords: Single molecule fluorescence; Confocal laser-scanning microscopy; Time-correlated single-photon counting; Wide-field imaging microscopy

Abbreviations

CLSM	Confocal laser-scanning microscope
FCS	Fluorescence correlation spectroscopy
FIDA	Fluorescence intensity distribution analysis
FIMDA	Fluorescence intensity multiple distribution analysis
MCP	Multi-channel plate
Mcps	Mega counts per second
MFD	Multi-parameter fluorescence detection
NA	Numerical aperture
NIR	Near-infrared region
PCH	Photon-counting histogramming
SMD	Single molecule detection
smWFM	Single-molecule wide-field microscopy
SMS	Single molecule spectroscopy
SNP	Single nucleotide polymorphism
TCSPC	Time-correlated single-photon counting
TIRFM	Total internal reflection fluorescence microscopy

7.1 Introduction

The first, indirect detection of a single molecule in solution was performed by Rotman when detecting the multiple reaction products of a single enzymatic molecule [1, 2]. The next step was the direct optical detection of a single although multiply labelled molecule by Hirschfeld [3, 4]. The importance of this work was to demonstrate the principal possibility of direct optical single molecule detection in a liquid, and it exercised a great influence on the further development of the field. Finally, the group of Richard A. Keller at Los Alamos National Laboratory was first able to detect individual rhodamine molecules in a fluid flow by 1996 [5, 6]. Since then, the research on single molecule detection and spectroscopy (in short SMS) in solution and on surfaces under ambient conditions has seen an explosive development. Recently, there have been several particularly exhaustive reviews concerning this quickly evolving field of research [7–13], and several books have been published [14, 15]. The present paper intends to give an overview of the actual state of the art of single molecule fluorescence spectroscopy and to present a few selected applications from our research group.

The scope of the review will be restricted to fluorescence SMS in liquids and on surfaces under ambient conditions (molecules embedded within solvents or bound to surfaces, measurements performed at room temperature), thus neglecting the broad research done on low-temperature spectroscopy (see e.g. [16]), and on atomic force and scanning tunnelling microscopy of single molecules. Moreover, near-field microscopy of single molecules [17–19] is also excluded from the considerations because it is too broad a field of research that would need the attention of a special overview by itself (but see the recent work of van Hulst's group [20–25]).

7.2 Photophysics, Probes and Markers

Quoting the words of Richard A. Keller, it is not difficult to detect the fluorescence of a single molecule but *not* to detect anything else. 'Anything else' means, first of all, elastic and inelastic scattering of light by the surrounding medium, and fluorescence from unwanted fluorescence sources (e.g. autofluorescence from biological samples). For successfully observing a single molecule by its fluorescence, it has to have suitable photophysical properties. These properties are: absorption and emission spectrum, optical absorption cross section, quantum yield of fluorescence, fluorescence lifetime, triplet-state dynamics, 'blinking' dynamics and photostability. In a very simplified form, a single fluorescing molecule can be described by the Jablonski scheme as depicted in Fig. 7.1: The non-excited molecule resides in its electronic ground state S_0. From there it can be excited by single- or multi-photon absorption into its first excited electronic state S_1, from where it can either (1) return to its ground state S_0, via emission of a fluorescence photon or by a non-radiative transition induced by the interaction with the surrounding medium, (2) switch into a triplet state (T_1) via intersystem crossing, or (3) be irreversibly destroyed by a photochemical reaction (photobleaching). In this simplified picture, vibrational modes within the electronic states were completely left out – a more detailed description of molecular photophysics can be found in many textbooks. Here, it is only important that, due to the vibrational mode substructure of the electronic states, fluorescence emission is red-shifted with respect to absorption (the so-called Stokes shift). This is crucially important for a successful implementation of any single molecule detection scheme: the Stokes shift allows for a spectral separation of the fluorescence emission from the excitation light which is elastically (Rayleigh) and inelastically (Raman) scattered back by the embedding medium. However, for most dyes dissolved within a solvent like water, Raman contributions will always overlap with the fluorescence emission spectrum.

The next important property is the absorption cross section: for a one-photon fluorescence excitation process, both the fluorescence signal and the intensity of the scattered light are linearly proportional (below optical saturation of the fluorescence) to the excitation intensity. Increasing the absorption cross section of the fluorescing molecules thus increases the ratio of detectable fluorescence to backscattered light, improving the signal-to-noise ratio of the detection. The absorption cross sections of many organic dyes are

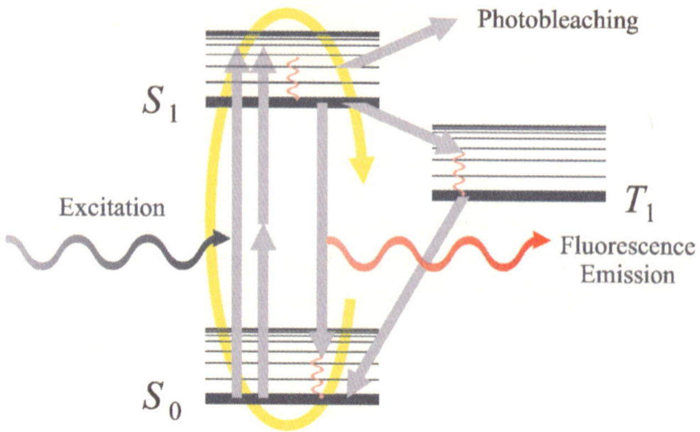

Fig. 7.1. Simplified three-state Jablonski diagram of a molecule's energetic states that are involved in its photophysics. The molecule can be excited, by a one- or multi-photon process, from its electronic ground state S_0 to the first excited state S_1. Its return to the ground state can be accompanied by the emission of a fluorescence photon. Repeated cycling of the molecule through its excited state generates a characteristic burst of detectable fluorescence photons. Competing with the S_1–S_0 transition is intersystem crossing into the first triplet state T_1, and irreversible photobleaching which eventually leads to the complete disappearance of the molecule from detection

typically between 10^{-16} and 10^{-15} cm^2, corresponding to optical extinctions between 26,000 and 260,000 l mol^{-1} cm^{-1}.

As just mentioned, the linear relation between excitation intensity and fluorescence signal is only correct for excitation intensities much below optical saturation: optical saturation occurs for excitation intensities where the molecule spends a significant part of the time not in the ground state but in the excited or the triplet states. When optical saturation starts, an increase in excitation power will not produce correspondingly more fluorescence photons per time unit, and thus the signal-to-noise ratio gets lower with increasing excitation intensity.

The next quantity to consider is the quantum yield of fluorescence. It refers to the probability that the return from the excited to the ground state is accompanied by the emission of a fluorescence photon (radiative transition). For many fluorescing molecules, this radiative transition is in competition with non-radiative transitions, where the energy of the excited state is transferred to the solvent collisions or processes such as electron transfer. A rule of thumb is that the fluorescence quantum yield decreases with increasing wavelength of the excitation/emission maximum of a dye. Thus, the standard laser dye rhodamine 6G, excitable around 520 nm, has a fluorescence quantum yield of nearly unity, whereas the popular dye Cy5 (excitation near 640 nm) shows a fluorescence quantum yield of 0.28 (in water).

Besides cycling between the ground and excited states, a molecule can undergo an intersystem crossing to its lowest triplet state T_1. The term intersystem crossing stems from the fact that the transition is accompanied by a spin flip of the excited electron and is thus symmetrically disfavoured.

Correspondingly, intersystem crossing rates are usually low (ca. one crossing for every 10^5–10^6 excitations). However, the average lifetime of the triplet state is usually much larger than the fluorescence lifetime (in the microsecond to millisecond range), leading to the repeated 'disappearing' of a single molecule from detection. Thus, triplet state dynamics often determines the saturation intensity. Even more problematic is the so-called blinking of single molecules [26–28], by which one denotes the optical disappearing of single molecules on even larger timescales (for seconds and more), thus seriously impeding their detection. The nature of this blinking behaviour is still an active area of research, and no final and universally accepted explanation has yet been put forward. Unfortunately, especially fluorescent proteins (cyan, green, yellow etc. fluorescent proteins), which are widely used and very popular in biological studies, show dramatic blinking [29–31], making it particularly difficult to use them as single molecule labels.

One of the most important photophysical characteristics is the photostability of a fluorescent molecule. It refers to the probability that a molecule undergoes an irreversible chemical reaction while it resides in its excited state. This photodestruction leads to the final disappearance of the molecule from detection and thus ultimately limits any SMS experiment. Typical fluorescent dye molecules survive ca. 10^5 to 10^6 excitation cycles until photodestruction occurs, although this number can vary widely and strongly depends on the nature of solvent. Especially in biological SMS applications, the photostability is the most limiting factor, defining the maximum number of photons that can be detected from a single molecule. Again, the popular fluorescent proteins show rather high photodestruction rates, the green fluorescent protein family being the most suitable one for SMS applications [31–33]. Recently, new kinds of fluorescing labels with high photostability and low triplet state/blinking dynamics were developed, such as fluorescing nano-crystals [34–37], solid-state colour centres [38], or metallic nano-particles [39–43].

An important strategy of background suppression is to use fluorescent dyes with long excitation and emission wavelengths in the near-infrared and infrared wavelength region [44]. As is well known, the intensity of non-resonant light scattering decreases roughly with the fourth power of the wavelength; thus, doubling the excitation wavelength decreases the scattering background by 16 times. Additionally, using long-wavelength dyes is also advantageous when performing SMS in biological samples containing autofluorescent biomolecules. Figure 7.2 shows the optical absorption of several classes of synthetic dyes and those of fluorescent biomolecules. In recent years, new classes of fluorescent dyes with photophysical parameters sufficient for single molecule detection in the near-infrared region (>600 nm) have become widely available, such as the popular cyanine dyes by Amersham or the Alexa dyes by Molecular Probes. Meanwhile, even dyes beyond 700 nm are available (e.g. the IRDye family by LI-COR), although not yet characterized on the single-molecule level.

A topic of considerable interest is the development of new near-infrared and infrared dyes with properties suitable for SMS (high fluorescence quantum yield, high photostability, see e.g. [45]). There are two principal reasons

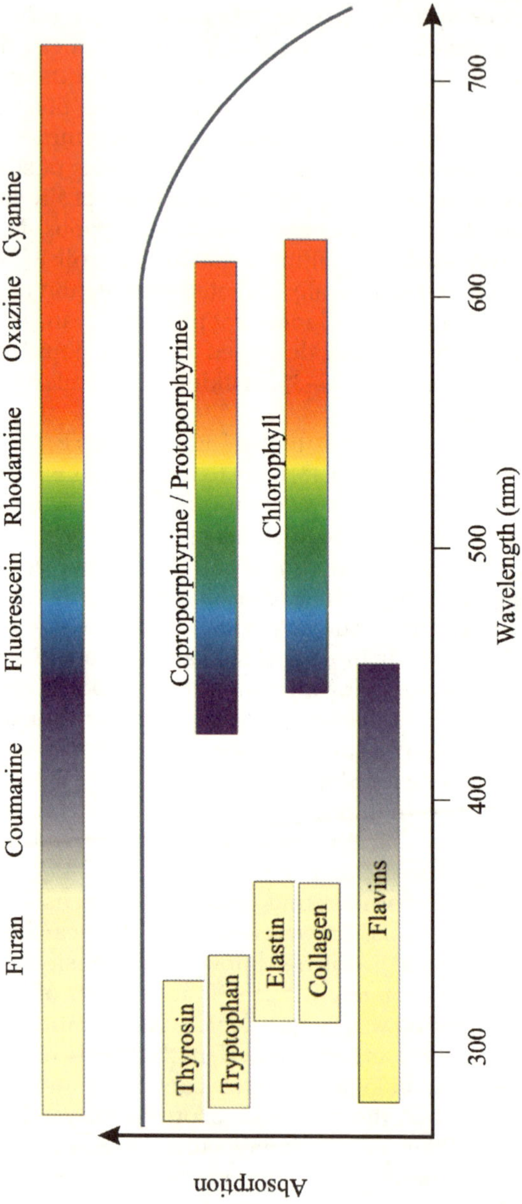

Fig. 7.2. Optical absorption regions of the fluorescence of biological molecules and several synthetic dye families

for being interested in such dyes: firstly, the scattering intensity of the excitation light, a major source of unwanted background in SMS, drops approximately off with the fourth power of the wavelength; secondly, using NIR and IR dyes prevents the unwanted excitation of autofluorescence in biological samples. This is of significance for many SMS applications in single-cell studies or single-molecule-sensitive screening of biological fluids.

A completely new and purely physical way of improving the photophysical properties of conventional dyes is to use the electrodynamic interaction of fluorescing dyes with metal for designing new composite fluorescing materials with strongly enhanced fluorescence properties. Lakowicz [46] coined the term 'radiative decay engineering' describing that modification of emission properties of fluorophores or chromophores in the presence of metals, leading to enhanced photostability and optical excitation efficiency. Detailed calculations of these effects are available for fluorescent metallic nano-cavities [47, 48], predicting remarkable enhancement values for photostability and fluorescence brightness, although the metal is completely enclosing the fluorescing entity. The calculations clearly show that the synthesis of such metal/fluorescent dye composites may lead to new fluorescing labels with extraordinary optical properties.

7.3 Physical Techniques

7.3.1 Modified Flow Cytometry, Microchannels and Microdroplets

Historically, the first successful detection of a single molecule in solution was realized with a modified flow cytometer using hydrodynamic focusing [6]. The principle of this detection scheme is depicted in Fig. 7.3. A sample stream containing the dissolved fluorescing molecules is injected into a surrounding sheath flow, providing transport of the injected molecules and simultaneous hydrodynamic focusing of the sample stream [49]. The molecules are transported through a focused laser beam (beam waist ca. 10 μm) which intersects the fluid flow at a right angle. Fluorescence is detected through an objective with its optical axis perpendicular to the flow direction *and* the exciting laser beam. Transit of a single molecule is detected as a burst in time of recorded photons. Due to the perpendicular arrangement of detection and excitation, the numerical aperture of the objectives is much smaller than that used in confocal epifluorescence setups (high NA objectives have a very small working distance of ca. 250 μm, leaving little space for arranging two objectives at 90 degrees), so that the detection volume of the flow cytometry setup is ca. 10^3 larger and the overall detection efficiency is lower than that of confocal systems. Thus, background suppression is a much more serious issue here. The main advantage of the flow cytometry setup is that all molecules pass the detection region in nearly the same way, so that the burst size directly reflects the fluorescence brightness of the molecules. By evaluating the number of detected photons per molecule it is possible to distinguish between different molecular species solely by their brightness (single

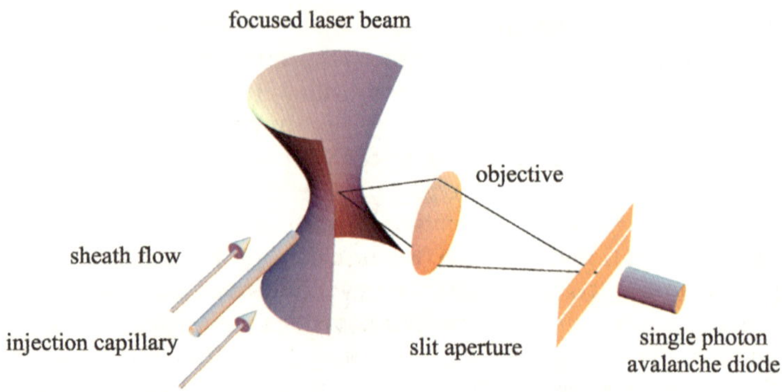

Fig. 7.3. Schematic of the modified flow cytometer for SMS application. A sample stream containing the dissolved fluorescing molecules is injected into a surrounding sheath flow, providing transport of the injected molecules and simultaneous hydrodynamic focusing of the sample stream. The molecules are transported through a focused laser beam, which intersects the fluid flow at a right angle. Fluorescence is detected through an objective with its optical axis perpendicular to the flow direction and the exciting laser beam. The fluorescence light is imaged onto a single-photon-sensitive light detector, such as a single-photon avalanche diode

molecule fluorescence intensity), without using any other fluorescence property [50]. A promising application of this technique is rapid and ultra-sensitive DNA fragment sizing, employing an intercalating dye for staining DNA fragments and relying on the assumption that dye load per fragment (and thus fluorescence brightness) is proportional to fragment size [51, 52]. The same working principle is also behind a new method of high-throughput detection of single nucleotide polymorphisms (SNPs) as described by Cai et al. [53]. The idea of 'guiding' molecules through the detection region can also be achieved by employing microcapillaries [54–59] or microchannels [60–63]. The advantage is its simplicity (no hydrodynamic focusing is necessary to ensure transition of the molecules through the detection region); the disadvantage may be a higher background signal.

Closely related to SMD in a fluid flow is SMD in gel capillary electrophoresis [64–66]. This is an important field of research because it combines the versatility of electrophoretic separation with the sensitivity of SMS. Most papers concerned with SMS in electrophoresis were dealing with DNA analysis (fragment sizing and/or specific sequence detection) [67–71].

One of the challenges of the flow cytometry approach is to prevent diffusion of the analyte molecules away from the flow axis. When using microcapillaries or microchannels, the molecules' motion is confined laterally, for the price of a heavily increased scattering background. A different solution to this problem is to include the analyte molecules into electrically charged micrometre sized droplets, which can then be handled with electric fields. A solution containing the analyte molecules is sprayed into a stream of microdroplets, which are subsequently led through the detection region by means of gravity and applied electric fields. The advantage of such an approach is the facility of perfectly guiding the microdroplets through the detection region while avoiding a high background signal stemming from any confining microstructures. Over the years, an impressive amount of literature was published by Mike Ramsey's group at Oak Ridge, describing the principles and numerous applications of that method [72–78].

A noteworthy variant for achieving extremely small detection volumes is to excite fluorescence by evanescent illumination together with high numerical aperture collection, achieving detection volumes as small as 0.1 μm^3 [79–82]. In a similar vein, Ruckstuhl et al. [83–86] use light collection above the critical angle of total internal reflection (supercritical angle fluorescence detection) for achieving comparably small detection volumes. In both cases, the extreme localization of the detection region at an interface (glass/water) is of considerable interest when trying to monitor binding kinetics on a single molecule level on the background of a rather high fluorophore concentration in solution.

7.3.2 Confocal Detection

As already mentioned, having a fluorescent dye with good photophysical properties is only one prerequisite for a successful detection of single molecules. Another important issue is the maximum suppression of Rayleigh and Raman scattering of the excitation light. One way, as discussed in connection with the Stokes shift of the fluorescence, is the spectral discrimination of fluorescence against scattered light by exploiting high-efficiency optical filters. In this way, it is possible to completely suppress all elastic scattering contributions (Rayleigh scattering at the excitation wavelength). However, Raman scattering often shows a significant spectral overlap with the fluorescence emission. Thus, a second and important way to minimize unwanted signal contributions due to Raman scattering is to minimize the detection volume, the volume of efficient fluorescence excitation *and* detection. Whereas the fluorescence signal of a single molecule does not change with decreasing detection volume (as long as it is contained within it), the intensity of the scattered light is directly proportional to its extent (number of scattering molecules). This is the basic motivation for all confocal detection schemes. The general scheme of such a system is shown in Fig. 7.4. A laser beam is coupled into a microscope objective with high numerical aperture, which focuses the beam into a diffraction-limited spot within the sample. The emerging fluorescence is collected by the same objective (epifluorescence detection) and separated from the excitation beam by the dichroic mirror. After passing additional spectral filters for background suppression, the collected fluorescence light is focused onto a confocal aperture, behind which it is detected with a single-photon sensitive photoelectric detector. For technical details of a single-molecule sensitive confocal epifluorescence microscope the reader is referred to Refs. [87, 88].

Two effects are employed for minimizing the detection volume: the diffraction-limited transversal extent of the laser beam at the focus (depending on wavelength ca. 300–500 nm), and the restriction of the axial extent of the detection volume by detecting through the confocal aperture. The principle of confocal detection is based on the fact that only light emerging near the focus region can

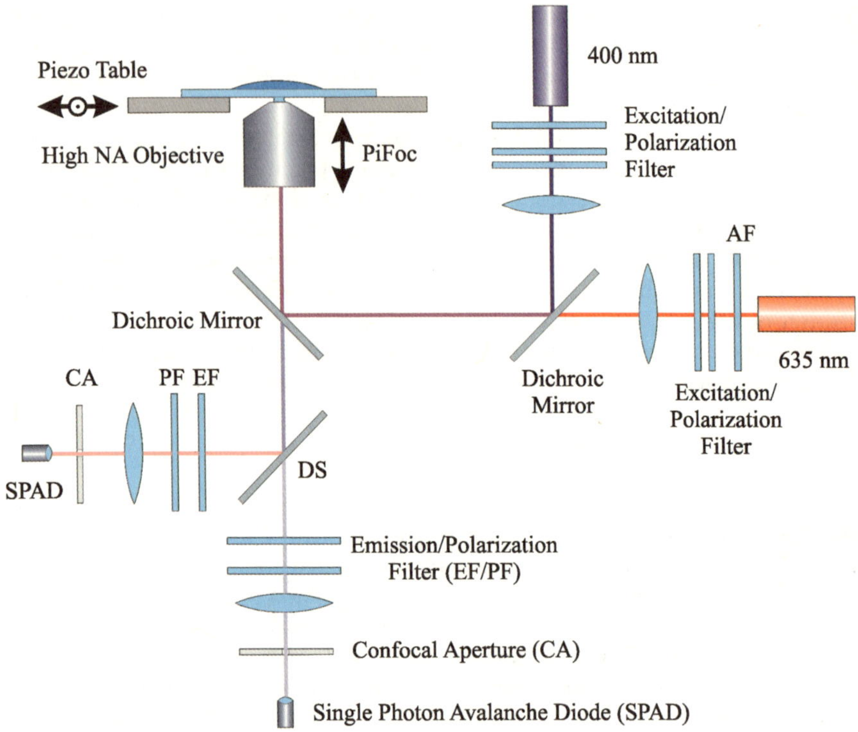

Fig. 7.4. Principal scheme of a confocal epifluorescence microscope for single molecule spectroscopy. Shown is a setup with two excitation lasers at different wavelengths for dual-colour excitation, and an additional lamp for wide-field illumination to observe the sample before starting the SMS measurement. Detection is carried out within two channels, either for two-colour or two-polarization detection, although there is no principal limitation on the total number of detection channels. The vertical position of the objective is changed by a piezo actuator (*PiFoc*); sample scanning is performed with a piezo table. For more details, see Refs. [87, 88]

completely pass through the aperture. Any light coming from points farther away from that region is largely blocked by the aperture because it does not converge within the aperture plane. The axial restriction of the detection volume that can be achieved by this technique is ca. 2–4 μm.

The simplest experiment one can perform with the confocal epifluorescence setup is to watch single molecules diffuse through the detection region. This supposes that one works at a sufficiently low concentration of fluorescing molecules: for a detection volume of 2 μm^3 and a concentration below ca. 10^{-9} M, one finds, on average, only single molecules (or none) within the detection region. Every transit of a single molecule through the detection region results in a characteristic burst of detected fluorescence photons.

For studying molecules that are immobilized on a surface, the focused laser is scanned over the surface, either to record the fluorescence signal along single lines or to record a complete fluorescence image by scanning in two directions.

Confocal imaging and scanning constitute the basic ingredients of a confocal laser-scanning microscope (CLSM). Scanning can also be extended into the third spatial dimension by scanning planes at different locations along the optical axis (so-called *z*-scanning), resulting in a thee-dimensional fluorescence image of a sample (e.g. a dye-tagged cell, or a transparent substrate with embedded fluorescent molecules).

Compared with SMD by highly sensitive CCD systems, the CLSM offers several advantages. Firstly, due to the small detection volume, the CLSM has an exceptionally high signal-to-background ratio. Secondly, uniform scanning provided, the CLSM assures a homogeneous sample illumination, which may be a challenge in conventional wide-field microscopy. Thirdly, the CLSM uses single-point detectors for light detection, thus offering the possibility to study extremely fast processes down to the picosecond timescale. This is especially interesting for sensitive fluorescence lifetime imaging, where the fluorescence decay time characteristics are measured for every point on the scanned surface. Also, the splitting of the detected light into several channels for measuring other spectral characteristics such as polarization or emission wavelength is straightforward with the CLSM. The main limitation of the CLSM, when compared with CCD imaging, is the limited scan speed with which an image can be recorded. This scan speed is mainly determined by the maximum number of photons per time interval that can be extracted from a single molecule (optical saturation limit). Typically, if one assumes that a single molecule occupies an image area of ca. 1 μm^2 and that a realistic detectable photon count intensity of a single molecule's fluorescence is of the order of 1 Mcps, an imaging speed of 0.1 $\mu m^2/ms$ will yield ca. 100 photons per detected molecule. This is a reasonable number for detecting and studying individual molecules. Thus, when imaging an area of 100×100 μm^2, it takes a minimum of 1 s, which is roughly three orders of magnitude slower than what is achievable with fast CCD systems.

7.3.3 Wide-Field Imaging

In the previous sections, SMD with point detectors (in most cases single-photo avalanche diodes) for detecting the fluorescence of single molecules was described. As already mentioned, this has the advantage of measuring processes on practically arbitrary short timescales, but at the price of relatively slow imaging speed. The straightforward solution to this problem is to use a parallel detection scheme, namely a charge-coupled detector, for imaging. The development of high-sensitivity, low-noise CCDs has finally made imaging of single fluorescing molecules possible. A single fluorophore attached to a protein molecule in aqueous solution was first observed in 1995 by using total internal reflection fluorescence microscopy (TIRFM) and conventional epifluorescence microscopy [89]. That it is possible to image single dye molecules even with a conventional video camera was demonstrated by Adachi et al. [90], where individual Cy3 molecules were used to image actin against heavy meromyosin sliding.

Two types of cameras are most suitable for single-molecule imaging: intensified CCD cameras and cooled, back-illuminated CCD cameras. Intensified

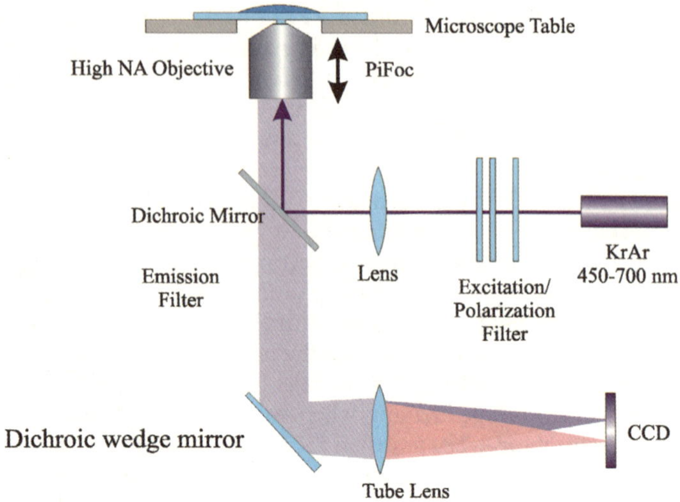

Fig. 7.5. Schematic of a single-molecule-sensitive wide-field microscope. The polarization and/or spectrally dividing optical element (a Wollaston prism and/or a wedged dichroic mirror as shown in the figure) splits the image into sub-images with different polarization and/or with different emission spectra, which are subsequently imaged at different positions on the same CCD chip. An essential element of the setup is the high-sensitivity CCD detector, in many cases a back-illuminated and cooled chip. In this way, multi-channel single-molecule imaging is realized; for details see Ref. [95]

cameras use a solid-state multi-channel plate (MCP) for amplifying the photoelectrons generated by photons on a photoelectric layer. However, they suffer from low spatial resolution (due to the coarse structure of MCPs compared with pixel size and packing on CCD chips) and high cost. Taking into account the recent emergence of affordable, high-sensitivity and Peltier-cooled CCD cameras on the market, the latter are much to be preferred for most single-molecule imaging applications. An exception may be applications where one desires to use the gating capability of MCPs on nanosecond timescales to allow, for example, time-resolved fluorescence imaging [91].

The basic schemes of single-molecule wide-field microscopy (smWFM) are shown in Fig. 7.5. Two principally distinct excitation methods are widely used: excitation by total internal reflection (so-called total internal reflection fluorescence microscopy – TIRFM), and direct wide-field illumination. For TIRF illumination, the collimated laser beam is coupled into the microscope objective off-axis so that it is converted into plane waves hitting the sample surface above the angle of total internal reflection. For achieving that, one has to use objectives with a numerical aperture larger than the sample's index of refraction [92]. Paige et al. [81] compared both TIRFM and direct epifluorescence illumination for SMS applications, and Vacha and Kotani [93] used alternate switching between both excitation modes for obtaining information about three-dimensional SM orientation. They used the fact that TIRF illumination produces a strong electric-field component perpendicular to the sample and, for a given in-plane angle o, the molecular dipole provides unambiguous orientation of the 3D molecular dipole moment. An important improvement in SM imaging was the ability to simultaneously record images at different wavelengths and polarizations, for dual-colour and/or polarization-resolved studies [94, 95]. Ma et al. [96] imaged complete spectra of individual molecules by placing a grating in front of the CCD.

The advantage of smWFM is the high imaging speed: imaging rates of ca. 1 kHz can be achieved for small regions of interest. Thus, smWFM is the ideal tool where molecular motion is of the order of several microns per second. A further important feature of single-molecule imaging is the possibility of localizing a single molecule with an accuracy of several dozen nanometres, much below the resolution limit of the imaging optics. The principle behind this is the knowledge that a single molecule is acting as a point emitter of light, generating a well-defined diffraction pattern on the CCD. By fitting the maximum position of that pattern, the lateral position of the imaged molecules can be determined with ca. 40-nm resolution [97]. The position along the optical axis can be determined with similar accuracy: when the molecule moves away from the focal plane of the objective, its image broadens in proportion to its distance from the focal plane [98, 99]. Pioneering work on imaging the diffusion of single molecules within synthetic lipid membranes as well as within cell membranes was done by the group of Hansgeorg Schindler [100–102]. The method allowed the observation of anomalous diffusion and membrane reorganization on a single molecule level. The basic limitation of this method is the finite number of photons that is emitted before irreversible photobleaching of the dye molecule, thus limiting the maximum length of the observable trajectory of a single molecule. A different application of the method was realized by the group of Toshio Yanagida: single-molecule imaging was used to follow the motion and working of individual motor proteins and ATPase. The observation of multiply labelled sliding actin filaments stood at the beginning of this exciting development [103]. One of the most fascinating applications of the single-molecule imaging technique was the direct visualization of turnovers of single ATP molecules by ATPase [89, 92, 104]. Another story of success was the direct imaging of the sliding motion of single kinesins along microtubules [105, 106]. A powerful extension of the method is its combination with optical tweezers for performing simultaneous imaging and force measurements on a single molecule level [107].

The ultimate goal for biological applications of smWFM is the observation of single molecules within living cells. The main obstacle on the route to achieving this goal is the autofluorescence of cellular compartments. The fluorescence brightness of native biomolecules (mostly NADH and FAD) is usually very low, but it becomes a severe problem at high concentrations of these molecules. There is no universal remedy to solve this problem, although using long-wavelength dyes together with optimized emission filters can be very helpful. There are several unique advantages that smWFM offers for single cell studies. Many cellular processes involve only a small number of molecules – single cells can be sensitive to single ligand–receptor binding events. Furthermore, as already mentioned, single molecules can be localized within the cell with much higher precision than the optical resolution limit of the used optics. Last but not least, following cyclic processes on a single molecule level does not necessitate an external synchronization, as is necessary when observing on a large ensemble of molecules. A number of exciting studies have been published showing both the feasibility and the great potential of smWFM for in vivo cells. Excellent reviews of this field can be found in Refs. [102] and [108].

Imaging techniques can be used to directly read the three-dimensional orientation of the *emission* dipole of single molecules. This can be achieved by introducing optical aberrations when imaging, or by simple defocused imaging. Then, every molecule is imaged onto the CCD detector not as a sharp diffraction-limited image, but as a complicated interference pattern containing information about the angular distribution of the imaged molecule. Aberrational imaging for determining molecular orientation was first demonstrated by Dickson et al. [109] and Bartko and Dickson [110, 111]. A similar approach is to purposely defocus the imaging of the molecules [112].

7.4 Data Acquisition and Evaluation

There are two principally different approaches for evaluating SMS data. The first set of methods can be called *fluctuation analysis* and processes the data with no explicit recourse to the single molecule nature of the experiment. These methods exploit the fluctuations of the measured light intensity that are caused by the fluctuating number of molecules within the detection volume. The most prominent representative of these methods is the so-called *fluorescence correlation spectroscopy* (FCS), but recently new and promising techniques such as *fluorescence intensity distribution analysis* (FIDA) or *photon-counting histogramming* (PCH) have been developed. These methods will be discussed in the next section.

The second set of methods explicitly uses the fact that the measured data are mostly generated by single-molecule transits through the detection volume. Here, every single-molecule transit is identified and isolated within the continuous data stream, and every transit is then individually analysed. These methods can be called *burst-by-burst* or *molecule-by-molecule analysis*.

7.4.1 Time-Tagged and Time-Correlated Photon Counting

Before discussing the above-mentioned different techniques of data analysis, a few principal remarks are necessary concerning the acquisition of SMS raw data. Besides recording the temporal behaviour of the *fluorescence intensity*, a very important and useful measurement parameter is the *fluorescence lifetime*. In SMS experiments, fluorescence lifetime is exclusively measured by time-correlated single-photon counting (TCSPC). Like FCS, TCSPC is also a correlation technique, but on a completely different timescale. In TCSPC, fluorescence excitation is performed by a pulsed laser with sufficiently short pulse width (usually in the femtosecond to picosecond time range) and high pulse repetition rate (several megahertz). For every detected photon, the time between the last exciting laser pulse and the photon arrival is measured. The time delays between laser pulses and photon detection events give direct information about the lifetime of the excited state of the molecule. For example, for a molecule following the simple photophysics depicted in Fig. 7.1, histogramming all delay times results in an exponentially decaying curve with a decay constant equal to the lifetime of the excited state or fluorescence lifetime (for a detailed introduction to TCSPC see the classical book by O'Connor and Phillips [113]). While TCSPC measurements were still a technical challenge some 20 years ago, the last decade has seen a revolutionary development of affordable and easy-to-use pulsed diode lasers and the arrival of the compact, self-contained high-speed electronics necessary for the timing of photon arrival times on a picosecond timescale.

The first successful TCSPC measurements on single molecules in solution were reported a decade ago [114, 115]. In SMS applications of TCSPC, data acquisition usually proceeds in an asynchronous way: data are collected and stored only when a photon is detected, tagging every photon with its arrival time on a picosecond timescale (TCSPC time) and on a 'macroscopic' timescale with temporal resolution of ca. 100 ns [87]. This assures that data are generated only if photons are detected, preventing the generation of large data flows as in synchronous data acquisition (synchronous counting of detected photons within evenly spaced subsequent time intervals). Subsequently, any desirable analysis of the data can be performed, e.g. FCS, TCSPC histogramming etc.

As already mentioned, TCSPC addresses the fluorescence lifetime of the detected molecules, which may be of interest as a photophysical parameter by itself. Besides that, TCSPC measurements offer the additional possibility of gating off photons that arrive directly with the laser pulse (detected within a few hundred picoseconds due to the finite width of the instrumental response function), which may be an effective way of additionally discriminating Raman scattering (occurring directly after the exciting laser pulse) against the desired single-molecule fluorescence (arriving at longer timescales after the laser pulse) [116].

One of the most interesting applications of TCSPC is to use the fluorescence lifetime information for distinguishing between different species of molecules. This can be done in combination with a statistical analysis such as FCS [117], or on a molecule-by-molecule basis [118–120]. In this sense, TCSPC is an elegant alternative or addition to multi-colour detection measurements [121].

7.4.2 Fluorescence Correlation Spectroscopy

Fluorescence correlation spectroscopy is a relatively old technique, originally introduced by Elson, Magde and Webb in the early 1970s [122–124]. However, it took nearly two decades, until the development of new lasers with high beam quality and temporal stability, low-noise single-photon detectors, and high-quality microscope objectives with nearly perfect imaging quality at high numerical aperture, for the technique to see a new renaissance in SMS. Achieving values of the detection volume within the range of a few cubic microns made the technique applicable to samples at reasonably high concentrations and short measurement times. An excellent and extensive description of this analysis method can be found in Ref. [125].

In FCS, the detected fluorescence intensity is correlated with a time-shifted replica of itself at different values of time shift. The result is the so-called autocorrelation function (second-order correlation function), which is calculated as

$$g(\tau) = \langle I(t) I(t+\tau) \rangle_t, \tag{7.1}$$

where $I(t)$ is the fluorescence intensity at time t, and the triangular brackets denote averaging over all time values t. The physical meaning of the autocorrelation as defined above is that it is directly proportional to the probability of detecting, on average, a photon at time τ if there was a detection event at time zero. This probability is composed of two basically different terms. The two photons detected at time zero and at time τ can originate from uncorrelated background or from different fluorescing molecules and therefore do not have any physical correlation (provided there is no interaction of the different fluorescing molecules). These events will contribute to a constant offset of $g(\tau)$ that is completely independent of τ. Alternatively, the two photons originate from one and the same molecule and are therefore physically correlated, leading to a time-dependent component of $g(\tau)$. Thus, the temporal behaviour of the autocorrelation function is solely determined by the photophysics and dynamics of individual molecules. In this sense, FCS is a true SMS technique, although the analysis is not explicitly identifying single molecule detection events. However, although FCS relies on the small number of fluctuations of molecules within the detection volume, any information about *individual* molecules is lost during the data analysis. The final result in FCS is the autocorrelation function $g(\tau)$ which is averaged over many single-molecule transits through the detection volume.

On different timescales, the temporal behaviour of the autocorrelation function is determined by different properties of the fluorescing molecules. On a nanosecond timescale, photon antibunching can be observed, reflecting the fact that directly after the emission of a photon the molecule needs to get re-excited again to be able to emit the next photon, leading to a steep decrease of $g(\tau)$ towards short times. On a microsecond timescale, $g(\tau)$ is dominated by triplet state dynamics. If excitation and/or detection are performed with polarization filters, the autocorrelation will also show contributions from rotational diffusion dynamics of the molecules. On a millisecond to second level, the autocorrelation function shows a typical decay due to the lateral diffusion of the molecules out of the detection region. The diffusion coefficient is also the parameter most frequently addressed by FCS measurements. In summary, the autocorrelation function contains information about:

- Translational and rotational diffusion of the fluorescent molecules
- Their average number within the detection volume (thus concentration)
- Triplet state dynamics
- Possible photoisomerization kinetics
- Reaction kinetics between different fluorescing molecular species.

The advantage of FCS is the relative simplicity of the analysis. Its drawback is that it works only within a very limited concentration range. If the concentration of fluorescing molecules becomes too large (typically $\gg 10^{-8}$ M), then the contribution from correlated photons from individual molecules, scaling with the number N of molecules within the detection volume, becomes very small compared with the contribution by uncorrelated photons from different molecules, scaling with N^2. If the concentration is too low (typically $<10^{-13}$ M), then the probability of finding a molecule within the detection region becomes extremely low. In both cases, the measurement time for obtaining a high-quality autocorrelation function gets prohibitively large. A remedy to that problem at very low concentrations is to rapidly scan the laser focus through the solution [126–128].

In contrast to single-channel FCS, where fluorescence is monitored at a single wavelength region only, multi-channel FCS uses two or more detection channels at different wavelength regions and subsequently applies a cross-correlation analysis [129]. Alternatively, two-photon excitation at a single wavelength is used together with multi-colour detection allowing, for example, monitoring of the binding dynamics of different molecular species with different emission spectra [130], or studies of the dissociation kinetics of bound molecules [131, 132]. The two challenges of multi-colour detection FCS are, firstly, to achieve an exact overlap of the detection volumes at different detection wavelengths, which may be difficult due to chromatic aberrations, and secondly, the necessity to use either several lasers for efficiently exciting molecules with different emission wavelengths, or a femtosecond pulsed laser for two-photon excitation of fluorescence, exploiting the broad two-photon absorption spectra of most molecules [133–135].

The technical challenge of multi-channel excitation/detection is to assure perfect overlap of the detection regions at different wavelengths, which is not a trivial task due to chromatic aberrations of most optical components. An alternative is to use the fluorescence lifetime instead of the fluorescence spectrum to distinguish between different molecules. By applying a sophisticated analysis that combines autocorrelation with fluorescence-lifetime measurements, a quasi-multi-channel FCS without the need of multiple excitation and/or detection channels can be realized [117].

A critical aspect of FCS is that it relies heavily on the correctness of the assumed models used for data evaluation, particularly the assumptions made about molecular diffusion and the geometry of the detection region. The final result of an FCS experiment is the autocorrelation function that is condensed out of a large file of recorded fluorescence intensities. When applying a model to that autocorrelation function, it is often difficult to distinguish whether something like anomalous diffusion of the observed molecules has caused a deviation of the autocorrelation from the ideal model, or whether some technical artefact was responsible for that deviation. For example, Balakrishnan [136] showed the influence of spatial curvature effects on FCS of molecules diffusing within membranes, and Chirico et al. [137] modelled a similar although small effect for dyes in solution that was earlier observed by Osborne et al. [138]. Egner et al. [139] made a detailed study of the aberration effects introduced by small refractive index mismatches, which may be a problem for correct FCS data evaluation when working in biological samples, although Ganic et al. [140] showed that such aberrational effects may be reduced when using two-photon excitation. Fradin et al. [141] showed that FCS measurements inside cells can lead to erroneous values of the diffusion coefficient if the influence of membranes is not recognized, and

Gennerich and Schild [142] demonstrated that FCS in small cytosolic compartments can lead to gross errors in diffusion coefficients, if confinement effects are not correctly taken into account.

FCS has been extensively used in numerous studies, but only some of those can be cited here. FCS was used for studying diffusion of molecules in homogeneous and heterogeneous environments [143–145], intermolecular binding and reaction kinetics [146–150], single molecule photophysics [151–159] and conformational dynamics [160]. For recent reviews see Refs. [161, 162] and the book [163].

7.4.3 Fluorescence Intensity Distribution Analysis and Related Techniques

A rather recent method of fluctuation analysis is fluorescence intensity distribution analysis (FIDA) [164, 165], or photon-counting histogramming [166, 167]. In FIDA, the detected photons are binned into time intervals of equal length (bin width usually in the range 10 µs up to 1 ms), and the obtained stream of photons per time bin is subsequently histogrammed (frequency of occurrence versus number of photons per time bin). The shape of the histograms depends, in a complicated way, on the intensity of the background, the fluorescence properties of the molecules, the molecules' concentration and, at large values of time-bin width, their diffusion. The molecular properties that are addressed by FIDA are concentration and fluorescence brightness. The latter is defined as the product of the absorption cross section and fluorescence quantum yield. In contrast to FCS, the exact analysis of the data, i.e. the connection between the shape of the experimentally obtained histograms and the fluorescence brightness, is very complicated.

The striking advantage of this method, when compared with FCS, is that it addresses molecular brightness instead of molecular diffusion, a parameter that changes much more from molecular species to molecular species than the diffusion coefficient. Thus, it is the method of choice when analysing mixtures of several different molecular species, as well as for studying processes such as dimerization. Moreover, molecular brightness is a more robust molecular parameter than the diffusion coefficient [33]. A potentially strong application of the FIDA or PCH method is the elucidation of the stoichiometry of macromolecular complexes or the number of binding sites on a macromolecule [168].

An extension of FIDA is fluorescence intensity multiple distribution analysis (FIMDA), where the data are histogrammed for different bin widths of the time intervals [169]. With increasing values of bin width, diffusion will have an increasing impact on the exact form of the histograms, so that diffusion coefficients can also be extracted from the analysis of the multiple histograms. As in the case of FCS, any combination of FIDA/FIMDA with multi-colour detection and/or fluorescence lifetime measurements is feasible.

7.4.4 Molecule-by-Molecule Analysis

Until now, only correlation or statistical analysis techniques of SMD data have been considered. FCS and FIDA perform a statistical analysis of photon detection intensities, whereas TCSPC correlates photon arrival times with the pulses of a pulsed laser. A different approach to analysing single molecule measurements is a molecule-by-molecule, or more correctly, a burst-by-burst analysis (a single photon burst may originate from more than only a single molecule present within the detection region). A first example of such a technique has already been mentioned above in connection with using TCSPC information to distinguish single molecules on a molecule-by-molecule basis. This approach can be generalized to other photophysical parameters such as the emission spectrum and polarization.

The starting point of all burst-by-burst methods is a suitable algorithm for identifying single bursts within the detected photon stream. One of the most powerful methods proved to be the so-called Lee filter as described in Ref. [170]. It identifies contiguous temporal regions where the photon detection intensities exceed a preset threshold. The algorithm takes into account that, due to shot noise in the photon detection and the stochastic character of the diffusing molecule's transit through the detection region, detection intensities may intermittently fall below the threshold during a single molecule's transit.

After having identified single molecule events, the photons belonging to each event can be analysed according to their TCSPC time [120, 121, 171–173], polarization [174] and colour [121, 173], in correspondence with parameters that were recorded by the measurement system. Finally, distributions of any desired parameters are obtained by histogramming the determined parameter values over all detected single molecule events. There are two fundamental and important differences of such burst-by-burst analysis when compared with a bulk analysis of the data (taking the information from all detected photons together without identifying single molecule events). First, cutting out single molecule events eliminates a large amount of background (low intensity scattering, detector and electronic noise) from the analysis, thus boosting the effective signal-to-background ratio of the analysis. The second difference becomes important when performing measurements on a mixture of different molecular species. In a bulk analysis, it is impossible to decide whether there is a single molecular species having a bimodal value distribution of some parameter (e.g. a two-exponential fluorescence decay), or whether there are two different molecular species with distinct average values of the recorded parameter (e.g. two species having a mono-exponential fluorescence decay time).

The technique of burst-by-burst analysis was perfected and pushed to its limit in a series of papers by the group of Claus Seidel [175–178] employing a specific burst analysis which they denote as multi-parameter fluorescence detection (MFD). While keeping the experimental advantage of monitoring single molecules diffusing through the microscopic open volume element of a confocal epi-illuminated setup as in FCS, MFD uses pulsed excitation and time-correlated single-photon counting to simultaneously monitor the evolution of four fluorescence parameters: intensity, lifetime, anisotropy and spectral signature. In an example that pushed the potential of the method to its limits, it was shown that one is able to distinguish between up to 16 different molecular species within one and the same sample, a task impossible to achieve by bulk analysis [179]. The method was applied to analysing conformational dynamics in DNA [175] and ATPase [180].

Acknowledgements
I thank my colleagues Jörg Fitter, Ingo Gregor, Benjamin Kaupp and Digambara Patra as well as Markus Sauer from the University of Bielefeld for many helpful discussions and suggestions. I am grateful to the Deutsche Forschungsgemeinschaft and the Volkswagenstiftung for their financial support of my work.

References

1. Rotman B (1961) Measurement of activity of single molecules of β-D-galactosidase. Proc Natl Acad Sci USA 47:1981–1991
2. Rotman B (1973) Measurement of single molecules of antibody by their ability to activate a deflective enzyme. In: Thaer AA, Sernetz M (eds) Fluorescence techniques in cell biology. Springer, Berlin Heidelberg New York, pp 333–337
3. Hirschfeld T (1976) Optical microscopic observation of single small molecules. Appl Opt 15:2965–2966
4. Hirschfeld T (1976) Quantum efficiency independence of the time integrated emission from a fluorescent molecule. Appl Opt 15:3135–3138
5. Dovichi NJ, Martin JC, Jett JH, Keller RA (1983) Attogram detection limit for aqueous dye samples by laser-induced fluorescence. Science 219:845–847
6. Shera EB, Seitzinger NK, Davis LM, Keller RA, Soper SA (1990) Detection of single fluorescent molecules. Chem Phys Lett 174:553–557
7. Meixner AJ (1998) Optical single-molecule detection at room temperature. Adv Photochem 24:1–59
8. Xie XS, Trautman JK (1998) Optical studies of single molecules at room temperature. Annu Rev Phys Chem 49:441–480
9. Enderlein J, Ambrose WP, Goodwin PM, Keller RA (1999) Fluorescence detection of single molecules applicable to small volume assays. In: Köhler M, Mejevaia T, Saluz HP (eds) Microsystem technology: A powerful tool for biomolecular studies. Birkhäuser, Basel, pp 311–329
10. Weiss S (1999) Fluorescence spectroscopy of single biomolecules. Science 283:1676–1683
11. Ambrose WP, Goodwin PM, Jett JH, van Orden A, Werner JH, Keller RA (1999) Single molecule fluorescence spectroscopy at ambient temperature. Chem Rev 99:2929–2956
12. Ishii Y, Yanagida T (2000) Single molecule detection in life science. Single Mol 1:5–14
13. Keller RA, Emory SR, Ambrose WP, Goodwin PM, Arias AA, Jett JJ, Cai H (2002) Analytical applications of single-molecule detection. Anal Chem 74:316–324A
14. Rigler R, Orrit M, Basché T (eds) (2001) Single molecule spectroscopy. Springer, Berlin Heidelberg New York
15. Zander C, Enderlein J, Keller RA (eds) (2002) Single molecule detection in solution. Wiley-VCH, Berlin
16. Orrit M, Bernard J, Brown R, Lounis B (1996) Optical spectroscopy of single molecule in solids. In: Wolf E (ed) Progress in optics, vol XXXV. Elsevier, Amsterdam, pp 63–144
17. Trautman JK, Macklin JJ, Brus LE, Betzig E (1994) Near-field spectroscopy of single molecules at room temperature. Nature 369:40–42
18. Trautman JK, Macklin JJ (1996) Time-resolved spectroscopy of single molecules using near-field and far-field optics. Chem Phys 205:221–229
19. Trautman JK, Ambrose WP (1997) Near-field optical imaging and spectroscopy of single molecules. In: Basché T, Moerner WE, Orrit M, Wild UP (eds) Single molecule optical detection, imaging and spectroscopy. VCH, Weinheim, pp 191–222
20. Ruiter AGT, Veerman JA, Garcia-Parajo MF, van Hulst NF (1997) Single-molecule rotational and translational diffusion observed by near-field scanning optical microscopy. J Phys Chem A 101:7318–7323
21. Garcia-Parajo MF, Veerman JA, van Noort SJT, Grooth BG de, Greve J, van Hulst NF (1998) Near-field optical microscopy for DNA studies at the single molecular level. Bioimaging 6:43–53

22. Garcia-Parajo MF, Veerman JA, Segers-Nolten GMJ, de Grooth BG, Greve J, van Hulst NF (1999) Visualising individual green fluorescent proteins with a near field optical microscope. Cytometry 36:239–246
23. Gersen H, Garcia-Parajo MF, Novotny L, Veerman JA, Kuipers L, van Hulst NF (2000) Influencing the angular emission of a single molecule. Phys Rev Lett 85:5312–5315
24. Gersen H, Garcia-Parajo MF, Novotny L, Veerman JA, Kuipers L, van Hulst NF (2001) Near-field effects in single molecule emission. J Microsc 202:374–378
25. van Hulst NF, Veerman JA, García-Parajó MF, Kuipers L (2000) Analysis of individual (macro)molecules and proteins using near-field optics. J Chem Phys 112:7790–7810
26. Göhde W, Fischer UC, Fuchs H, Tittel J, Basché T, Bräuchle C, Herrmann A, Müllen K (1998) Fluorescence blinking and photobleaching of single terrylenediimide molecules studied with a confocal microscope. J Phys Chem A 102:9109–9116
27. Panzer O, Göhde W, Fischer UC, Fuchs H, Müllen K (1998) Influence of oxygen on single molecule blinking. Adv Mat 10:1469–1472
28. Weston KD, Buratto SK (1998) Millisecond intensity fluctuations of single molecules at room temperature. J Phys Chem A 102:3635–3638
29. Garcia-Parajo MF, Segers-Nolten GMJ, Veerman JA, Greve J, van Hulst NF (2000) Real-time light-driven dynamics of the fluorescence emission in single green fluorescent protein molecules. Proc Natl Acad Sci USA 97:7237–7242
30. Dickson RM, Cubitt AB, Tsien RY, Moerner WE (1997) On/off blinking and switching behavior of single molecules of green fluorescent protein. Nature 388:355–358
31. Moerner WE (2002) Single-molecule optical spectroscopy of autofluorescent proteins. J Chem Phys 117:10925–10937
32. Chen Y, Müller JD, Ruan QQ, Gratton E (2002) Molecular brightness characterization of EGFP in vivo by fluorescence fluctuation spectroscopy. Biophys J 82:133–144
33. Kubitscheck U, Kuckmann O, Kues T, Peters R (2000) Imaging and tracking of single GFP molecules in solution. Biophys J 78:2170–2179
34. Chan WCW, Nie S (1998) Quantum dot bioconjugates for ultrasensitive nonisotopic detection. Science 281:2016–2019
35. Han M, Gao X, Su JZ, Nie S (2001) Quantum-dot-tagged microbeads for multiplexed optical coding of biomolecules. Nat Biotechnol 19:631–635
36. Dahan M, Laurence T, Pinaud F, Chemla DS, Alivisatos AP, Sauer M, Weiss S (2001) Time-gated biological imaging by use of colloidal quantum dots. Opt Lett 26:825–827
37. Dahan M, Levi S, Luccardini C, Rostaing P, Riveau B, Triller A (2003) Diffusion dynamics of glycine receptors revealed by single-quantum dot tracking. Science 302:442–445
38. Jelezko F, Tietz C, Gruber A, Popa I, Nizovtsev A, Kilin S, Wrachtrup J (2001) Spectroscopy of single N-V centers in diamond. Single Mol 2:255–260
39. Lee TH, Gonzalez JI, Dickson RM (2002) Strongly enhanced field-dependent single-molecule electroluminescence. Proc Natl Acad Sci USA 99:10272–10275
40. Peyser LA, Vinson AE, Bartko AP, Dickson RM (2001) Photoactivated fluorescence from individual silver nanoclusters. Science 291:103–106
41. Boyer D, Tamarat P, Maali A, Orrit M, Lounis B (2003) Imaging single metal nanoparticles in scattering media by photothermal interference contrast. Physica E 17:537–540
42. Maali A, Cardinalb T, Treguer-Delapierre M (2003) Intrinsic fluorescence from individual silver nanoparticles. Physica E 17:559–560
43. Cognet L, Tardin C, Boyer D, Choquet D, Tamarat P, Lounis B (2003) Single metallic nanoparticle imaging for protein detection in cells. Proc Natl Acad Sci USA 100:11350–11355
44. Soper SA, Wabuyele M, Owens CV, Hammer RP (2002) Single molecule detection in the near-infrared. In: Zander C, Enderlein J, Keller RA (eds) Single-molecule detection in solution – methods and applications. Wiley-VCH, Berlin, pp 323-362
45. Arden-Jacob J, Frantzeskos J, Kemnitzer NU, Zilles A, Drexhage KH (2001) New fluorescent markers for the red region. Spectrochim Acta A 57:2271–2283

46. Lakowicz JR (2001) Radiative decay engineering: biophysical and biomedical applications. Anal Biochem 298:1–24
47. Enderlein J (2002) Theoretical study of single molecule fluorescence in a metallic nanocavity. Appl Phys Lett 80:315–317
48. Enderlein J (2002) Spectral properties of a fluorescing molecule within a spherical metallic nanocavity. Phys Chem Chem Phys 4:2780–2786
49. Kachel V, Fellner-Feldegg H, Menke E (1990) Hydrodynamic properties of flow cytometry instruments. In: Melamed MR, Lindmo T, Mendelsohn ML (eds) Flow cytometry and sorting. Wiley-Liss, New York, pp 27–44
50. van Orden A, Machara NP, Goodwin PM, Keller RA (1998) Single molecule identification in flowing sample streams by fluorescence burst size and intraburst fluorescence decay rate. Anal Chem 70:1444–1451
51. Ambrose WP, Cai H, Goodwin PM, Jett JH, Habbersett RC, Larson EJ, Grace WK, Werner JH, Marrone BL, Keller RA (2000) Flow fragment sizing of DNA fragments. In: Lakowicz JR (ed) Topics in fluorescence, vol 7. Plenum, New York
52. Larson EJ, Hakovirta JR, Cai H, Jett JH, Burde S, Keller RA, Marrone BL (2000) Rapid DNA fingerprinting of pathogens by flow cytometry. Cytometry 41:203–208
53. Cai H, White PS, Torney D, Deshpande A, Wang Z, Marrone B, Nolan JP (2000) Flow cytometry-based minisequencing: a new platform for high-throughput single-nucleotide polymorphism scoring. Genomics 66:135–143
54. Lee YH, Maus RG, Smith BW, Winefordner JD (1994) Laser-induced fluorescence detection of a single molecule in a capillary. Anal Chem 66:4142–4149
55. Guenard RD, King LA, Smith BW, Winefordner JD (1997) 2-Channel sequential single-molecule measurement. Anal Chem 69:2426–2433
56. Zander C, Drexhage KH, Han KT, Wolfrum J, Sauer M (1998) Single-molecule counting and identification in a microcapillary. Chem Phys Lett 286:457–465
57. Neumann M, Ehrfeld W, Han KT, Jacob P, Konrad R, Siebert S, Wolfrum J, Sauer M (1999) Electrophoretic sizing of DNA fragments and detection of single mononucleotide molecules in microstructures. Biomed Chromatogr 13:109–110
58. Sauer M, Angerer B, Han K-T, Zander C (1999) Detection and identification of single dye-labeled mononucleotide molecules released from an optical fiber in a microcapillary: first steps towards a new single molecule DNA sequencing technique. Phys Chem Chem Phys 1:2471–2477
59. Becker W, Hickl H, Zander C, Drexhage KH, Sauer M, Siebert S, Wolfrum J (1999) Time-resolved detection and identification of single analyte molecules in microcapillaries by time-correlated single-photon counting (TCSPC). Rev Sci Instrum 70:1835–1841
60. Fister JC, Jacobson SC, Davis LM, Ramsey JM (1998) Counting single chromophore molecules for ultrasensitive analysis and separations on microchip devices. Anal Chem 70:431–437
61. Effenhauser CS, Bruin GJM, Paulus A, Ehrat M (1997) Integrated capillary electrophoresis on flexible silicone microdevices: analysis of DNA restriction fragments and detection of single DNA molecules on microchips. Anal Chem 69:3451–3457
62. Chou HP, Spence C, Scherer A, Quake S (1999) A microfabricated device for sizing and sorting DNA molecules. Proc Natl Acad Sci USA 96:11–13
63. Foquet M, Korlach J, Zipfel W, Webb WW, Craighead HG (2002) DNA fragment sizing by single molecule detection in submicrometer-sized closed fluidic channels. Anal Chem 74:1415–1422
64. Chen DY, Adelhelm K, Cheng XL, Dovichi NJ (1994) Simple laser-induced fluorescence detector for sulforhodamine 101 in a capillary electrophoresis system – detection limits of 10 yoctomoles or six molecules. Analyst 119:349–352
65. Castro A, Shera EB (1995) Single-molecule electrophoresis. Anal Chem 67:3181–3186
66. Castro A, Shera EB (1995) Single-molecule detection: applications to ultrasensitive biochemical analysis. Appl Opt 34:3218–3222
67. Castro A, Williams JGK (1997) Single-molecule detection of specific nucleic-acid sequences in unamplified genomic DNA. Anal Chem 69:3915–3920

68. Haab BB, Mathies RA (1995) Single molecule fluorescence burst detection of DNA fragments separated by capillary electrophoresis. Anal Chem 67:3253–3260
69. Haab BB, Mathies RA (1997) Optimization of single-molecule fluorescence burst detection of ds-DNA – application to capillary electrophoresis separations of 100–1,000 basepair fragments. Appl Spetrosc 51:1579–1584
70. Shortreed MR, Li H, Huang WH, Yeung ES (2000) High-throughput single-molecule DNA screening based on electrophoresis. Anal Chem 72:2879–2885
71. Castro A (2002) Single molecule detection of specific nucleic acid sequences. In: Zander C, Enderlein J, Keller RA (eds) Single-molecule detection in solution – methods and applications. Wiley-VCH, Berlin, pp 303–321
72. Whitten WB, Arnold JM, Ramsey S, Bronk BV (1991) Single-molecule detection limits in levitated microdroplets. Anal Chem 63:1027–1031
73. Barnes MD, Ng KC, Whitten WB, Ramsey JM (1993) Detection of single rhodamine 6G molecules in levitated microdroplets. Anal Chem 65:2360–2365
74. Barnes MD, Lermer N, Kung CY, Whitten WB, Ramsey JM, Hill SC (1997) Real-time observation of single-molecule fluorescence in microdroplet streams. Opt Lett 22: 1265–1267
75. Lermer N, Barnes MD, Kung CY, Whitten WB, Ramsey JM (1997) High-efficiency molecular counting in solution: single-molecule detection in electrodynamically focused microdroplet streams. Anal Chem 69:2115–2121
76. Kung CY, Barnes MD, Lermer N, Whitten WB, Ramsey JM (1998) Confinement and manipulation of individual molecules in attoliter volumes. Anal Chem 70:658–661
77. Barnes MD, Kung C-Y, Whitten WB, Ramsey JM, Arnold S, Holler S (1996) Fluorescence of oriented molecules in a microcavity. Phys Rev Lett 76:3931–3934
78. Kung CY, Barnes MD, Lermer N, Whitten WB, Ramsey JM (1999) Single-molecule analysis of ultradilute solutions with guided streams of 1-µm water droplets. Appl Opt 38:1481–1487
79. Hansen RL, Harris JM (1998) Measuring reversible adsorption kinetics of small molecules at solid/liquid interfaces by total internal reflection fluorescence correlation spectroscopy. Anal Chem 70:4247–4256
80. Hansen RL, Harris JM (1998) Total internal reflection fluorescence correlation spectroscopy for counting molecules at liquid/solid interfaces. Anal Chem 70:2565–2575
81. Paige MF, Bierneld EJ, Moerner WE (2001) A comparison of through-the-objective total internal reflection microscopy and epifluorescence microscopy for single-molecule fluorescence imaging. Single Mol 2:191–201
82. Ambrose WP, Goodwin PM, Nolan JP (1999) Single-molecule detection with total internal reflection excitation: comparing signal-to-background and total signals in different geometries. Cytometry 36:224–231
83. Ruckstuhl T, Enderlein J, Jung S, Seeger S (2000) Forbidden light detection from single molecules. Anal Chem 72:2117–2123
84. Ruckstuhl T, Rankl M, Seeger S (2003) Highly sensitive biosensing using a supercritical angle fluorescence (SAF) instrument. Biosens Bioelectron 18:1193–1199
85. Krieg A, Laib S, Ruckstuhl T, Seeger S (2003) Real-time detection of nucleotide incorporation during complementary DNA strand synthesis. Chembiochem 4:589–592
86. Ruckstuhl T, Seeger S (2003) Confocal total-internal-reflection fluorescence microscopy with a high-aperture parabolic mirror lens. Appl Opt 42:3277–3283
87. Böhmer M, Pampaloni F, Wahl M, Rahn HJ, Erdmann R, Enderlein J (2001) Advanced time-resolved confocal scanning device for ultrasensitive fluorescence detection. Rev Sci Instrum 72:4145–4152
88. Böhmer M, Enderlein J (2002) Single molecule detection on surfaces with the confocal laser scanning microscope. In: Zander C, Enderlein J, Keller RA (eds) Single-molecule detection in solution – methods and applications. Wiley-VCH, Berlin, pp 145–183
89. Funatsu T, Harada Y, Tokunaga M, Saito K, Yanagida T (1995) Imaging of single fluorescent molecules and individual ATP turnovers by single myosin molecules in aqueous solution. Nature 374:555–559

90. Adachi K, Kinosita K, Ando T (1999) Single-fluorophore imaging with an unmodified epifluorescence microscope and conventional video camera. J Microsc 195:125–132
91. Straub M, Hell SW (1998) Fluorescence lifetime three-dimensional microscopy with picosecond precision using a multifocal multiphoton microscope. Appl Phys Lett 73:1769–1771
92. Tokunaga M, Kitamura K, Saito K, Iwane AH, Yanagida T (1997) Single-molecule imaging of fluorophores and enzymatic-reactions achieved by objective-type total internal-reflection fluorescence microscopy. Biochem Biophys Res Commun 235:47–53
93. Vacha M, Kotani M (2003) Three-dimensional orientation of single molecules observed by far- and near-field fluorescence microscopy. J Chem Phys 118:5279–5282
94. Mörtelmaier M, Kögler EJ, Hesse J, Sonnleitner M, Huber LA, Schütz GJ (2002) Single molecule microscopy in living cells: subtraction of autofluorescence based on two-color recording. Single Mol 3:225–231
95. Cognet L, Harms GS, Blab GA, Lommerse PHM, Schmidt T (2000) Simultaneous dual-color and dual-polarization imaging of single molecules. Appl Phys Lett 77:4052–4054
96. Ma Y, Shortreed MR, Yeung ES (2000) High-throughput single-molecule spectroscopy in free solution. Anal Chem 72:4640–4645
97. Schütz GJ, Trabesinger W, Schmidt T (1998) Direct observation of ligand colocalization on individual receptor molecules. Biophys J 74:2223–2226
98. Schütz GJ, Hesse J, Freudenthaler G, Pastushenkoo VP, Knaus HG, Pragl B, Schindler H (2000) 3D mapping of individual ion channels on living cells. Single Mol 1:153–157
99. Schütz GJ, Axmann M, Schindler H (2001) Imaging single molecules in three dimensions. Single Mol 2:69–74
100. Schütz GJ, Schindler H, Schmidt T (1997) Single-molecule microscopy on model membranes reveals anomalous diffusion. Biophys J 73:1073–1080
101. Schütz GJ, Kada G, Pastushenko VP, Schindler HG (2000) Properties of lipid microdomains in a muscle cell membrane visualized by single molecule microscopy. EMBO J 19:892–901
102. Schütz GJ, Schindler H (2002) Single dye tracing for ultrasensitive microscopy on living cells. In: Zander C, Enderlein J, Keller RA (eds) Single-molecule detection in solution – methods and applications. Wiley-VCH, Berlin, pp 231–245
103. Yanagida T, Nakase M, Nishiyama K, Oosawa F (1984) Direct observation of motion of single F-actin filaments in the presence of myosin. Nature 307:58–60
104. Oiwa K, Eccleston JF, Anson M, Kikumoto M, Davis CT, Reid GP, Ferenczi MA, Corrie JET, Yamada A, Nakayama H, Trentham DR (2000) Comparative single-molecule and ensemble myosin enzymology: sulfoindocyanine ATP and ADP derivatives. Biophys J 78:3048– 3071
105. Okada Y, Hirokawa N (1999) A processive single-headed motor: kinesin superfamily protein KIF1A. Science 283:1152–1157
106. Inoue Y, Iwane AH, Miyai T, Muto E, Yanagida T (2001) Motility of single one-headed kinesin molecules along microtubules. Biophys J 81:2838–28350
107. Ishii Y, Iwane AH, Yokota H, Inoue Y, Wazawa T, Nishiyama M, Tanaka H, Kitamura K, Yanagida T (2002) Studying molecular motors on the single molecule level. In: Zander C, Enderlein J, Keller RA (eds) Single-molecule detection in solution – methods and applications. Wiley-VCH, Berlin, pp 273–292
108. Kubitschek U (2002) Single protein molecules visualized and tracked in the interior of eukaryotic cells. Single Mol 3:267–274
109. Dickson RM, Norris DJ, Moerner WE (1998) Simultaneous imaging of individual molecules aligned both parallel and perpendicular to the optic axis. Phys Rev Lett 81: 5322–5325
110. Bartko AP, Dickson RM (1999) Imaging three-dimensional singe molecule orientations. J Phys Chem B 103:11237–11241
111. Bartko AP, Xu K, Dickson RM (2002) Three-dimensional single molecule rotational diffusion in glassy state polymer films. Phys Rev Lett 89:026101
112. Böhmer M, Enderlein J (2003) Orientation imaging of single molecules by wide-field epifluorescence microscopy. J Opt Soc B 20:554–559

113. O'Connor DV, Phillips D (1984) Time-correlated single photon counting. Academic, London
114. Soper SA, Davis LM, Shera EB (1992) Detection and identification of single molecules in solution. J Opt Soc Am B 9:1761–1769
115. Wilkerson CW, Goodwin PM, Ambrose WP, Martin JC, Keller RA (1993) Detection and lifetime measurement of single molecules in flowing sample streams by laser-induced fluorescence. Appl Phys Lett 62:2030–2032
116. Li L-Q, Davis LM (1993) Single photon avalanche diode for single molecule detection. Rev Sci Instrum 64:1524–1529
117. Böhmer M, Wahl M, Rahn HJ, Erdmann R, Enderlein J (2002) Time-resolved fluorescence correlation spectroscopy. Chem Phys Lett 353:439–445
118. Köllner M, Fischer P, Arden-Jacob J, Drexhage K-H, Müller R, Seeger S, Wolfrum J (1996) Fluorescence pattern recognition for ultrasensitive molecule identification: comparison of experimental data and theoretical predictions. Chem Phys Lett 250:355–360
119. Enderlein J, Goodwin PM, van Orden A, Ambrose WP, Erdmann R, Keller RA (1997) A maximum likelihood estimator to distinguish single molecules by their fluorescence decays. Chem Phys Lett 270:464–470
120. Enderlein J, Sauer M (2001) Optimal algorithm for single molecule identification with time-correlated single photon counting. J Phys Chem A 105:48–53
121. Herten DP, Tinnefeld P, Sauer M (2000) Identification of single fluorescently labeled mononucleotide molecules in solution by spectrally resolved time-correlated single photon counting. Appl Phys B 71:765–771
122. Magde D, Elson E, Webb WW (1972) Thermodynamic fluctuations in a reacting system – measurement by fluorescence correlation spectroscopy. Phys Rev Lett 29:705–708
123. Elson EL, Magde D (1974) Fluorescence correlation spectroscopy. I. Conceptual basis and theory. Biopolymers 13:1–27
124. Magde D, Elson E, Webb WW (1974) Fluorescence correlation spectroscopy. II. An experimental realization. Biopolymers 13:29–61
125. Widengren J, Mets Ü (2002) Conceptual basis of fluorescence correlation spectroscopy and related techniques as tools in bioscience. In: Zander C, Enderlein J, Keller RA (eds) Single-molecule detection in solution – methods and applications. Wiley-VCH, Berlin, pp 69–95
126. Petersen NO (1986) Scanning fluorescence correlation spectroscopy. I. Theory and simulation of aggregation measurements. Biophys J 49:809–815
127. Petersen NO, Johnson DC, Schlesinger MJ (1986) Scanning fluorescence correlation spectroscopy. II. Application to virus glycoprotein aggregation. Biophys J 49:817–820
128. Amediek A, Haustein E, Scherfeld D, Schwille P (2002) Scanning dual-color cross-correlation analysis for dynamic co-localization studies of immobile molecules. Single Mol 3:201–210
129. Schwille P, Meyer-Almes FJ, Rigler R (1997) Dual-color fluorescence cross-correlation spectroscopy for multicomponent diffusional analysis in solution. Biophys J 72:1878–1886
130. Bacia K, Majoul IV, Schwille P (2002) Probing the endocytic pathway in live cells using dual-color fluorescence cross-correlation analysis. Biophys J 83:1184–1193
131. Kettling U, Koltermann A, Schwille P, Eigen M (1998) Real-time enzyme kinetics monitored by dual-color fluorescence cross-correlation spectroscopy. Proc Natl Acad Sci USA 95:1416–1420
132. Koltermann A, Kettling U, Bieschke J, Winkler T, Eigen M (1998) Rapid assay processing by integration of dual-color fluorescence cross-correlation spectroscopy: high-throughput screening for enzyme activity. Proc Natl Acad Sci USA 95:1421–1426
133. Heinze K, Koltermann A, Schwille P (2000) Simultaneous two-photon excitation of distinct labels for dual-color fluorescence cross-correlation analysis. Proc Natl Acad Sci USA 97:10377–10382
134. Rarbach M, Kettling U, Koltermann A, Eigen M (2001) Dual-color fluorescence cross-correlation spectroscopy for monitoring the kinetics of enzyme-catalyzed reactions. Methods 24:104–116

135. Heinze KG, Rarbach M, Jahnz M, Schwille P (2002) Two-photon fluorescence coincidence analysis: rapid measurements of enzyme kinetics. Biophys J 83:1671–1681
136. Balakrishnan J (2000) Spatial curvature effects on molecular transport by diffusion. Phys Rev E 61:4648–46451
137. Chirico G, Fumagalli C, Baldini G (2002) Trapped Brownian motion in single- and two-photon excitation fluorescence correlation experiments. J Phys Chem B 106:2508– 2519
138. Osborne MA, Balasubramanian S, Furey WS, Klenerman D (1998) Optically biased diffusion of single molecules studied by confocal fluorescence microscopy. J Phys B 102: 3160–3167
139. Egner A, Schrader M, Hell SW (1998) Refractive index mismatch induced intensity and phase variations in fluorescence confocal, multiphoton and 4Pi-microscopy. Opt Commun 153:211–217
140. Ganic D, Gan X, Gu M (2000) Reduced effects of spherical aberration on penetration depth under two-photon excitation. Appl Opt 39:3945–3947
141. Fradin C, Abu-Arish A, Granek R, Elbaum M (2003) Fluorescence correlation spectroscopy close to a fluctuating membrane. Biophys J 84:2005–2020
142. Gennerich A, Schild D (2000) Fluorescence correlation spectroscopy in small cytosolic compartments depends critically on the diffusion model used. Biophys J 79:3294– 3306
143. Korlach J, Schwille P, Webb WW, Feigenson GW (1999) Characterization of lipid bilayer phases by confocal microscopy and fluorescence correlation spectroscopy. Proc Natl Acad Sci USA 69:8461–8466
144. Gell C, Brockwell DJ, Beddard GS, Radford SE, Kalverda AP, Smith DA (2001) Accurate use of single molecule fluorescence correlation spectroscopy to determine molecular diffusion times. Single Mol 2:177–181
145. Yoshida N, Tamura M, Kinjo M (2000) Fluorescence correlation spectroscopy: a new tool for probing the microenvironment of the internal space of organelles. Single Mol 1:279–283
146. Widengren J, Rigler R (1998) Fluorescence correlation spectroscopy as a tool to investigate chemical reactions in solution and on cell surfaces. Cell Mol Biol 44:857–879
147. Björling S, Kinjo M, Földes-Papp Z, Hagman E, Thyberg P, Rigler R (1998) Fluorescence correlation spectroscopy of enzymatic DNA polymerization. Biochemistry 37:12971–12978
148. Häsler K, Pänke O, Junge W (1999) On the stator of rotary ATP synthase: the binding strength of subunit δ to $\alpha\beta_3$ as determined by fluorescence correlation spectroscopy. Biochemistry 38:13759–13765
149. Schwille P, Bieschke J, Oehlenschlager F (1997) Kinetic investigations by fluorescence correlation spectroscopy: the analytical and diagnostic potential of diffusion studies. Biophys Chem 66:211–228
150. Wohland T, Friedrich K, Hovius R, Vogel H (1999) Study of ligand–receptor interactions by fluorescence correlation spectroscopy with different fluorophores: evidence that the homopentameric 5-hydroxytryptamine Type 3 as receptor binds only one ligand. Biochemistry 38:8671–8681
151. Widengren J, Mets Ü, Rigler R (1999) Photodynamic properties of green fluorescent proteins investigated by fluorescence correlation spectroscopy. Chem Phys 250:171–186
152. Haupts U, Maiti S, Schwille P, Webb WW (1998) Dynamics of fluorescence fluctuations in green fluorescent protein observed by fluorescence correlation spectroscopy. Proc Natl Acad Sci USA 95:13573–13578
153. Heikal AA, Hess ST, Baird GS, Tsien RY, Webb WW (2000) Molecular spectroscopy and dynamics of intrinsically fluorescent proteins: Coral red (dsRed) and yellow (Citrine). Proc Natl Acad Sci USA 97:11996–12001
154. Brock R, Vàmosi G, Vereb G, Jovin TM (1999) Rapid characterization of green fluorescent protein fusion proteins on the molecular and cellular level by fluorescence correlation microscopy. Proc Natl Acad Sci USA 96:10123–10128
155. Heikal AA, Hess ST, Webb WW (2001) Multiphoton molecular spectroscopy and excited-state dynamics of enhanced green fluorescent protein (EGFP): acid-base specificity. Chem Phys 274:37–55

156. Schwille P, Kummer S, Heikal AA, Moerner WE, Webb WW (2000) Fluorescence correlation spectroscopy reveals fast optical excitation-driven intramolecular dynamics of yellow fluorescent proteins. Proc Natl Acad Sci USA 97:151–156
157. Widengren J, Seidel CAM (2000) Manipulation and characterization of photo-induced transient states of Merocyanine 540 by fluorescence correlation spectroscopy. Phys Chem Chem Phys 2:3435–3441
158. Huang S, Heikal AA, Webb WW (2002) Two-photon fluorescence spectroscopy and microscopy of NAD(P)H and flavoprotein. Biophys J 82:2811–2825
159. Malvezzi-Campeggi F, Jahnz M, Heinze KG, Dittrich P, Schwille P (2001) Light-induced flickering of dsRed provides evidence for distinct and interconvertible fluorescent states. Biophys J 81:1776–1785
160. Widengren J, Schwille P (2000) Characterization of photoinduced isomerization and back-isomerization of the cyanine dye Cy5 by fluorescence correlation spectroscopy. J Phys Chem A 104:6416–6428
161. Schwille P (2001) Fluorescence correlation spectroscopy and its potential for intracellular applications. Cell Biochem Biophys 34:383–408
162. Hess ST, Huang S, Heikal AA, Webb WW (2002) Biological and chemical applications of fluorescence correlation spectroscopy: a review. Biochemistry 41:697–705
163. Rigler R, Elson E (eds) (2001) Fluorescence correlation spectroscopy. Springer, Berlin Heidelberg New York
164. Kask P, Palo K, Ullmann D, Gall K (1999) Fluorescence-intensity distribution analysis and its application in biomolecular detection technology. Proc Natl Acad Sci USA 96: 13756–13761
165. Kask P, Palo K, Fay N, Brand L, Mets U, Ullmann D, Jungmann J, Pschorr J, Gall K (2000) Two-dimensional fluorescence intensity distribution analysis: theory and applications. Biophys J 78:1703–1713
166. Chen Y, Müller JD, So PTC, Gratton E (1999) The photon counting histogram in fluorescence fluctuation spectroscopy. Biophys J 77:553–567
167. Chen Y, Müller JD, Berland KM, Gratton E (1999) Fluorescence fluctuation spectroscopy. In: Conn PM (ed) Methods: a companion to methods in enzymology, vol 19. Academic, London, pp 234–252
168. Chen Y, Müller JD, Tetin SY, Tyner JD, Gratton E (2000) Probing ligand protein binding equilibria with fluorescence fluctuation spectroscopy. Biophys J 79:1074–1084
169. Palo K, Mets U, Jager S, Kask P, Gall K (2000) Fluorescence intensity multiple distributions analysis: concurrent determination of diffusion times and molecular brightness. Biophys J 79:2858–2866
170. Enderlein J, Robbins DL, Ambrose WP, Keller RA (1998) Molecular shot noise, burst size distributions, and single molecule detection in fluid flow: effects of multiple occupancy. J Phys Chem A 102:6089–6094
171. Fries JR, Brand L, Eggeling C, Köllner M, Seidel CAM (1998) Quantitative identification of different single molecules by selective time-resolved confocal fluorescence spectroscopy. J Phys Chem A 102:6601–6613
172. Maus M, Cotlet M, Hofkens J, Gensch T, De Schryver FC, Schaffer J, Seidel CAM (2001) An experimental comparison of the maximum likelihood estimation and nonlinear least-squares fluorescence lifetime analysis of single molecules. Anal Chem 73:2078–2086
173. Knemeyer J-P, Herten D-P, Sauer M (2003) Detection and identification of single molecules in living cells using spectrally resolved fluorescence lifetime imaging microscopy. Anal Chem 75:2147–53
174. Dahan M, Deniz AA, Ha T, Chemla DS (1999) Ratiometric measurement and identification of single diffusing molecules. Chem Phys 247:85–106
175. Eggeling C, Fries JR, Brand L, Günther R, Seidel CAM (1998) Monitoring conformational dynamics of a single molecule by selective fluorescence spectroscopy. Proc Natl Acad Sci USA 95:1556–1561

176. Eggeling C, Schaffer J, Seidel CAM, Korte J, Brehm G, Schneider S, Schrof W (2001) Homogeneity, transport, and signal properties of single Ag particles studied by single-molecule surface-enhanced resonance Raman scattering. J Phys Chem A 105:3673–3679
177. Schaffer J, Volkmer A, Eggeling C, Subramaniam V, Striker G, Seidel CAM (1999) Identification of single molecules in aqueous solution by time-resolved fluorescence anisotropy. Phys Chem A 103:331–336
178. Kühnemuth R, Seidel CAM (2001) Principles of single molecule multiparameter fluorescence spectroscopy. Single Mol 2:251–254
179. Seidel C, personal communication
180. Börsch M, Turina P, Eggeling C, Fries JR, Seidel CAM, Labahn A, Gräber P (1998) Conformational changes of the H+-ATPase from *Escherichia coli* upon nucleotide binding detected by single molecule fluorescence. FEBS Lett 437:251–254

Part 2

Application of Fluorescence Spectroscopy to Biological Membranes

8 Raft Microdomains in Model Membranes as Revealed by Fluorescence Quenching

G. Duportail

Keywords: Detergent-resistant membranes; Lipid rafts; Fluorescence quenching; Phase separation

Abbreviations

Chol	Cholesterol
DOPC	Dioleoylphosphatidylcholine
DPH	Diphenylhexatriene
DPPC	Dipalmitoylphosphatidylcholine
DRMs	Detergent-resistant membranes
FCS	Fluorescence correlation spectroscopy
FRET	Fluorescence resonance energy transfer
GPI	Glycosyl phosphatidylinositol
Lα	Liquid-crystalline phase
LcTMADPH	22-Diphenylhexatrienyldocosyltrimethylammonium
Lo	Liquid-ordered phase
NBD	7-Nitrobenz-2-oxa-1,3-diazol-4-yl
PC	Phosphatidylcholine
PE	Phosphatidylethanolamine
Py	Pyrene
12SLPC	1-Palmitoyl-2-(12-doxyl)stearoylphosphatidylcholine
SM	Sphingomyelin
TempoPC	1,2-Dioleoylphosphatidyl{4-[*N,N*-dimethyl-*N*-(2-hydroxyethyl)ammonium]-2,2,6,6,-tetramethylpiperidine-1-oxyl}
T_m	Transition temperature

8.1 Introduction

Since Singer and Nicholson presented their fluid mosaic model in 1972 [1], our view of biological membranes as two-dimensional fluids in which molecules are free to diffuse laterally has been altered by the demonstration of various kinds of inhomogeneities in cell membranes. In the prevailing view of cellular membrane structure, the lipid bilayer functions as a neutral two-dimensional solvent, having little influence on membrane protein function. But biophysicists found that lipids exist in several phases in model lipid bilayers. In the gel state, the lipids form a tightly packed ordered array that undergoes little motion, and in the liquid-like disordered fluid state, the lipids undergo considerable motion. In the presence of cholesterol, these states were replaced by an intermediate state named the liquid-ordered state (Lo). In this Lo phase, phospholipids with saturated hydrocarbon chains pack tightly with cholesterol but nevertheless remain mobile in the plane of the membrane [2]. Despite a detailed biophysical characterization of model membranes, it has been difficult to show that lipids exist in these different phases in the complex environment of the cell.

A turning point came when the lipid raft hypothesis was formulated more than 10 years ago. It originated from cell biology studies, and its main postulate was the existence of microdomains consisting of dynamic assemblies of cholesterol and sphingolipids in the exoplasmic leaflet of the bilayer. The high content of saturated hydrocarbon chains in sphingolipids results in a tightly packed membrane structure characteristic of the liquid-ordered phase, with cholesterol intercalated between the saturated fatty acid chains. In contrast, the surrounding membrane, which has higher concentrations of phospholipids with unsaturated, kinked fatty acid chains, is in a more fluid, liquid-disordered phase. Simons and Ikonen [3] coined the term "lipid rafts" to describe these liquid-ordered microdomains moving within a fluid matrix of unsaturated glycerolipids. The name rafts was subsequently adopted for all sphingolipid-rich domains.

The raft concept has long been controversial, largely because it has been difficult to prove definitively that rafts exist in living cells. This reserve has been overcome by the development of methods leading to detergent-resistant membranes (DRMs). Membrane fragments that are insoluble in non-ionic detergents, like Triton X-100 at 4 °C, can be isolated from most mammalian cells [4]. These fragments appear to be derived from rafts, are rich in cholesterol and sphingolipids and are in the Lo phase when isolated from cells. Furthermore, Lo-phase lipid vesicles are also detergent-insoluble under the conditions used to extract cells. Thus there is obviously a close relation between rafts and DRMs.

So far, despite the wide use of detergent extraction to isolate DRMs from cells and model membranes, this method may introduce artefacts into the analysis, such as altering the lipid composition of the domains or even, in some cases, inducing their formation from lipid systems where they do not exist. For this reason, it was important to develop other non-detergent-based techniques to detect formation of a Lo-like phase in complex lipid mixtures. With this aim in mind, fluorescence, and especially fluorescence quenching methods, have contributed to our current understanding of the conditions under which membrane domains are formed. Quenching appeared to be particularly suited to phase separation studies because it can detect relatively small domains. In addition, unlike other spectroscopic assays, it does not require that the two phases have different physical properties. Only a difference in lipid composition in each phase is necessary to detect phase separation. This review will focus on some selected and recent papers devoted to this topic, in order to highlight the interest of such an approach.

8.2 Identification of Lipid Compositions Forming Rafts

Short-range quenchers, like bromines and nitroxide groups, completely quench the fluorescence of directly neighbouring fluorophores. The basic quenching experiment involves the use of mixed lipid bilayers in which one lipid carries the quencher group. Phase separation results in the formation of one set of domains enriched and a second set of domains depleted in the quencher-carrying lipid. When a fluorescent probe is embedded into bilayers exhibiting such a phase separation, the fluorescence intensity is very different from that measured in bilayers made up of a single lipid phase.

The quenching behaviour of a binary lipid system can be modelled using the equations of London and Feigenson [5]. The fluorescence quenching of a fluorescent probe in a uniform phase with a random lateral distribution of quencher mixed with an unlabelled lipid is assumed to be given by:

$$F/F_0 = \exp\left[-\pi \left(R_c^2 + z^2\right) C\right] \tag{8.1}$$

where R_c is the critical quenching distance, z the difference in depth of fluorophore and quencher, and C is the quencher concentration in units of molecules per square angström [6]. This expression for F/F_0 applies to mixtures forming a uniform single phase over some range of compositions. In mixtures where two phases are present, F/F_0 is calculated from:

$$F/F_0 = F/F_0' + (F/F_0'' - F/F_0')\,[X/(K_p - K_pX + X)] \tag{8.2}$$

where F/F_0' is the fluorescence at the boundary between the one-phase and two-phase regions at which the quencher-enriched phase saturated with the non-quenching lipid is predominant; F/F_0'' is the fluorescence at the other boundary between the one-phase and two-phase regions at which the quencher-depleted phase saturated with the quenching lipid is predominant; K_p is the partition coefficient of the fluorescent probe between the two phases; and X is the fraction of the total bilayer that is in the quencher-depleted phase (see [6, 7] for details).

Practically, a series of liposomes containing graded mixtures of two or more lipids is prepared. In a binary lipid system, one lipid is linked to the fluorescence quencher, and the identity of the other unlabelled lipid is varied as desired. A hydrophobic fluorescent probe is incorporated into the samples. The fluorescence intensity of each sample is measured and normalized to that of a sample without quencher. If both lipids mix to form a single uniform phase, then the fluorescence intensity, according to Eq. (8.1), has a smooth and nearly exponential dependence on the membrane concentration of the quencher-labelled lipid (solid line, Fig. 8.1a). However, some lipids can exhibit lipid–lipid immiscibility when combined. If such lipids are usually miscible when the concentration of either component is low, they undergo phase separation at other concentrations. As the two phases have different lipid compositions, one phase is enriched and the other is relatively depleted in quencher. As a result, probe molecules that partition in the quencher-enriched phase present a greater quenching than for those partitioning into the quencher-depleted phase, and consequently there can be striking changes in the fluorescence dependence on quencher concentration when two phases appear. The boundaries of the two-phase region are given by the lipid composition at which the quenching diverges from that for a lipid mixture forming a single phase (Fig. 8.1a).

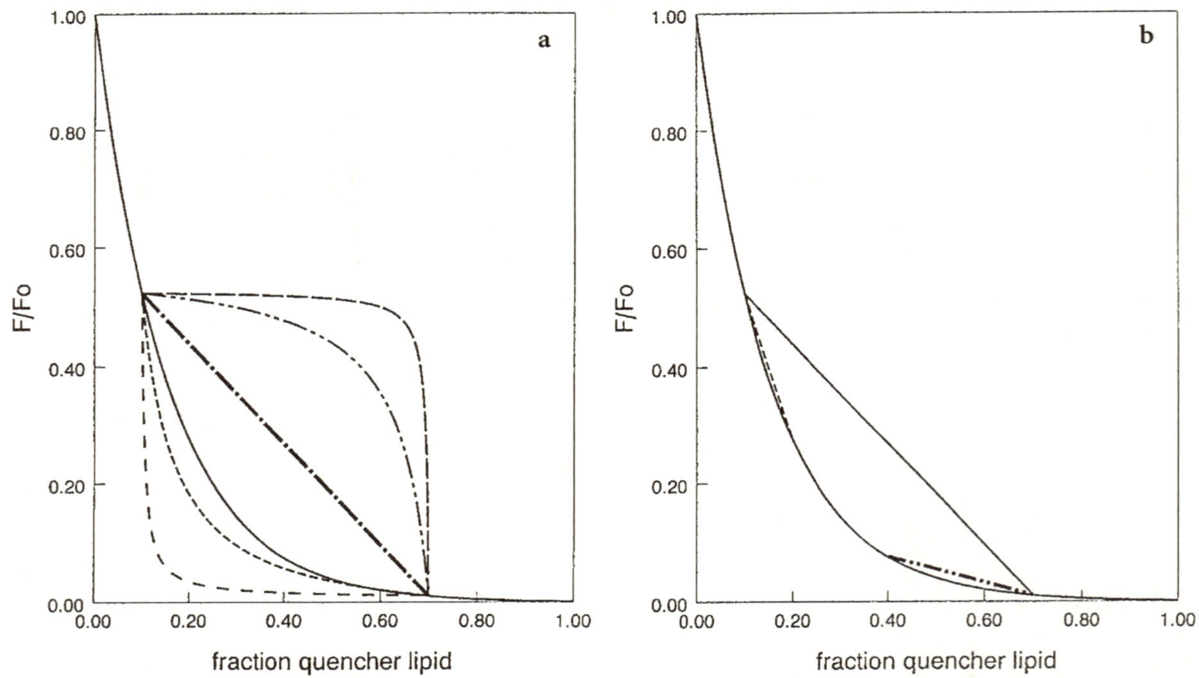

Fig. 8.1a, b. Theoretical quenching patterns showing the effect of probe partition coefficient and phase separation compositions on quenching curves; (y-axis): ratio of fluorescence in the presence of quencher to that in its absence; (x-axis): fraction of quencher lipid in a binary mixture with a non-quenching lipid. **a** Fluorescence vs mol fraction quencher lipid for a mixture forming a uniform phase at all compositions (—) or mixtures forming two phases between 10% quencher lipid (fraction 0.1) and 70% quencher lipid (fraction 0.7), with the fluorescent probe having a partition coefficient (K_p) of (– – –) 100, (- - -) 10, (–·–) 1, (–··–) 0.1 and (——) 0.01. **b** Effect of phase separation points on fluorescence. Data are calculated for a fluorescent probe with K_p=1 and phase boundaries at (- - -) 10% and 20% quencher lipid, (—) 10% and 70% quencher lipid, and (*curved solid line*) for a uniform phase at all compositions. Figure reproduced from [7] with permission of the American Chemical Society and Prof. E. London

The exact shape of the quenching curve in the two-phase region depends on the partitioning of the fluorescent probe between the two phases. For example, if the probe partitions preferentially into the quencher-enriched phase, fluorescence will decrease more rapidly than in a completely miscible mixture ($K_p = 10$ and 100 curves). Conversely, if the probe partitions preferentially into the quencher-depleted phase, then fluorescence will decrease more slowly ($K_p = 0.1$ and 0.01 curves). If the probe partitions evenly between the two phases, then fluorescence will decrease linearly with increasing quencher concentration ($K_p = 1$ curve). The usefulness of quenching for detecting phase separation is strongly influenced by the range of lipid composition over which the phase separation occurs. When it occurs over a wide range of concentrations, for example between 10 and 70% quencher lipid, the phase separation is easily detected (Fig. 8.1b, solid lines). Conversely, the divergence is much more difficult to discern when phase separation occurs over only a narrow range of lipid compositions, for example between 40 and 70% quencher or between 10 and 20% quencher (Fig. 8.1b, –···–/–––). For this reason, it is best to compare the quenching curve of lipid mixtures in which phase separation is suspected to occur with a control curve, showing quenching in mixtures known to be fully miscible throughout the concentration range.

Such an approach was applied by London and collaborators to examine the phase behaviour of lipid mixtures [7]. Diphenylhexatriene (DPH) was chosen as fluorescent probe because it partitions nearly equally between different phases. DOPC was chosen as a lipid that remains in a fluid phase under all experimental conditions. The quenching lipid used was generally 12SLPC, a phosphatidylcholine with a doxyl ring on carbon-12 of the second position fatty acyl chain. Because this ring inhibits close packing with other lipids, 12SLPC phase behaviour is similar to that of an unsaturated lipid. As examples, Fig. 8.2 shows the DPH quenching curves obtained at 23 and 37 °C in mixtures of fully miscible 12SLPC and DOPC (circles) compared with the quenching curves (triangles) in mixtures of 12SLPC and DPPC or sphingomyelin (SM). The experiments were performed in systems containing either no cholesterol (Chol) or 33 mol%. At 23 °C, in the absence of Chol (Fig. 8.2a,c), each pair of curves coincides at both ends of the concentration range, under conditions in which DPPC/12SLPC or SM/12SLPC are in either a single gel phase or a single fluid phase. However, discontinuities in both curves indicative of phase separation are seen at about 20 and 80% DPPC or SM. In the two-phase region between these points, fluorescence is less quenched than in the control single-phase DOPC/12SLPC mixture. The similar phase separation behaviour of both DPPC and SM probably reflects their similar transition temperature (T_m) values. Presumably, the two lipids share a tendency for tight acyl chain packing. At 37 °C, both DPPC/12SLPC (Fig. 8.2b) and SM/12SLPC (Fig. 8.2d) mixtures do not exhibit a discernable two-phase region, as indicated by the coincidence with the DOPC/12SLPC curve. This indicates a quite complete miscibility of DPPC or SM and 12SLPC at 37 °C and is not surprising since at this temperature, these two lipids are close to their T_m.

The most striking result of this study is that addition of 33 mol% of Chol to a mixture promoted separation of a DPPC- or SM-enriched phase from the

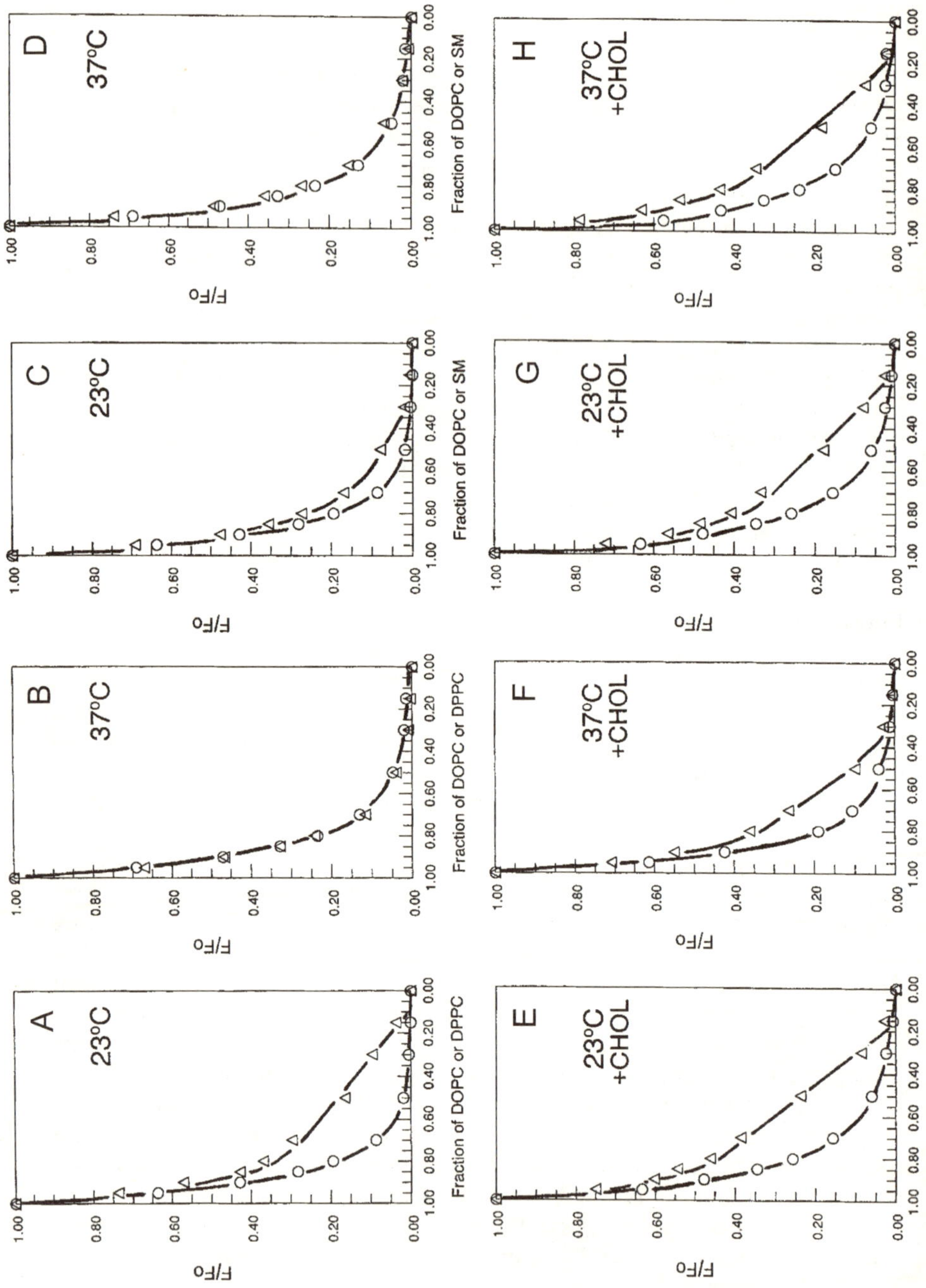

Fig. 8.2 a–h. Comparison of quenching curves for mixtures of DOPC and 12SLPC with those for mixtures of DPPC or SM and 12SLPC. Data are shown for (○) DOPC/12SLPC mixtures and (△) DPPC/12SLPC (*left curves*) or SM/12SLPC (*right curves*) mixtures at 23 °C or 37 °C. *Upper curves*: no cholesterol; *lower curves*: 33 mol% cholesterol. The x-axis shows DOPC, DPPC or SM as a mol fraction of total phospholipid. In the cholesterol-containing samples, phospholipids are 67% of total lipid including cholesterol. Figure reproduced from [7] with permission of the American Chemical Society and Prof. E. London

fluid phase at 37 °C (Fig. 8.2f, h). In contrast, at 23 °C, which is significantly below T_m for both lipids, Chol either has no effect (for DPPC, Fig. 8.2e) or more moderately promotes phase separation (for SM, Fig. 8.2g). It may be a general rule that cholesterol has little or no effect on phase separation at temperatures far below T_m, in accordance with previous results [8]. This suggested that rafts could form at physiological sphingolipid and cholesterol concentration, and that SM could effectively form rafts in the absence of glycosphingolipids, as was confirmed by studies in cells [9]. In addition, it was observed that the lipid compositions for which quenching detected raft formation were the same as those for which detergent insolubility was observed, showing that the isolation of DRMs usually does not reflect some detergent-induced artefacts, but rather pre-existing domains. This study, combined with a previous one [10], showed that cholesterol allows Lo-state domains to form under conditions in which no domain formation would occur in its absence.

By using a bromolipid as quencher and indolyl-labelled phospholipids as fluorescent probes, Wang and Silvius [11] explored the capacity of the types of lipids found in the inner leaflet of plasma membranes to form rafts. The variation of the quenching experiments consisted in the use of two fluorescent lipids differing in their acyl chain unsaturation, and thus with different abilities (different K_p values) to bind to ordered domains. It was found that phosphatidylethanolamine, phosphatidylserine and their mixtures, with acyl chains of similar unsaturation level to those of natural lipids, fail to form separated domains in the presence of 33 mol% Chol, which suggests that these lipids are not able to form rafts in the inner leaflet. Nevertheless, a small amount of SM could raise the possibility of raft formation in a complex mixture of lipids.

8.3 Temperature Dependence in Domain/Raft Formation

A second application of fluorescence quenching is to study the stability of ordered domain formation. Stability can be assessed from the temperature dependence of domain formation. Varying the temperature is particularly useful when comparing the effect of changing lipid structure on phase separation (e.g. examining lipids having different acyl chains or headgroups). The stronger the tendency of a particulate mixture to phase separate, the higher the temperature above which ordered domains melt (T_{mix}) and are replaced by a homogeneous disordered lipid phase [12].

Let us consider a quenching experiment in which the temperature dependence of DPH fluorescence is measured in a sample containing the quenching lipid 12SLPC and either a saturated lipid like DPPC or a sphingolipid. The way to determine the conditions under which two phases are present in this sample, and the value of T_{mix}, consists of comparing the temperature dependence of quenching in the DPPC- or sphingolipid-containing "test" sample with that in a standard sample that contains the same amount of quencher lipid, but exists in a single fluid phase at all temperatures. If two phases are present in the test sample, it will show a weaker quenching, a higher F/F_0, than in the

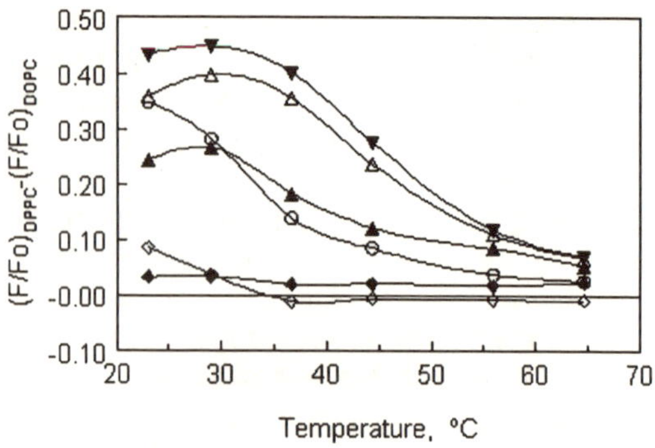

Fig. 8.3. The effect of temperature on domain formation in 1:1 (mol:mol) DPPC/12SLPC mixtures with and without 15 mol% sterols assayed by DPH fluorescence quenching. Samples contained 1 mol% DPH embedded in multilamellar vesicles at a total lipid concentration of 50 μM. The y-axis ($\Delta F/F_0$) is the difference between the fraction of DPH fluorescence unquenched in samples containing a 1:1 DPPC/12SLPC mixture and those containing a 1:1 DOPC/12SLPC mixture. (▼): ergosterol; (△): 7-dehydrocholesterol; (○): stigmasterol; (▲): cholesterol; (◇): no sterol; (◆): C4-sterol. Figure reproduced from [14] with permission of the American Society for Biochemistry and Molecular Biology and Prof. E. London

standard sample. While quenching should be relatively temperature independent in a bilayer containing a single fluid phase, it is strongly temperature dependent in a bilayer containing two phases.

This approach was used to explore the effect of a variety of sterols having different abilities to pack tightly with lipids [13, 14]. Domain formation in mixtures of phospholipids and sphingolipids was examined in the presence of several natural sterols, like ergosterol and 7-dehydroxycholesterol having a double bond not found in cholesterol, or plant sterols such as stigmasterol and sitosterol having the same ring structure as cholesterol but an additional ethyl group on the aliphatic side chain. To compare sterol effects on domain formation, the effect of 15 and 33 mol% sterol in mixtures of various lipids with the quencher lipid 12SLPC was measured as a function of temperature. An example is given in Fig. 8.3. In such experiments, thermal transitions are easily detected when domains are present. Above this transition domain formation is abolished. The transition temperature, T_{mix}, can be defined as the temperature at which the domain mixing (or melting) is half-realized and considered as a measure of the stability of the domains. It can be approximated by the inflection point in the curve of $\Delta F/F_0$ versus temperature. It was shown that T_{mix} defined this way is easier to determine accurately than when defined as the temperature at which domains disappear completely [12]. Another way to assess domain formation is to present the value of $\Delta F/F_0$ at a fixed temperature (Fig. 8.4). In general, $\Delta F/F_0$ reflects both the fraction of the bilayer in the form of sterol-containing saturated lipid-rich domains and the degree of enrichment of these domains in saturated lipid and sterol [7, 13]. For simplicity, we could refer to a decreased level of quenching in the presence of domains, i.e. a high value of $\Delta F/F_0$, as indicative of a high extent of domain formation.

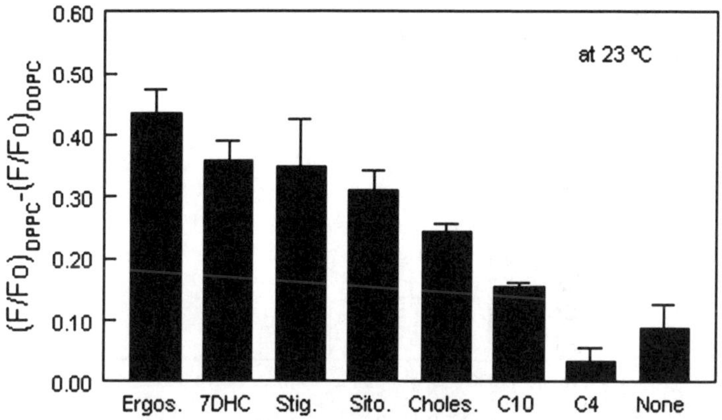

Fig. 8.4. $\Delta F/F_0$ at 23 °C in 1:1 (mol:mol) DPPC/12SLPC mixtures with and without 15 mol% sterols. Samples were prepared as described in Fig. 8.3 with the same sterols (with the addition of C10-sterol). Figure reproduced from [14] with permission of the American Society for Biochemistry and Molecular Biology and Prof. E. London

Both Figs. 8.3 and 8.4 show that there is less quenching in the presence of the natural sterols than in their absence, indicating that all the natural sterols tested promote domain formation. The domains formed in the presence of sterol can be abolished by increasing the temperature to above 60 °C. As the temperature above which domains disappear (T_{mix}) is lowest in the absence of sterol, it shows that all the natural sterols stabilize domain formation. Overall, T_{mix} and $\Delta F/F_0$ give a similar rank order to the effect of sterol structure on domain formation. Interestingly, a good correlation was found between the stability of ordered domains and the ability of the sterols to pack tightly with lipids, as judged by their resistance to solubilization by detergent and the molecular order measured by fluorescence anisotropy [13]. Also, some sterols with structures that do not pack well with saturated lipids show a behaviour opposite to that of cholesterol, tending to abolish rather than to promote domain formation.

The order of sterols in terms of most-to-least strongly promoting domain formation is similar at 15 and 33 mol% sterol. Thus it appears that the effect of sterol structure on domain formation is not strongly dependent on sterol concentration. Experiments at 33 mol% were repeated using LcTMADPH, a DPH derivative attached to a long, saturated alkyl chain ending with a trimethylammonium head. In contrast to DPH, which partitions roughly equally between ordered and fluid domains, LcTMADPH partitions preferentially into ordered domains [15]. Thus, formation of an ordered domain depleted in quencher lipid should reduce quenching of LcTMADPH fluorescence more strongly than that of DPH. As a result, LcTMADPH should be a more sensitive probe of domain formation than DPH if the quencher-depleted domains are in a more ordered phase than the rest of the bilayer. The greater sensitivity of LcTMADPH to the formation of DPPC-rich domains is confirmed by the increased $\Delta F/F_0$ as compared to DPH (Fig. 8.5). This indicates that the

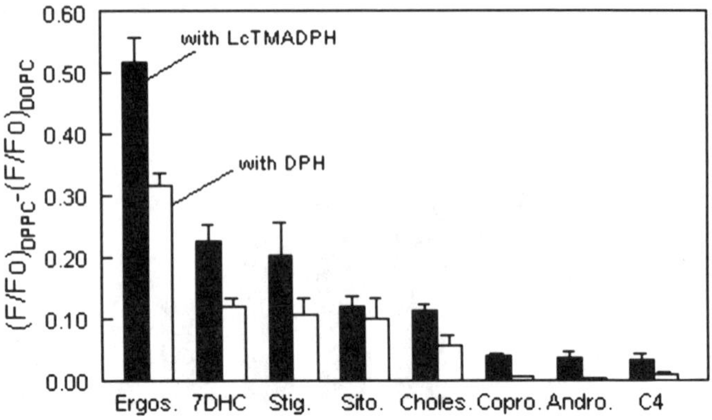

Fig. 8.5. $\Delta F/F_0$ at 23 °C in 1:1 (mol:mol) DPPC/12SLPC mixtures with and without 33 mol% sterols. Samples were prepared as in Fig. 8.3 except at 65 °C in the presence of 0.4 mol% DPH or 0.2 mol% LcTMADPH. *Open bars* correspond to values obtained with DPH, *closed bars* with LcTMADPH. Coprostanol and androstenol are new sterols compared to Figs. 8.3 and 8.4. Figure reproduced from [14] with permission of the American Society for Biochemistry and Molecular Biology and Prof. E. London

DPPC-rich domains formed in the presence of most sterols are more highly ordered that the remainder of the bilayer which is enriched in 12SLPC. Interestingly, the DPPC-rich domains formed in the presence of sitosterol do not exhibit a $\Delta F/F_0$ with LcTMADPH that is significantly higher than that with DPH. This suggests that the DPPC-rich domains in the presence of sitosterol are not highly ordered, consistent with their relatively loose packing as judged by solubility in Triton X-100. It should also be noted that in control samples lacking domains, i.e. 12SLPC–DOPC mixtures, quenching of LcTMADPH and DPH is very similar. This shows that the difference in quenching of both probes in samples containing domains is not an artefact of the inherently lesser ability of 12SLPC to quench LcTMADPH relative to DPH. Such probes, presenting a high partitioning preference versus one phase, should be confirmed as powerful tools for the study of rafts in the future (London, personal communication).

It was also found that the formation of DPPC-rich and of SM-rich domains have similar sensitivities to sterol structure. However, the stability of ordered domain formation by cerebrosides was relatively insensitive to sterol presence and structure. Also, it was found that a small amount of ceramide is able to strongly stabilize domain formation, which may have important physiological consequences. In another study of sphingolipid domain formation, it was shown that SM with a dihydrosphingosine core forms ordered domains with cholesterol that are more stable than those formed by SM with a sphingosine core [16]. This property was correlated to the higher T_m of bilayers composed of dihydro-SM relative to those composed of ordinary SM.

8.4 Affinity of Lipids and Proteins for Rafts as Detected by Quenching

A third application of fluorescence quenching is to determine the affinity of lipids and proteins for rafts. Under conditions in which quencher-rich and quencher-poor domains coexist, fluorescence intensity depends on the relative affinity of a fluorescently labelled molecule for one type of domain over the other [5]. Under favourable conditions, it is possible to derive the partition coefficients (K_p) between the different domains.

Silvius and collaborators developed methods for determining relative K_p values for ternary mixtures [17, 18]. They compared the distribution of various bimane-labelled lipids and lipidated peptides (at a concentration of 0.6 mol% vs lipids) between Lo and liquid-disordered domains in representative mixtures combining saturated sphingo- or glycerophospholipids with 12SLPC or Tempo-PC and physiological proportions (33 mol%) of cholesterol. The experimental quenching curves were plotted in the modified scaled form:

$$(F/F_0)_{cor} = (F_N - F_{100\%Q})/(F_{0\%Q} - F_{100\%Q}) \tag{8.3}$$

where F_N is the normalized fluorescence measured for the labelled lipid in vesicles of a given composition and $F_{0\%Q}$ and $F_{100\%Q}$ are the normalized fluorescence values measured in vesicles containing, respectively, 0% or 100% 12SLPC in the non-sterol fraction. These quenching curves are then fitted on an empirical basis (see [17, 18] for details).

First were measured the relative abilities of different bimane-labelled diacyl phospholipids (Fig. 8.6) to partition into Lo domains [17]. The results generally support previous conclusions, based on detergent extraction assays [19], that longer-chain disaturated species are relatively enriched in such domains while multiply unsaturated species show very little partitioning into the Lo phase. These findings lend further support to the idea that GPI-anchored proteins of the cell surface may associate with Lo-phase lipid rafts, at least in part, because their lipid anchors frequently carry long saturated acyl chains. However, the present results add to this conclusion the nuance that monounsaturated species, or species bearing saturated chains as short as C_{14} (myristoyl), may also be able to partition to a significant degree into Lo domains, albeit with significantly lower affinities than longer-chain disaturated lipids.

A second series of experiments were devoted to examining the association of a variety of fluorescent lipidated peptides with rafts [18]. Their structure is represented in Fig. 8.7. They incorporate lipid-modified sequence motifs found in different classes of cellular proteins. As an example, the quenching curves determined for DPPC/Tempo-PC/33 mol% Chol bilayers incorporating these peptides are represented in Fig. 8.8. The presented fluorescence quenching data provide direct confirmation that certain lipidated motifs found in proteins, including the combination of a cholesteryl and an *N*-palmitoyl group as well as motifs combining multiple saturated acyl chains, confer significant affinities for Lo membrane domains. Interestingly, however, a given combina-

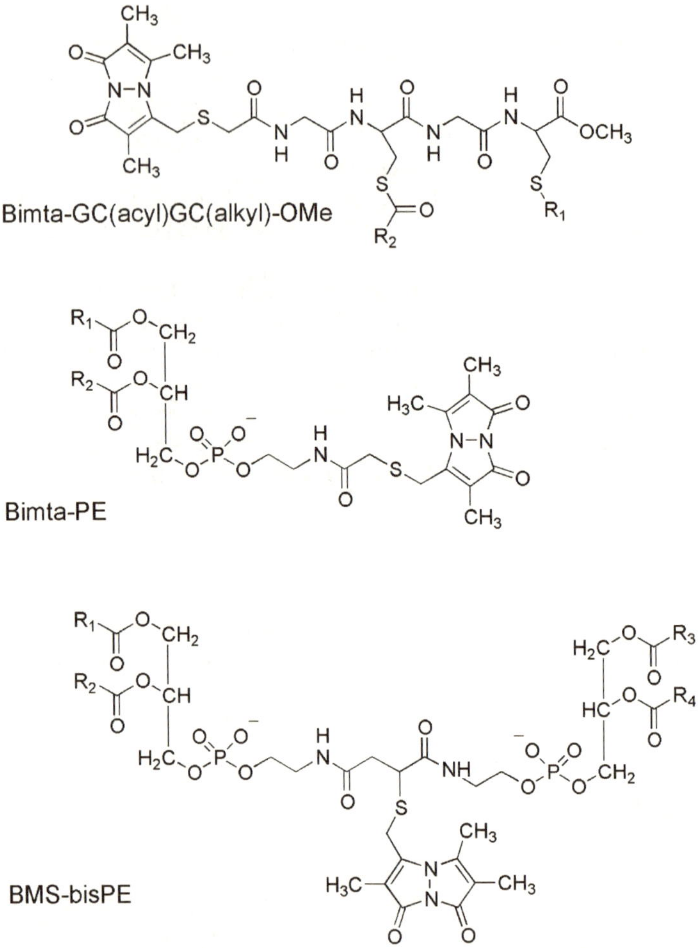

Fig. 8.6. Representative structures of bimane-labelled diacyl phospholipids. Figure reproduced from [17] with permission of the Biophysical Society and Prof. J. Silvius

tion of lipid residues can confer markedly different affinities for Lo domains, depending on the precise manner in which the lipid chains are attached to the peptide backbone. This is strikingly illustrated by comparing the affinities for Lo domains measured for peptides bearing the motif palmitoylGlyCys-(palmitoyl)-, -Cys(palmitoyl)Cys-(palmitoyl)- and palmitoylCys(palmitoyl). While all of these peptides exhibit significant affinities for Lo domains, the first shows an affinity that exceeds by more than tenfold those observed for the latter two species. The differing behaviour of these peptides, in which the acyl groups are attached by the same chemical linkages, suggests that in addition to their structures, the relative spacing and/or orientation of the coupled acyl chain can be an important determinant of affinity for rafts. This possibility has been previously suggested [20] and raises the interesting prospect

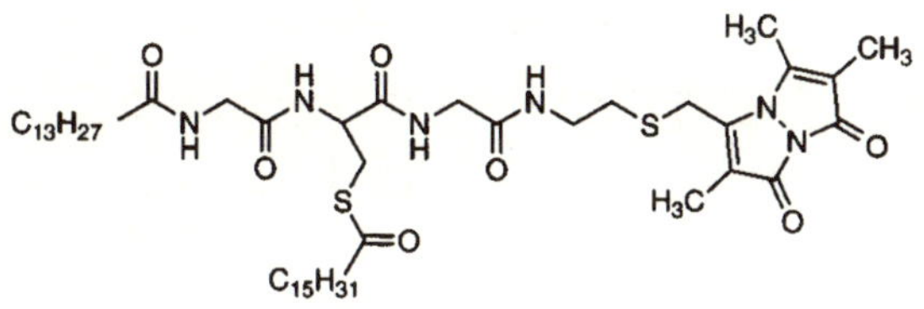

myristoylGlyCys(palmitoyl)Gly-caBim

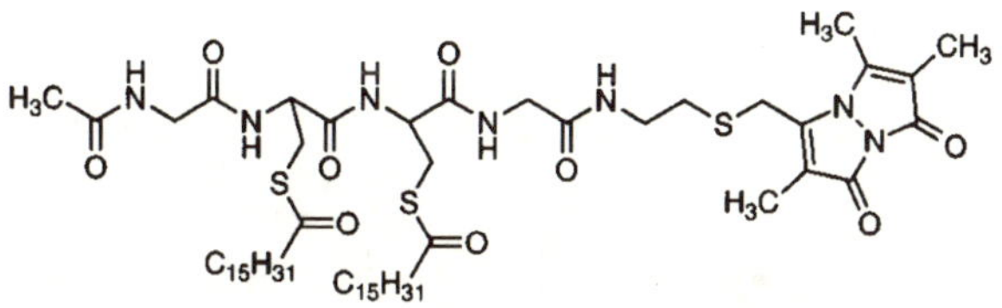

acetylGlyCys(palmitoyl)Cys(palmitoyl)Gly-caBim

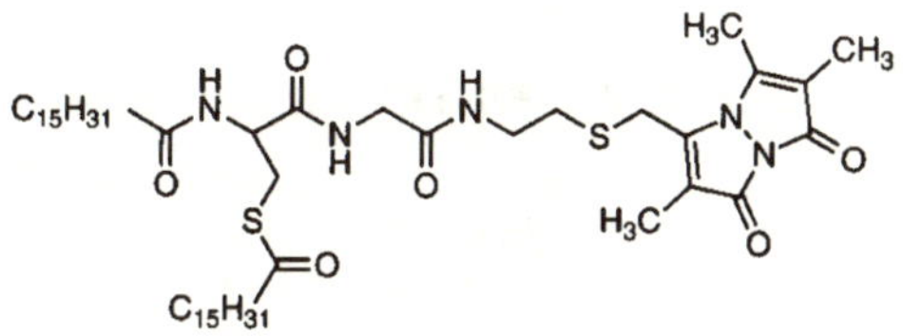

palmitoylCys(palmitoyl)Gly-caBim

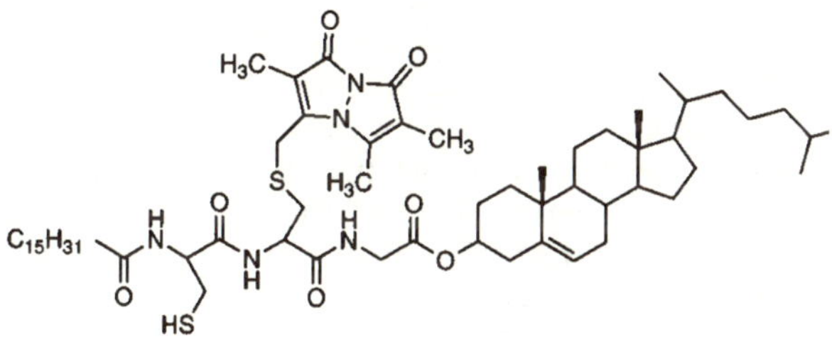

palmitoylCysCys(bimanyl)Gly-cholesterol

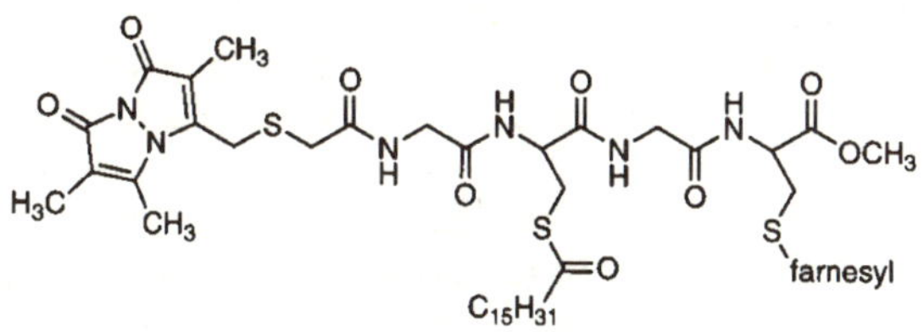

Bimta-GlyCys(palmitoyl)GlyCys(farnesyl)-OMe

Fig. 8.7. Representative structures of bimane-labelled lipidated peptides. Figure reproduced from [18] with permission of the American Chemical Society and Prof. J. Silvius

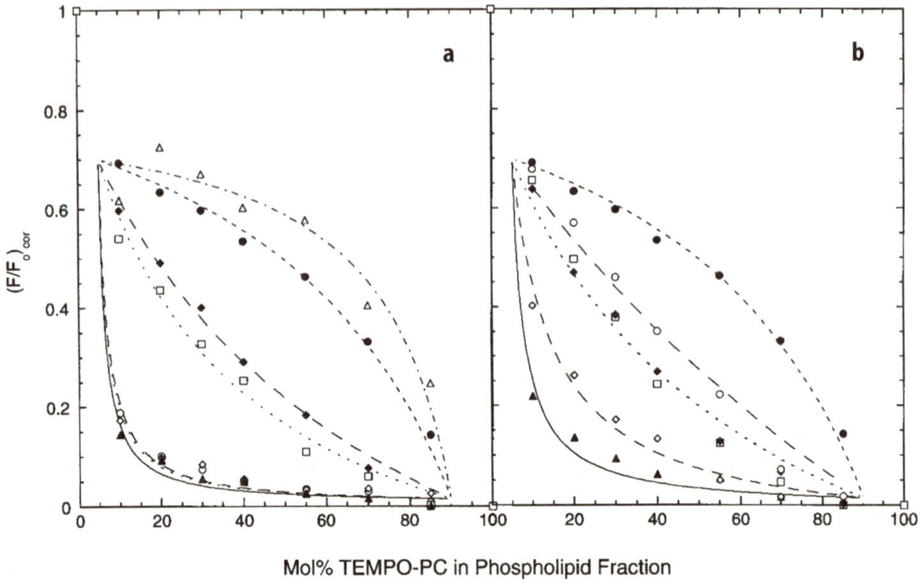

Fig. 8.8 a, b. Quenching curves determined at 25 °C for DPPC/TEMPO-DOPC/33 mol% cholesterol bilayers incorporating various bimane-labelled lipidated peptides. Figure reproduced from [18] with permission of the American Chemical Society and Prof. J. Silvius

that the affinities of lipidated proteins for rafts could be modulated by factors that modify the local conformation of the lipidated sequence.

The same approach was used to determine relative K_p values for partitioning between raft and non-raft domains for a series of lipids containing fluorescently labelled acyl chains [21]. Inclusion of a DPH-labelled acyl chain allows sphingolipids to maintain a significant degree of association with Lo domains. Presumably, the rigid, rod-like DPH group introduces only few perturbations in the ability of the lipid to pack tightly as compared to a molecule with a saturated chain. It was also found that cerebroside derivatives partition into rafts more strongly than SM, whereas lipids containing one DPH-labelled chain and one unsaturated chain partition into rafts more weakly. In contrast to the behaviour of DPH-labelled lipids, BODIPY- and NBD-labelled sphingolipids were found to partition into rafts relatively poorly. These relatively bulky groups presumably perturb the tight acyl chain packing, and we must be aware that these labels may not always faithfully reflect the behaviour of the parent lipid.

More recently, the same authors characterized the partitioning of indolyl-labelled phospho- and sphingolipids between gel, Lo and Lα domains [22]. In both Chol-free and Chol-containing lipid mixtures, sphingolipids with varying polar headgroups show a net preference for partitioning into ordered domains. The affinities of different sphingolipids for Lo domains do not vary in a consistent manner with the size or other structural properties of the polar headgroup. For example, ganglioside GM1 partitions in a very similar manner to SM. Ceramide exhibits a dramatically higher affinity for Lo domains than do other sphingolipids. These findings confirm that sphingolipids with a variety of headgroup structures will be enriched by substantial factors in Lo versus Lα domains of biomembranes, with only little dependence on polar headgroup structure. The very high affinity observed for partitioning of ceramide into gel or Lo domains is consistent with previous reports that this lipid species promotes the formation of such domains [14, 23], and that ceramide-mediated signalling may be localized in these domains [24].

8.5 Alternative Fluorescence Methods for the Detection of Rafts

Short-range quenching is only one of several methods used for detecting rafts in model membranes. Alternative fluorescence methods include energy transfer and fluorescence microscopy. Among these methods, only short-range quenching is sensitive to relatively small domains. Energy transfer can usually sense only somewhat larger domains and fluorescence microscopy can detect only relatively large domains. Unfortunately, short-range quenching requires relatively high levels of quencher lipid, which limits the types of lipid mixtures that can be investigated by this method, and makes it difficult to adapt it to studies of cells because such quencher lipids cannot be introduced into cells at high levels.

The potential use of energy transfer in model membranes is illustrated by a study showing the exclusion of rhodopsin from rafts [25]. Bovine rhodopsin was reconstituted into mixtures of di22:6-PC, di16:0-PC, (16:0,22:6-PC) and cholesterol. Fluorescence resonance energy transfer (FRET), using lipids labelled at the headgroup with pyrene (Py) as donor and rhodopsin retinal group as acceptor of fluorescence, was used to study rhodopsin association with lipids. Higher FRET efficiencies detected for di22:6-PE-Py compared to di16:0-PE-Py, in mixed di22:6-PC/Di16:0-PC/Chol bilayers, indicate preferential segregation of rhodopsin with polyunsaturated lipids. It was shown that the effective range of the rhodopsin–lipid interactions facilitating raft formation exceeds two adjacent lipid layers. In similar mixed bilayers containing no cholesterol, raft formation is absent at temperatures above the lipid phase transition, confirming the crucial role of cholesterol in microdomain formation. Very recently, Silvius described a FRET method able to detect inhomogeneities in lipid bilayers on distance scales of the order of tens of nanometres or greater [26]. This approach compares the efficiency of energy transfer between two fluorescent lipid donors, differing in their affinities for ordered versus disordered regions of the bilayer, and an acceptor lipid that distributes preferentially into disordered regions. Interestingly, the present FRET method reports clear evidence of inhomogeneity in mixtures combining SM or saturated phospholipids with unsaturated phospholipids and Chol, even at temperatures where these systems appear homogeneous by fluorescence microscopy. This work showed that under physiological conditions, lipid mixtures mimicking the lipid composition of the outer leaflet of the plasma membrane can form domains on a spatial scale comparable to that inferred for rafts in biomembranes.

However, when the raft size is large, light microscopy [27], and especially fluorescence confocal microscopy on giant vesicles [28–30], remains the most valuable fluorescence method because of its ability to probe many different domain properties, including domain shape, heterogeneity, connectivity, lipid and protein lateral diffusion coefficients and time-dependent domain–domain interactions. Fluorescence microscopy is also valuable because it can be applied to both bilayers and supported model membrane monolayers [31]. Noteworthy is the work of Samsonov et al. [32]. The formation of rafts was studied by using planar bilayer membranes that contained rhodamine-PE as a fluorescent probe, and wide-field fluorescence microscopy was used to detect phase separation of this probe. It was shown that SM–cholesterol domains did not form at high temperature but spontaneously formed when the temperature was lowered below the T_m of SM. The domains were observed to be circular (resolution ≥0.5 μm); they quickly reassumed their circular shape after deformation and merged with each other to form larger domains, all phenomena consistent with Lo rather than solid-ordered domains. A saturated PC like distearoyl-PC could substitute for SM to complex with cholesterol into Lo domains. But in the presence of Chol, saturated phosphatidylethanolamines or phosphatidylserines yielded solid-ordered domains of irregular shape, demonstrating that Lo domain formation is dependent on the polar headgroup of the lipid. An individual raft always extended through both monolayers. Degrading cholesterol in one monolayer with cholesterol oxidase first caused the boundary of the raft to become irregular, then the raft gradually disappeared. The fluid nature of rafts, demonstrated in this study, may be important for allowing dynamic interactions between proteins localized within rafts.

Fluorescence correlation spectroscopy (FCS) also appears to be a promising tool in the study of rafts. In a very recent work [33], Kahya et al., by applying this spectroscopy to giant vesicles, proved that FCS is able to evaluate lipid composition by taking the molecular diffusion coefficient of a lipophilic fluorescent probe as a fingerprint of membrane phase compositions. Moreover, FCS data were used to build a ternary phase diagram showing areas of phase coexistence, transition points and, importantly, how lipid dynamics varies between and within phase regions.

An alternative method to detect the preferential association of molecules with different lipid environments should also be noted. For molecules for which exchange can be catalysed by carriers such as cyclodextrins, the equilibrium distribution, and thus K_p, of a molecule between populations of vesicles with different compositions can be determined [34–36].

Acknowledgements

This review is mainly making use and taking the best account of recent works from Profs. J. Silvius (McGill University at Montreal) and E. London (State University of New York at Stony Brook). Both are gratefully acknowledged for their courtesy. The author also wants to mention previous reviews from Prof. London and his collaborators [12, 37], giving him starting elements and a conducting wire to write the present paper: "*Thank you, Erwin, for your permanent availability, and also for giving me the opportunity of a collaboration* [14] *which raised up my interest for rafts in biomembranes*".

References

1. Singer SJ, Nicholson GL (1972) Science 175:720
2. Brown DA, London E (1998) Annu Rev Cell Dev Biol 14:111
3. Simons K, Ikonen E (1997) Nature 387:569
4. Brown DA, Rose JK (1992) Cell 68:533
5. London E, Feigenson GW (1981) Biochim Biophys Acta 649:89
6. Chattopadhyay A, London E (1987) Biochemistry 26:39
7. Ahmed SN, Brown DA, London E (1997) Biochemistry 36:10944
8. Silvius JR (1992) Biochemistry 31:3398
9. Ostermeyer AG, Beckrich BT, Ivarson KA, Grove KE, Brown DA (1999) J Biol Chem 274:34459
10. Silvius JR, del Giudice D, Lafleur M (1996) Biochemistry 35:15198
11. Wang TY, Silvius JR (2001) Biophys J 81:2762
12. London E, Brown DA, Xu X (2000) Methods Enzymol 312:272
13. Xu X, London E (2000) Biochemistry 39:843
14. Xu X, Bittman R, Duportail G, Heissler D, Vilcheze C, London E (2001) J Biol Chem 276: 33540
15. Beck A, Heissler D, Duportail G (1993) Chem Phys Lipids 66:135
16. Kuikka M, Ramstedt B, Ohvo-Rekilä H, Tuuf J, Slotte JP (2001) Biophys J 80:2327
17. Wang TY, Leventis R, Silvius JR (2000) Biophys J 79:919
18. Wang TY, Leventis R, Silvius JR (2001) Biochemistry 40:13031
19. Schroeder RJ, London E, Brown DA (1994) Proc Natl Acad Sci USA 91:12130
20. Mesquita RM, Melo RS, Thompson TE, Vaz WL (2000) Biophys J 78:3019
21. Wang TY, Silvius JR (2000) Biophys J 79:1478
22. Wang TY, Silvius JR (2003) Biophys J 84:367
23. Massey JB (2001) Biochim Biophys Acta 1510:167
24. Liu P, Anderson RG (1995) J Biol Chem 270:27179
25. Polozova A, Litman BJ (2000) Biophys J 79:2
26. Silvius JR (2003) Biophys J 85:1034
27. Jacobson K, Dietrich C (1999) Trends Cell Biol 9:87
28. Korlach J, Schwille P, Webb WW, Feigenson GW (1999) Proc Natl Acad Sci USA 96:8461
29. Feigenson GW, Buboltz JT (2001) Biophys J 80:2775
30. Dietrich C, Bagatolli LA, Volovyk ZN, Thompson NL, Levi M, Jacobson K, Gratton E (2001) Biophys J 80:1417
31. Dietrich C, Volvyk ZN, Levi M, Thompson NL, Jacobson K (2001) Proc Natl Acad Sci USA 98:10643
32. Samsonov AV, Mihalyov I, Cohen FS (2001) Biophys J 81:1486
33. Kahya N, Scherfeld D, Bacia K, Poolman B, Schwille P (2003) J Biol Chem 278:28109
34. Ramstedt B, Slotte JP (1999) Biophys J 76:908
35. Leventis R, Silvius JR (2001) Biophys J 81:2257
36. Niu SL, Litman BJ (2002) Biophys J 83:3408
37. London E (2002) Curr Opin Struct Biol 12:480

9 The Lateral Structure of Lipid Membranes as Seen by Fluorescence Microscopy

L. A. Bagatolli

Keywords: Lipid domain; Laurdan; Fluorescent probes; Phase coexistence; Giant vesicles

Abbreviations

AFM	Atomic force microscopy
BLES	Bovine lipid extract surfactant
DANCA	2′-(*N*,*N*-dimethylamino)-6-naphthoyl-4-*trans*-cyclohexanoic acid
DMPC	Dimyristoylphosphatidylcholine
DSPC	Disteroylphosphatidylcholine
DSC	Differential scanning calorimetry
GP	Generalized polarization
GUV	Giant unilamellar vesicle
Laurdan	6-Lauroyl-2-(*N*,*N*-dimethylamino) naphthalene
LUV	Large unilamellar vesicle
MLV	Multilamellar vesicle
PE	Phosphatidylethanolamine
Prodan	6-Propionyl-2-(*N*,*N*-dimethylamino) naphthalene
SUV	Small unilamellar vesicle

9.1 Introduction

The importance of lateral heterogeneity of lipid membranes had not been widely acknowledged, particularly within the biology field, until the postulation of the raft hypothesis in recent years [1–3]. Even though the presence of phase coexistence in cell membranes is still a matter of controversy, experimental evidence is being accumulated favoring the presence of lateral organized domains [1–7]. It is important to notice, however, that the coexistence of stable lipid domains in membranes was discovered in 1977 [8], 20 years before the raft hypothesis. This last phenomenon that involves lipid phase separation and is characterized by the coexistence of ordered (tightly packed) and disordered (loosely packed) lipid phases in the lipid bilayer plays a central role in the stabilization of multicomponent vesicles and in the fission of small vesicles after budding [9]. The lateral structure of lipid membranes strongly depends on composition and is extremely sensitive to environmental parameters such as temperature. In this sense, several phase diagrams have been constructed in the last 30 years by studying the thermotropic behavior of different lipid mixtures using several experimental (and also theoretical) approaches [10–27]. Although important thermodynamic information is presently available for different lipid mixtures, the physical characteristics of the different lipid domains, such as shape, size, and their time evolution, are not. In general, these experimental approaches detect average physical changes in solutions of multilamellar (MLVs), small unilamellar (SUVs), or large unilamellar vesicles (LUVs) without offering the possibility of direct observation of the morphological and topological features of the membrane lateral structure. In recent years a new experimental strategy, based on the direct visualization of free-standing lipid bilayers using giant unilamellar vesicle (GUV) technology and fluorescence microscopy techniques, has opened the possibility of correlating structural and dynamical information between molecular and supramolecular levels of lipid membranes [28–32]. This correlation provides very interesting and new insights about the lateral structure of membranes. The aim of this chapter is to revisit the experimental methodology and discuss the impact of the new results on the membrane field.

9.2 Giant Vesicles

In recent years, several papers appeared which described the use of giant vesicles as model systems to address biophysical aspects of lipid-lipid, lipid-DNA, and lipid-protein interactions [28–41]. One of the reasons why giant vesicles represent a suitable membrane model system is their size, on the order of a few tens of micrometers, similar to the size of the plasma membrane of cells. Due to their size, *single* vesicles can be directly observed using microscopy techniques (see below). Additionally, because experiments are performed at the level of single vesicles, heterogeneities in shape, size, and presence of multilamellar vesicles are ruled out.

There are different protocols for preparing giant vesicles such as the gentle hydration method [42] and the electroformation method [43]. The latter method is getting popular, mainly because it is not time consuming and the yield of giant vesicles observed after preparation is high (see ref. [44] for comparison of different methods of preparing GUVs). Many different experimental strategies were reported using giant vesicles, such as manipulation of single vesicles, microinjection, etc. The reader can find additional information about the giant vesicle field in an excellent review by Menger and Keiper [45] and a book completely devoted to giant vesicles edited by P. L. Luisi and P. Walde [46].

9.3 Domains in Membranes

Knowledge of the correlation between the morphology of single vesicles and particular events that occur in the lipid membrane at molecular and supramolecular levels is crucial in order to understand the different mechanisms that occur through this important biological component. Studies involving fluorescence spectroscopy and microscopy are suitable for investigating such a correlation, in particular for studying lipid-lipid and lipid-protein interactions. As mentioned above, the combination of giant vesicle technology, fluorescence

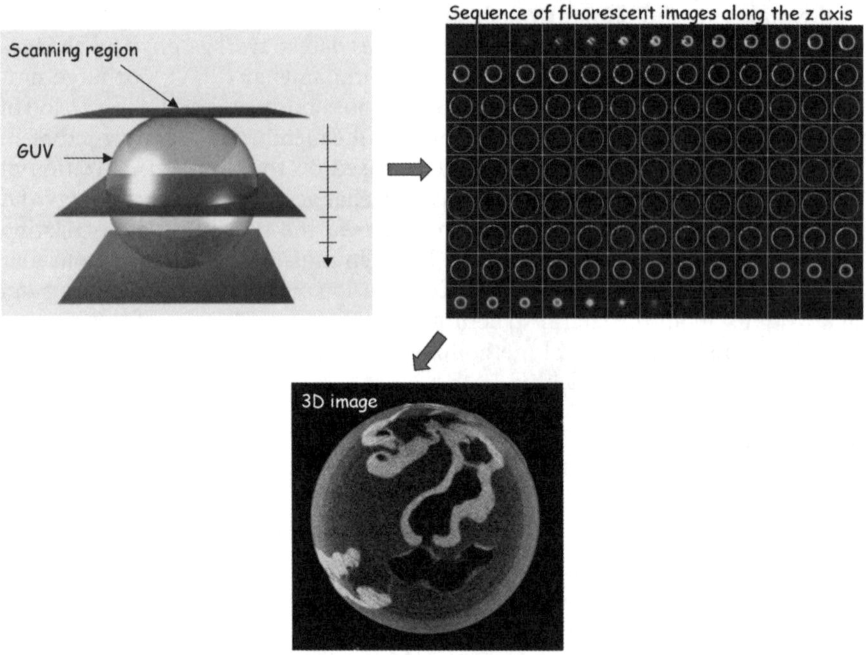

Fig. 9.1. Sketch of a 3D reconstruction of fluorescent label GUV using confocal fluorescence microscopy. Giant vesicle and the movement of the scanning region along the z-axis (*top left*); sequence of fluorescent images obtained along the z-axis (*top right*); 3D fluorescent image of a GUV. The GUV is composed of (POPC:ceramide)/ergosterol mixture (5:1 mol/ 26 mol%) at 25 °C. Three different regions can be observed in the GUV using the fluorescent probe $DiIC_{18}$

microscopy (epifluorescence, confocal, or two-photon excitation), and the particular spectral and partition properties of different fluorescent probes is now used to study the behavior of simple and complex lipid mixtures [28–36] and their interaction with proteins and peptides [47]. Figure 9.1 summarizes the procedure used to obtain a 3D image of a giant vesicle in a confocal or two-photon excitation microscope. As can be seen in the figure, not just the size and shape of the coexisting lipid domain are accessible but the fraction contributed by each area can be visualized and computed directly from the images. In studies regarding the temperature behavior of lipid mixtures one can obtain this last piece of information at different temperatures. In such cases it is possible to add a new dimension to the classical phase diagrams, i.e., the morphological information of lipid domains, as is represented in Fig. 9.2. The experimental data presented in this figure are connected with information obtained by other experimental techniques (NMR in this case, [48]).

A key goal in the single vesicle experiments is to identify the phase state of the different areas observed in the vesicles. There are many ways to achieve this goal [28, 29], a crucial issue being careful selection of the fluorescent probes.

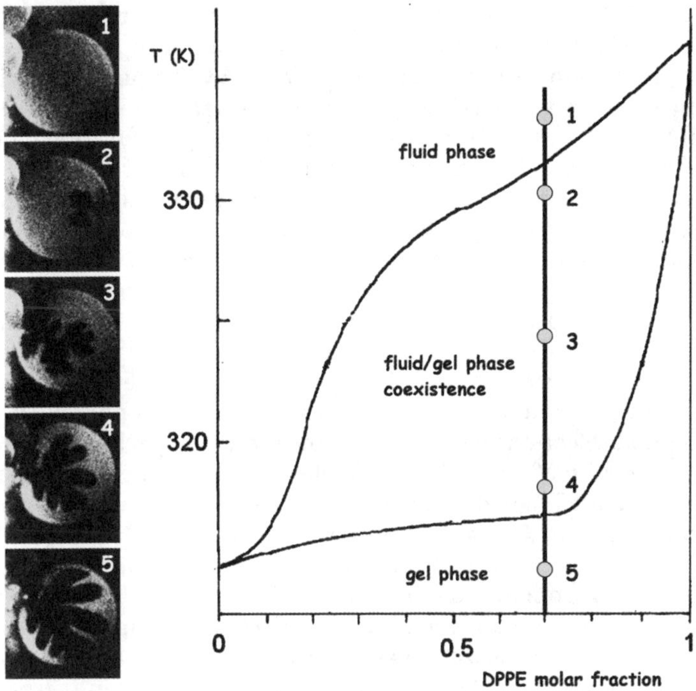

Fig. 9.2. Phase diagram of DPPE/DPPC (adapted from ref. [48]) and fluorescence images of GUVs composed of DPPE/DPPC 7:3 mol obtained at different temperatures. A new dimension (domain shape and size) can be added to the phase diagram (see text)

9.4 Fluorescent Probes: Advantages and Disadvantages

There is a group of membrane probes that show different partition properties in lipid bilayers displaying phase coexistence [28, 31, 32, 35, 36]. The fluorescence intensity images obtained with such probes display the shape of the lipid domains but additional experimental strategies (to measure the diffusion coefficients of the probes in each phase, for example, see refs. [28] and [36]) are necessary to further evaluate the local physical characteristics of the observed lipid domains. However, there is a tendency to associate the partition properties of the fluorescent molecules with the phase state of lipid domains. Nevertheless, as was demonstrated previously, the partition properties of the probe depend on the composition of the lipid membrane [49 and references therein]. The message here is that it is not wise to generalize the fluorescent molecule's affinity for the different lipid phases without a careful characterization of the probe.

A very interesting fluorescent molecule to study the lateral separation phenomenon is Laurdan. This probe belongs to the family of polarity-sensitive fluorescent probes – that includes 6-propionyl-2-(*N*,*N*-dimethylamino)naphthalene (Prodan) and 2′-(*N*,*N*-dimethylamino)-6-naphthoyl-4-*trans*-cyclohexanoic acid (DANCA) – first designed and synthesized by Gregorio Weber to study the phenomenon of dipolar relaxation of fluorophores in solvents, bound to proteins and associated with lipids [50–53]. The advantages in using Laurdan to explore phase coexistence in bilayers are based on three fundamental characteristics of this molecule: (1) the Laurdan transition moment in lipid vesicles is aligned parallel to the hydrophobic lipid chains [49]; (2) the phase-dependent emission spectral shift, i.e., Laurdan's emission is blue in the gel phase and greenish in the fluid phase [49]; and (3) homogeneous probe distribution on lipid membranes that displays phase coexistence [49]. Since Laurdan is similarly distributed between the lipid phases and the lateral packing of the lipid domains determines the emission wavelength of the probe, it is relatively easy to differentiate among lipid phases simply by the position of the emission band. In addition, due to the particular position of the Laurdan transition moment relative to the lipids, it is possible to use the so-called photoselection effect to determine the lateral packing of the bilayer. The photoselection effect is dictated by the fact that only those fluorophores are excited that are aligned parallel, or nearly so, to the plane of polarization of the excitation light. Using circular polarized light and observing the top or bottom part of a lipid vesicle displaying gel/fluid phase coexistence, only fluorescence coming from the fluid part of the bilayer can be observed (Fig. 9.3a). In the fluid phase a component of Laurdan's transition dipole is always parallel to the excitation polarization because of the relatively low lipid order, i.e, the wobbling movement of the Laurdan molecule increases with respect to the gel phase where no fluorescence intensity is observed (even though Laurdan molecules are present in this region of the bilayer). The photoselection effect exists only at the top and bottom surfaces of the vesicle where the Laurdan molecules are located along the z-axis (the excitation light polarization plane is defined as the x–y plane). As shown in Fig. 9.3b, the excitation efficiency in the center of the GUV is equal in both phases. The photoselection effect provides another way, in addition to the position of Laurdan's emission band, to characterize the phase state of a particular membrane area.

To ascertain the phase state in a quantitative fashion, the Laurdan generalized polarization (GP) function is used in the single vesicle experiments (for a review see [49, 54]). The Laurdan GP function is simply related with the position of the probe's emission spectrum in the membrane. The position of the emission spectra is related to the extent of water dipolar relaxation processes in the membrane. This relaxation is more pronounced in the fluid phase compared with that observed in the gel phase. Consequently, the characteristic values of the Laurdan GP function can be associated with different lipid phases [55, 56]. In addition, the compositional differences between the coexisting lipid phases can be evaluated using the GP function [30]. For a complete description of Laurdan GP in membranes the reader is encouraged to explore the following review articles [49, 54–56]. The important message here is that an image of a GUV labeled with Laurdan provides *simultaneous* information about the morphology and phase state of lipid domains.

Laurdan belongs to the family of membrane probes that are excited in the ultraviolet (UV) region. Using epifluorescence or confocal microscopy it is

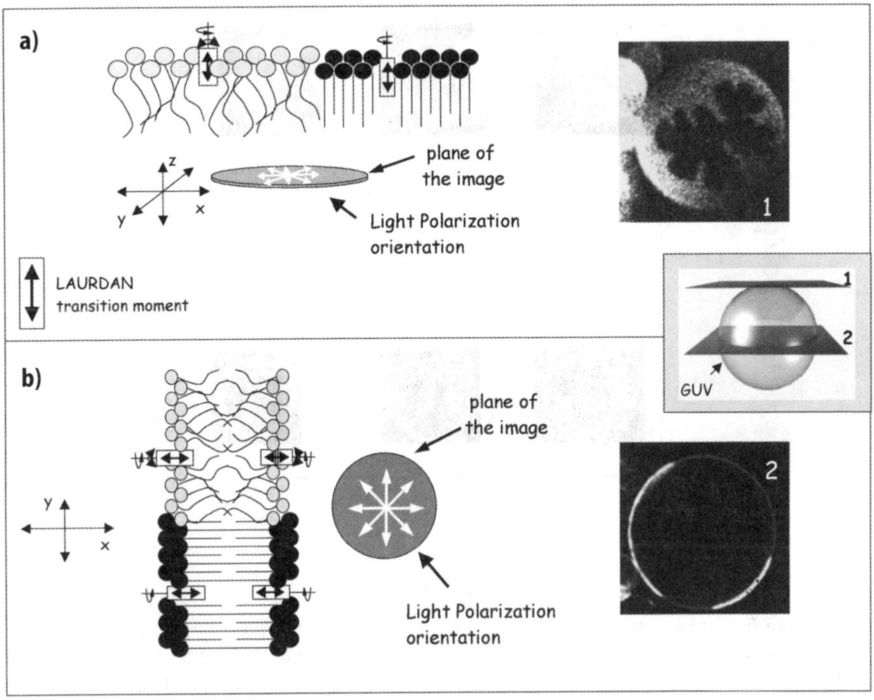

Fig. 9.3 a, b. Laurdan photoselection effect operating in the GUV's polar region (**a**) and center cross section (**b**). The composition of GUV is DPPE/DPPC (7:3 mol) at a temperature corresponding to the gel/fluid phase coexistence (50 °C)

extremely difficult to obtain Laurdan images, simply because of the high extent of photobleaching. Instead, Laurdan images can be easily obtained by using multiphoton excitation techniques, as was demonstrated previously [49, 54 and references therein].

9.5 Correlation with Other Experimental Techniques

Differential scanning calorimetry (DSC) experiments provide extremely useful thermodynamic parameters to characterize the temperature behavior of lipid mixtures. However, as the number of components in the lipid mixture is increased the data analysis becomes very difficult. For example, a calorimetry profile of a natural complex lipid mixture from bovine lipid extract surfactant (BLES) is shown in Fig. 9.4. Although thermodynamic information from the DSC experiment can be extracted from the system, no detailed information about the physical characteristics of lipid lateral structure at different temperatures can be obtained using this technique. As observed in Fig. 9.4, from the GUV data three different temperature regimes with particular membrane lateral structure (phase state) are observed and characterized using Laurdan [34]. The information

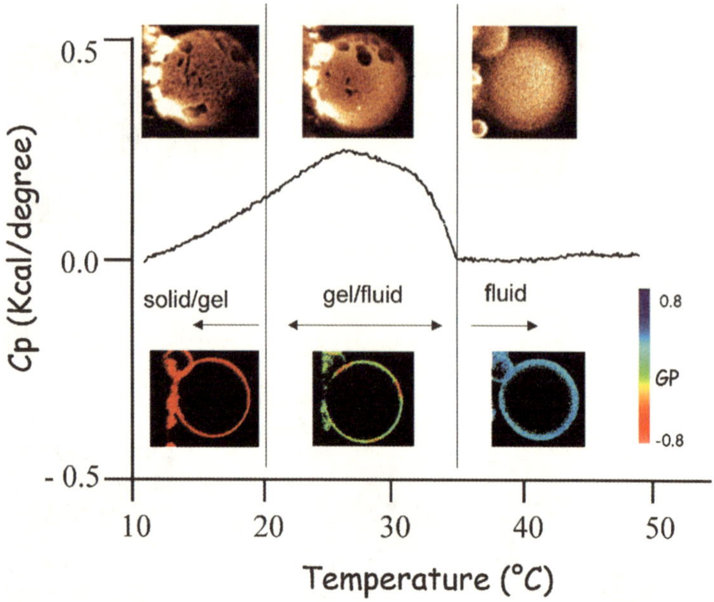

Fig. 9.4 Bovine lipid extract surfactant (BLES) preparations. The thermogram obtained from DSC experiments indicates a very broad and complex phase transition temperature region. Laurdan intensity and Laurdan GP images reveal additional information showing three different phase transitions (fluid → fluid/gel, fluid/gel → solid/gel) in the same temperature range used in the DSC experiments [34]

obtained from these two techniques is rather complementary and very useful to characterize the lateral structure of lipid mixtures.

Other techniques can provide information that complement that obtained from GUV experiments. Atomic force microscopy (AFM), for instance, allows bridging of the micrometer size scale with the nanometer size scale. As shown in Fig. 9.5, the nanoscopic details of a binary mixture (DMPC/DSPC) can be obtained using a probe-free technique [57, 58]. In this case the phase information obtained with the probe Laurdan (that is related with water dipolar relaxation processes in the membrane) is expanded with the nanoscopic information (lateral packing structure and height) obtained by using AFM. In this case the highly laterally packed areas (gel phase) observed using Laurdan fluorescence have a nanometer size structure corresponding to a ripple phase as observed with AFM. The similarities in shape and size are remarkable in both model systems. In addition, Fig. 9.5 also addresses a very important point regarding the fluorescence experiments. As seen very often in the literature, a major criticism of fluorescent techniques is the potential invasive nature of fluorescent probes. For example Aussenac et al. [59] pointed out that fluorescent probes (such as Laurdan or rhodamine PE) may perturb the system by favoring the formation or aggregation of lipid microdomains by intrinsic cross-linking. Besides the many control experiments already reported in the literature (such as the similar lipid domain scenario observed using probes

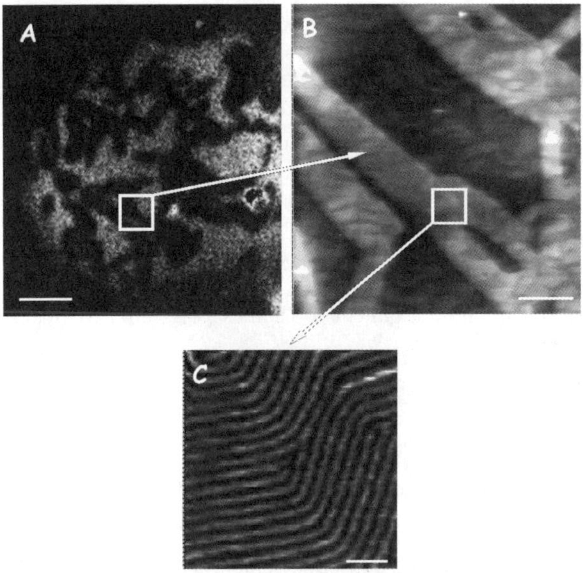

Fig. 9.5. **a** Two-photon excitation fluorescence image of a single Laurdan-labeled GUV composed of DMPC/DSPC 1:1 mol mixture. The *dark areas* correspond to the gel phase (photoselection effect). **b**, **c** AFM images of double planar bilayers composed of DSPC/DMPC 1:1 binary mixture. The *brighter areas* in **b** correspond to gel areas. The nanoscopic details inside these areas reveal the presence of a metastable ripple phase [57, 58]. The *bars* are **a** 4 μm, **b** 1 μm, **c** 0.1 μm

with completely different chemical structure, see for example ref. [29]), the similar shape and size of lipid domain shape observed with fluorescent probes (GUVs using fluorescence microscopy) and in probe-free membranes (planar-supported bilayers using AFM) clearly rule out major effects of fluorescence probes on the lateral structure (Fig. 9.5) of the lipid membranes. In addition, the transition temperatures observed for different lipid mixtures using different probes [29, 30] are in line with those reported in phase diagrams obtained using experimental techniques such as NMR and DSC, both probe-free techniques.

9.6 Concluding Remarks and Future Directions

This article demonstrates the usefulness of the combination of giant vesicle technology and fluorescence microscopy techniques, an exquisite experimental tool to explore the concerted phenomena occurring at different complexity levels of biological membranes (molecular and supramolecular). Using the approach outlined above we were able to characterize the lateral structure of many different lipid mixtures ranging from simple artificial lipid mixtures to natural lipid mixtures [29–31, 33, 34; for reviews see refs. 49 and 54]. After 30 years of studies of the physical and chemical aspects of lipid–lipid interactions (in particular the lipid lateral separation) in model systems, the importance of the relationship between membrane morphology and membrane physical characteristics is now well recognized. The challenge now is to explore membrane systems with native compositions to ultimately correlate membrane lateral structure with membrane function.

Acknowledgements
Research in the laboratory of L. A. B. is founded by a grant from SNF, Denmark (21-03-0569) and the Danish National Research Foundation (which supports MEMPHYS-Center for Biomembrane Physics). The author is grateful to Thomas Kaasgaard and Chad Leidy for providing the AFM images included in Fig. 9.5.

References

1. Simons K, Ikonen E (1997) Nature 387:569
2. Brown DA, London E (1998) Annu Rev Cell Dev Biol 14:111
3. Edidin M (2003) Annu Rev Biophys Biomol Struct 32:257
4. Tocanne JF (1992) Commun Mol Cell Biophys 8:53
5. Jacobson K, Dietrich C (1999) Curr Opin Cell Biol 9:84
6. Oliveira RG, Maggio B (2000) Neurochem Res 25:77
7. Gauss KE, Gratton EP, Kable W, Jones AS, Gelissen I, Kritharides L, Jessup W (2003) Proc Natl Acad Sci USA 100:15554
8. Gebhardt CH, Gruler CH, Sackmann E (1997) Z Naturforsch 32C:581
9. Sackmann E, Feder T (1995) Mol Membr Biol 12:21
10. Ipsen JH, Mouritsen OG (1988) Biochim Biophys Acta 944:121
11. Jørgensen K, Mouritsen OG (1995) Biophys J 69:942
12. Lee AG (1975) Biochim Biophys Acta 413:11
13. Lentz BR, Barenholtz Y, Thompson TE (1976) Biochemistry 15:4529
14. Mabrey S, Sturtevant JM (1976) Proc Natl Acad Sci USA 73:3862
15. Van Dijck PWM, Kaper AJ, Oonk HAJ, De Gier J (1977) Biochim Biophys Acta 470:58
16. Arnold K, Lösche A, Gawrisch K (1981) Biochim Biophys Acta 645:143
17. Blume A, Wittebort RJ, Das Gupta SK, Griffin RG (1982) Biochemistry 21:6243
18. Caffrey M, Hing FS (1987) Biophys J 51:37
19. Shimshick EJ, McConnell HM (1973) Biochemistry 12:2351
20. Maggio B (1985) Biochim Biophys Acta 815:245
21. Maggio B, Fidelio GD, Cumar FA, Yu RK (1986) Chem Phys Lipids 42:49
22. Bagatolli LA, Maggio B, Aguilar F, Sotomayor CP, Fidelio GD (1997) Biochim Biophys Acta 1325:80
23. Vaz WLC, Melo ECC, Thompson TE (1989) Biophys J 56:869
24. Vaz WLC, Melo ECC, Thompson TE (1990) Biophys J 58:273
25. Bultmann T, Vaz WLC, Melo ECC, Sisk RB, Thompson TE (1991) Biochemistry 30:5573
26. Almeida PFF, Vaz WLC, Thompson TE (1992) Biochemistry 31:7198
27. Schram V, Lin HN, Thompson TE (1996) Biophys J 71:1811
28. Korlach J, Schwille P, Webb WW, Feigenson GW (1999) Proc Natl Acad Sci USA 96:8461
29. Bagatolli LA, Gratton E (2000) Biophys J 78:290
30. Bagatolli LA, Gratton E (2000) Biophys J 79:434
31. Dietrich C, Bagatolli LA, Volovyk ZN, Thompson NL, Levi M, Jacobson K, Gratton E (2001) Biophys J 80:1417
32. Feigenson GW, Buboltz JT (2001) Biophys J 80:2775
33. Bagatolli LA, Gratton E, Khan TK, Chong PLG (2000) Biophys J 79:416
34. Nag K, Pao JS, Harbottle RR, Possmayer F, Petersen NO, Bagatolli LA (2002) Biophys J 82:2041

35. Veatch SL, Keller SL (2003) Biophys J 85:3074
36. Scherfeld D, Kahya N, Schwille P (2003) Biophys J 85:3758
37. Wick R, Angelova MI, Walde P, Luisi PL (1996) Chem Biol 3:105
38. Longo ML, Waring AJ, Gordon LM, Hammer DA (1998) Langmuir 14:2385
39. Bucher P, Fischer A, Luisi PL, Oberholzer T, Walde P (1998) Langmuir 14:2712
40. Angelova MI, Hristova N, Tsoneva I (1999) Eur Biophys J 28:142
41. Holopainen JM, Angelova MI, Kinnunen PKJ (2000) Biophys J 78:830
42. Reeves JP, Dowben RM (1969) J Cell Physiol 73:49
43. Angelova MI, Soléau S, Meléard PH, Faucon JF, Bothorel P (1992) Colloid Polym Sci 89:127
44. Bagatolli LA, Parasassi T, Gratton E (2000) Chem Phys Lipids 105:135
45. Menger FM, Keiper JS (1998) Curr Opin Chem Biol 2:726
46. Luisi PL, Walde P (eds) (2000) Giant vesicles, 1st edn. Wiley, Chichester
47. Fahsel S, Pospiech EM, Zein M, Hazlet TL, Gratton E, Winter R (2002) Biophys J 83:334
48. Arnold K, Lösche A, Gawrisch K (1981) Biochim Biophys Acta 645:143
49. Bagatolli LA, Gratton E (2001) J Fluoresc 11:141
50. Weber G, Farris FJ (1979) Biochemistry 18:3075
51. Macgregor RB, Weber G (1986) Nature 319:70
52. Parasassi T, Conti F, Gratton E (1986) Cell Mol Biol 32:103–108
53. Lasagna M, Vargas V, Jameson DM, Brunet JE (1996) Biochemistry 35:973
54. Bagatolli LA, Sanchez S, Hazlett T, Gratton E (2003) Meth Enzymol 360:481
55. Parasassi T, Gratton E (1995) J Fluoresc 5:59
56. Parasassi T, Krasnowska E, Bagatolli LA, Gratton E (1998) J Fluoresc 8:365
57. Leidy C, Kaasgaard T, Crowe JH, Mouritsen OG, Jørgensen K (2002) Biophys J 83:2625
58. Kaasgaard T, Leidy C, Crowe JH, Mouritsen OG, Jorgensen K (2003) Biophys J 85:350
59. Aussenac F, Tavares M, Dufourc EJ (2003) Biochemistry 42:1383

Part 3

Application of Fluorescence Spectroscopy to Protein Studies

10 Protein Dynamics and Protein Folding Dynamics Revealed by Time-Resolved Fluorescence

Anoop Saxena, J. B. Udgaonkar and G. Krishnamoorthy

Keywords: Protein folding dynamics; Time-resolved fluorescence; Barstar; Tryptophan motional dynamics; Fluorescence resonance energy transfer

Abbreviations

CD	Circular dichroism
FRET	Fluorescence resonance energy transfer
Gdn-HCl	Guanidine hydrochloride
MEM	Maximum entropy method
NATA	*N*-acetyltryptophanamide
TCSPC	Time-correlated single photon counting
TNB	Thionitrobenzoate
tr-FRET	Time-resolved fluorescence resonance energy transfer

10.1 Introduction

The concept that dynamics along with structure form the basis of the activity of biomolecular systems has gained experimental support in recent years [1, 2]. By virtue of its sensitivity, selectivity, and large temporal range, fluorescence spectroscopy has become one of the most revealing windows on biomolecular dynamics [3–6]. Of the several experimental methods available for studying macromolecular dynamics, fluorescence-based methods have the following advantages: (1) dynamics of a specific group or a segment of a massive macromolecular system can be observed without interference from the rest of the system; (2) the timescale of observable dynamics covers a wide range from femtoseconds to seconds; and (3) observations can be made at very low concentrations.

The dynamic nature of protein structure, which is essential for its function, leads to structural heterogeneity. Due to its origin in dynamics, the apparent level of heterogeneity depends upon the time window used for observation. Large time windows result in an averaging of the structural parameters, whereas shorter windows produce an instantaneous snapshot of the structural variants populating a distribution. The time window set by fluorescence-based methods is linked to the excited state lifetime of the fluorophore, which lies generally in the range of 10 ps to 10 ns. Since the timescale of large-scale and high-amplitude protein dynamics is generally in the range of nanoseconds and beyond [7], fluorescence methods capture essentially a snapshot of the various structural forms present. The ensemble of structural forms of proteins could vary in their level of heterogeneity. While the spread in structural parameters is expected to be quite small for native and stable structures, partially structured intermediates encountered along protein folding pathways are expected to have broader distributions of their structural characteristics. Furthermore, the amplitudes of local and segmental dynamics are expected to be nonuniform throughout the structure and probably related to functional domains of the protein. Hence the level of structural heterogeneity could also follow a nonuniform pattern. Such considerations demand that observations be made on fluorescence probes at specific locations, guided by information on the function of the system. Similarly, the location of the probe is quite critical while using it as a probe of folding dynamics. In the sections to follow we shall see examples where the dynamic fluorescence of the intrinsic fluorophore, tryptophan (trp), has been used effectively to shed light on protein dynamics and protein folding dynamics.

10.2
Dynamic Fluorescence of Tryptophan

The main parameters derivable from time-domain fluorescence measurements, fluorescence lifetime and fluorescence depolarization time, yield information on the microenvironment around the fluorophore [3–6]. While both these parameters could, in principle, reflect structural heterogeneity, fluorescence lifetime is the preferred parameter due to the relatively higher accuracy with which it can be determined. The very high signal-to-noise ratios (*S*/*N*) achievable in intensity decay measurements enable reliable recovery of multiple and distributed lifetimes [6, 8]. However, this benefit is often offset by the ultrasensitivity of fluorescence lifetime to a variety of environmental factors, which makes the interpretation of lifetimes and changes in them ambiguous. However, when the dominant source of nonradiative decay can be identified with a factor, such as fluorescence resonance energy transfer (FRET) [9, 10], the excited state lifetime becomes a very useful structural signature [9, 10]. In contrast to fluorescence lifetime, fluorescence depolarization time, albeit with low accuracy, is a direct indicator of molecular dynamics [3–6] and hence could offer relatively less ambiguous conclusions.

While the high sensitivity of the excited state lifetime of tryptophan in proteins has resulted in a wealth of information on protein dynamics (e.g., see Vol. 6 in ref. 4), the complex intensity decay kinetics remain a subject of continuing interest. Although there have been a variety of models for explaining the multiexponential nature of decay kinetics [11–13], the rotamer-based ground state heterogeneity model [14, 15] seems to explain most of the observations. A simple demonstration of this model is provided by the data shown in Fig. 10.1. Fluorescence lifetime distributions obtained by analyzing intensity decay curves by the maximum entropy method (MEM) [15–17] clearly show that *N*-acetyltryptophanamide (NATA), a model compound for tryptophan, is associated with a single exponential decay kinetics. In contrast, the di- and tripeptides showed multiple peaks in their lifetime distributions. The

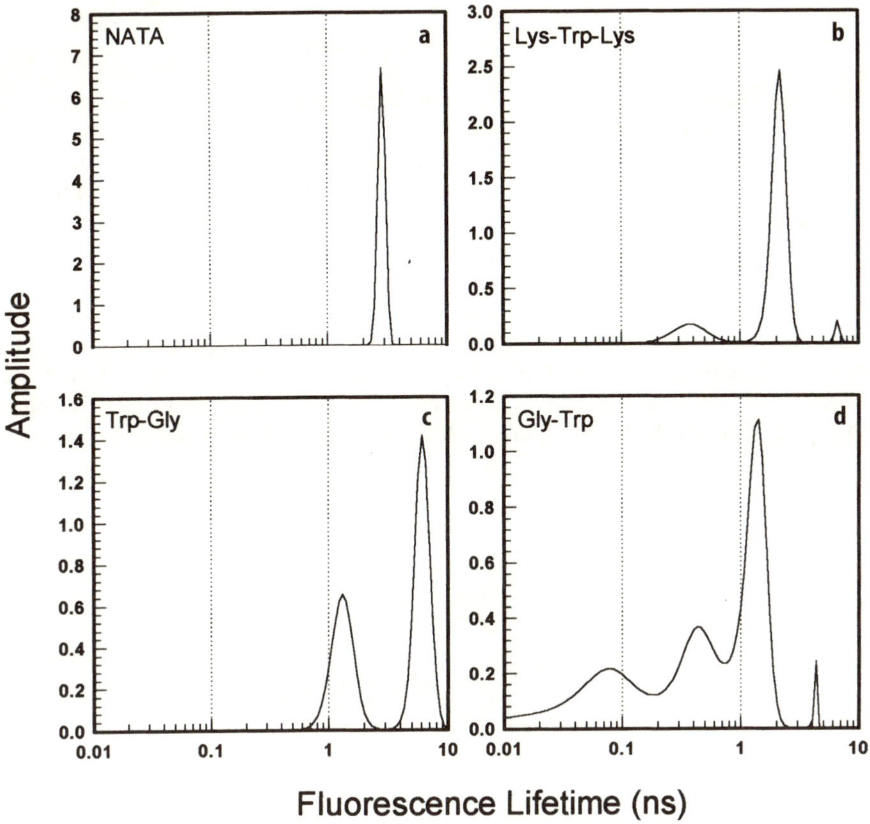

Fig. 10.1a–d. Fluorescence lifetime distributions analyzed by MEM. **a** *N*-acetyltryptophan-amide (NATA); **b** Lys-Trp-Lys; **c** Trp-Gly; **d** Gly-Trp

monoexponential nature of the decay of NATA could be explained by the fact that the distance between the indole chromophore and the carbonyl carbon, which is presumed to be the quenching group, is similar in all three rotamers along the C_α–C_β bond. In contrast, any peptide is associated with dissimilar distances between the indole and the quenching groups in the rotamers. Multi-exponential lifetimes have been observed in many single tryptophan proteins in their native state [18, references in 15]. However, there have been situations where either single or nearly single lifetimes are seen, such as in the case of the single tryptophan (Trp53) mutant of barstar [19]. This unique situation is associated with the rigidity of the trp side chain, as shown by the lack of internal motion in the native (N) state of this protein [19, 20]. The same protein shows three lifetimes along with the presence of internal rotational dynamics of the tryptophan in molten globule-like states [19, 20]. These observations constitute strong support for the ground state heterogeneity model based on rotamers to account for the origin of the multiexponential nature of tryptophan decay kinetics in general.

10.2.1
Tryptophan Motional Dynamics and Protein Surface Hydration

In spite of the lack of general agreement on the origin of the multiexponential nature of the fluorescence decay kinetics of tryptophan, its rotational dynamics have found remarkable use in providing insights into protein dynamics. The dynamics of surface-located side chains are expected to influence the binding of other molecules. Solvent-exposed trp side chains can be used for probing these dynamics. The motional dynamics of such trp side chains have been used to demonstrate preferential hydration at native protein surfaces [21]. It was found that, in native proteins, the correlation time (φ) associated with the rotational dynamics of surface tryptophans did not scale linearly with bulk viscosity (η). In contrast, φ scaled linearly with η in denatured proteins similar to free tryptophan in solution. The latter behavior is expected from Stokes–Einstein hydrodynamics, $\varphi = \eta V/kT$, where V, k, and T are molecular volume, Boltzmann constant, and

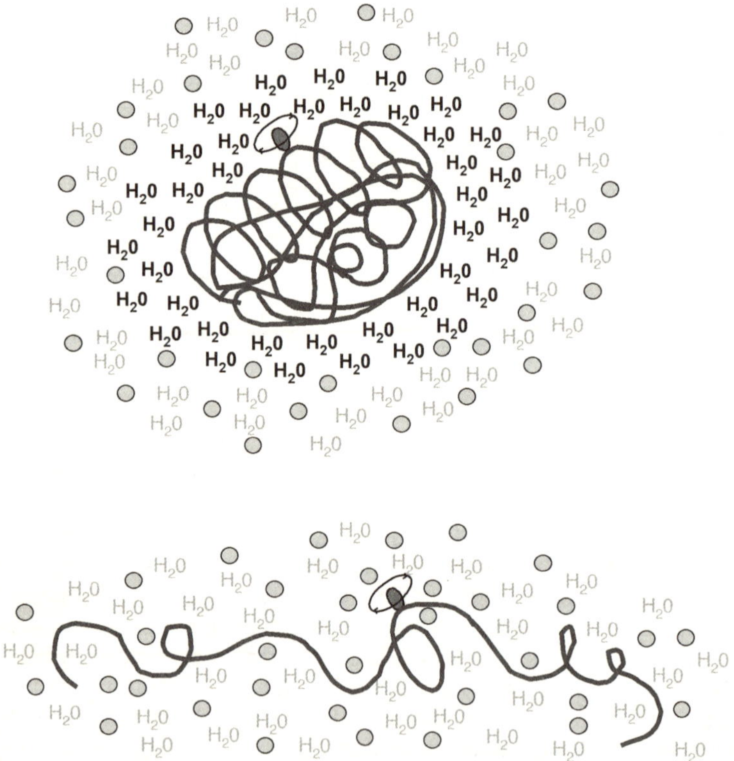

Fig. 10.2. Schematic diagram illustrating the control of dynamics of surface-located side chains (marked by an *ellipsoid*) by bound water (marked by ***bold*** $\mathbf{H_2O}$) in native proteins (*top*). In denatured proteins (*bottom*), the network of bound water is absent and hence the dynamics are controlled by bulk solvent comprising water (marked by H_2O) and the cosolvent, glycerol (marked by *small circles*)

absolute temperature, respectively. These results were used to construct a model wherein surface dynamics in native proteins are controlled by a bound water layer (Fig. 10.2). Hydration of polar side chains at the surface of native proteins could lead to a shield of bound water. Such a contiguous layer of bound water is unlikely to be present in the denatured state, and hence the dynamics of side chains are expected to be controlled by bulk solvent as observed. Such information is relevant to understanding the finer aspects of protein-protein, protein-ligand, and protein-nucleic acid interactions.

An alternative way of gaining insight into dynamics at the protein–water interface is to monitor the relaxation of bound water subsequent to excitation of surface-located fluorescence probes [22]. Measured with femtosecond time resolution, these solvent relaxation studies have shown that the relaxation of bound ("biological") water is slower than that of bulk water [22], similar to observations in heterogeneous media [23]. Building a comprehensive model for surface dynamics requires information on the dynamics of side chains themselves and of the bound layer of water.

10.2.2 Motional Dynamics of Trp53 in Stable Structural Forms of Barstar

Barstar, an 89-residue protein, has served as an effective model system in folding and unfolding studies [9, 19, 20, 24–27]. The motional dynamics of Trp53, which is located at the core of the protein, are very sensitive to the overall structural characteristics of the protein. In the native (N) state of the protein, the decay of fluorescence anisotropy of Trp53 follows a single correlation time (φ) of ~5.1 ns [20, 27]. This corresponds to the overall tumbling of the protein which has a molecular weight of ~10 kD. The absence of any local dynamics within the observable time resolution of ~20 ps is an indication of the high rigidity of the protein core where Trp53 is located. As mentioned earlier, this rigidity is probably the origin of the unique single exponential decay of fluorescence intensity. In contrast, when the protein is unfolded (U), Trp53 is associated with two φ values, ~3.5 and ~0.7 ns, with almost equal amplitudes [20]. Since the model compound NATA shows a value of φ <0.1 ns in aqueous solutions, the two correlation times seen in the U form should reflect the segmental dynamics of the polypeptide. In some of the models proposed for protein folding [28, 29], the initial step is nucleation of folding in a specific segment of the chain. Thus, the segmental dynamics seen in the U form could be used as a locator of the nucleation site. The two correlation times observed in the U form could also have arisen due to the two forms of U differentiated by the *cis–trans* isomer status of Pro48 (see below).

Apart from the N and U forms, barstar is known to exist in several partially folded states. The A form seen at pH 3 [30, 31] has properties similar to those of a molten globule which is a generic form of folding intermediate [32]. The rotational dynamics of Trp53 in the A form show two φ values, ~1 and >50 ns (Table 10.1). This indicates that the core is flexible (φ ~1 ns) and the protein is extensively aggregated (φ >50 ns) in the A form [31, 33]. Another molten globule-like structural form was observed in salt-stabilized, high-pH (12) denatured

Table 10.1. Rotational correlation times associated with the core Trp53 in various stable structural forms of barstar

Structural form of barstar	Rotational correlation times φ (ns), (amplitude)	
1. Native (N) state	5.1 (1.0)	–
2 Unfolded (U) state	0.76 (0.51)	3.7 (0.49)
3. Low-pH molten globule-like form (A form)	1.1 (0.26)	>50 (0.74)
4. High-pH (12) denatured form (D state)	0.26 (0.50)	2.6 (0.50)
5. High-pH (12) salt-stabilized compact form (P state)	7.4 (1.0)	–
6. Denaturation transition zone form	1.5 (0.53)	11.9 (0.47)

barstar [20]. The rotational dynamics of Trp53 indicated a flexible interior and enhanced overall volume of the protein as hallmarks of this form.

The titration of barstar with chemical denaturants such as urea or guanidine hydrochloride (Gdn-HCl), when monitored by steady-state probes such as fluorescence intensity or circular dichroism (CD), could be fitted to two-state models similar to observations for many single-domain small proteins [34]. However, folding intermediates have been observed during kinetic studies [35, 36], thus creating an apparent contradiction. A likely scenario is that either such intermediates are present at very low concentrations or these intermediates are invisible through the probes normally used during equilibrium titration. Studies of the dynamics of Trp53 come in handy for clarifying the picture. In the transition zone (between the N and U states) the fluorescence anisotropy decay of Trp53 showed φ ~12 ns, which is different from that of either the N or U forms [19]. This long correlation time indicates partially folded and subcompact intermediate structure(s). Table 10.1, which summarizes the φ values obtained in various structural forms of barstar, demonstrates the use of the rotational dynamics of tryptophan in gaining structural information.

10.2.3 Tryptophan Dynamics and "Double Kinetics" in Protein Folding

Studies of the motional dynamics of tryptophan, observed through time-resolved fluorescence anisotropy measurements, provide a revealing insight into the complexities of protein folding dynamics [27, 37]. The ability to observe the dynamics in an ensemble of heterogeneous molecules gives fluorescence-based methods an edge over high-resolution techniques such as NMR and X-ray crystallography. Protein folding and unfolding processes generally occur in the timescale of microseconds to seconds [38] and can be monitored through measurements of the fluorescence intensity of appropriately located probes. Although integrated fluorescence intensity could be collected with sufficiently high *S/N* with a time resolution of a few microseconds, collection of picosecond time-resolved fluorescence usually requires data acquisition windows of several tens

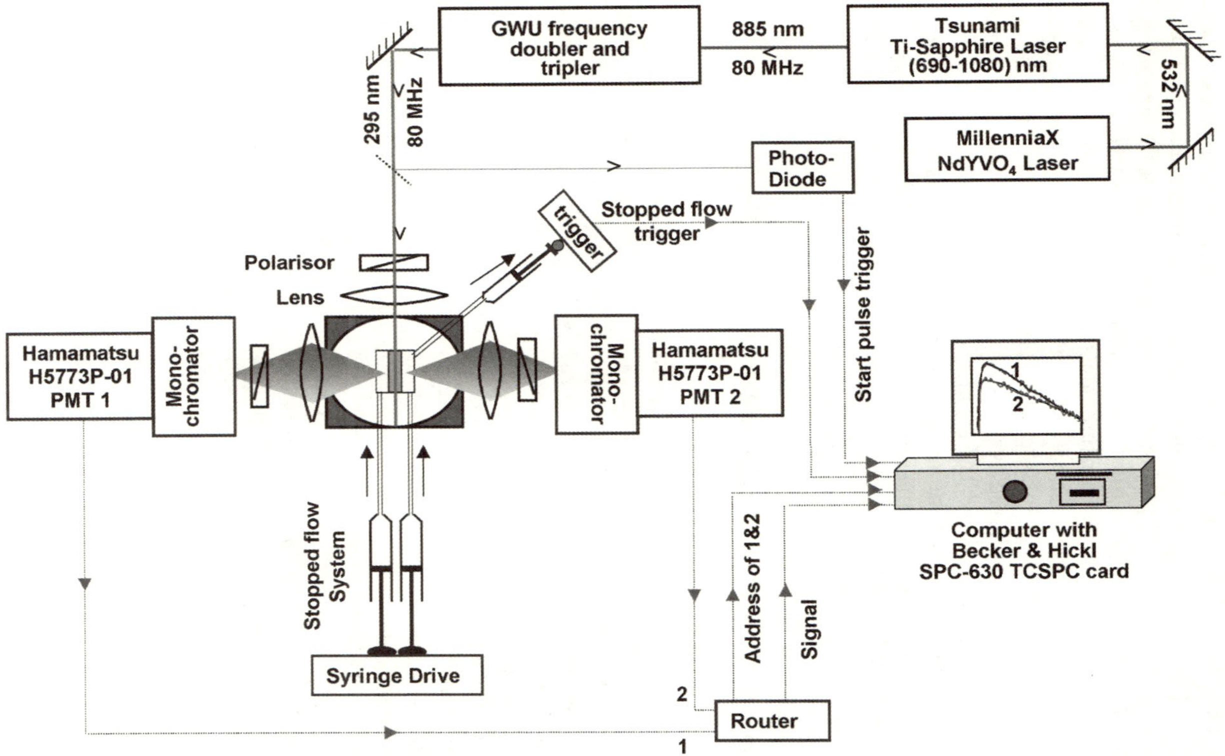

Fig. 10.3. Experimental setup for monitoring the time evolution of time-resolved fluorescence ("double kinetics") during folding or unfolding of proteins. The setup consists of three main parts: (1) picosecond high repetition laser system; (2) stopped-flow sample handling system; and (3) dual-channel high repetition rate time-correlated single photon counting (TCSPC) system. The two detection channels can be set either at two polarizations (for fluorescence anisotropy measurements) or at two wavelengths when needed

of seconds. The necessity to collect only a single photon after every pulse of excitation in the time-correlated single photon counting (TCSPC) method [39], and time-spreading of the collected photons, are the main causes of the normal requirement of long (~10 s) acquisition times. (Although a part of the problem is overcome in streak camera-based measurements [40], these measurements suffer from low dynamic range when compared to the TCSPC method.) Nevertheless, a time resolution of ~20 ms in such "double kinetics" experimental studies of protein folding reactions monitored by picosecond-resolved rotational dynamics was demonstrated in a specialized TCSPC method [37]. Similar time resolution can be achieved with the help of newer generation TCSPC cards, which operate in the reverse start–stop mode at the full repetition rate (~80 MHz) of the pulsed laser. A schematic of the experimental setup based on such a card, and used in some of the experiments described in later sections, is shown in Fig. 10.3.

10.2.4 Motional Dynamics of Trp53 During Folding of Barstar

Trp53, which is located in the core of the protein, is very sensitive to the level of structure acquisition during folding of the protein. Upon initiation of folding by changing the solvent from 8 M urea to 2 M urea, barstar folds in two phases when observed through measurement of fluorescence intensity of Trp53 [27, 41]. The two phases arise due to the presence of two forms of the unfolded protein, namely U_F and U_S, which are differentiated by *cis–trans* isomers of Pro48. The two forms exist in the ratio of ~3:7 under the unfolding conditions. Folding of U_F is fast and is complete within a second whereas the major (~70%) fraction, U_S, folds through a series of intermediates and takes several tens of minutes to fold [27, 41]. Thus, the slow phase of the folding process offers the possibility of monitoring the time evolution of dynamic fluorescence of Trp53. Upon initiation of folding, the description of the rotational dynamics of Trp53 changes from a combination of local and segmental motion in the U form (φ ~0.7 and ~3 ns) to a combination of local (φ ~1 ns) and global (φ ~7 ns) motion [27]. This then evolves during the slow phase of folding to the N state, which is characterized by a single value of φ ~5 ns reflecting the global tumbling dynamics. This process is modeled [27] as an initial collapse of the unfolded polypeptide into an intermediate form (labeled I_N), which has a flexible core (φ ~1 ns) and an enlarged volume (φ ~7 ns which is higher than that of N). Thus, this kinetic intermediate form appears similar to the equilibrium structure seen in the salt-stabilized, high-pH (12) form [20] discussed above. Subsequent to the initial collapse, I_N evolves to N in two steps:

$$U \rightarrow I_N \rightarrow I'_N \rightarrow N \tag{10.1}$$

In the step leading to I'_N, the core becomes rigid as revealed by the disappearance of φ ~1 ns. During the next (slower) step the overall size becomes compact (global φ decreases from ~7 to ~5 ns) and secondary and tertiary interactions consolidate [27]. Thus, these studies of the motional dynamics of tryptophan demonstrated that tight packing of the core is an early event in the overall process of folding.

10.2.5 Evolution of Core Dynamics During Unfolding of Barstar

The question whether information retrieved from unfolding kinetics can be used to shed light on the folding process depends upon the protein folding model in a very fundamental manner [42]. Early models assumed definite pathways of folding [43] and hence, according to microscopic reversibility, information obtained from either folding or unfolding kinetics could be used to infer the complementary process. In contrast, energy landscape models [44, 45], which are favored by both theorists and experimentalists, predict that the folding process is heterogeneous at the microscopic level and hence the concept of folding pathway and intermediates becomes less relevant. In such a scenario, the folding process is unlikely to be a direct reversal of unfolding [42]. However, intermediates during folding and unfolding have been detected in many systems [35, 36]. Hence kinetic studies on the unfolding process are expected to provide useful information on the pathway connecting the N and U states of the protein.

Unfolding of barstar by urea, for example, shows two kinetically distinct phases when observed through the fluorescence intensity of Trp53 [46]. This has been modeled by a two-step process, $N \rightarrow U_C \rightarrow U_T$, where U_C is the unfolded form in which Pro48 is in the *cis* conformation (similar to N where Pro48 is in the *cis* conformation) and U_T is the unfolded form with Pro48 in the *trans* conformation [46]. While U_T could be identified with U_S, the slow-folding unfolded form mentioned in Sect. 10.2.4, the identity of U_C could still be debated. If U_C is similar to U_F, the fast-folding form mentioned, we would expect the overall properties of the protein at the end of the fast phase to be similar to those seen at the end of the slow phase of unfolding except, of course, for the local isomer status at Pro48. On the other hand, if U_C is more structured when compared to U_T one could expect to demonstrate this through the motional dynamics of Trp53. Another motivation for monitoring the time evolution of rotational dynamics is to see whether the unfolding process is discrete (having only two states, N and U) or is associated with a continuous opening of the structure. With such motivations in mind we monitored the double kinetics of time evolution of the rotational dynamics of Trp53 during unfolding by using the setup described in Fig. 10.3.

Figure 10.4b–g shows the typical decay of fluorescence anisotropy of Trp53 in the native protein and the decays captured at successive time points when unfolding was initiated by change of solvent from 0 to 5.2 M urea. Figure 10.4a shows the time evolution of steady-state fluorescence intensity. Rotational correlation times (φ) obtained by analysis [19, 20, 27] of anisotropy decay profiles are listed in Table 10.2. It can be seen that φ evolves from ~5 ns (corresponding to N) to a combination of ~2 and ~0.5 ns (corresponding to U) during the unfolding process. The initial jump of φ from ~5 to ~7 ns could be mainly due to the increase in the solvent viscosity by a factor of 1.3 when the solvent was changed from 0 to 5.2 M urea. Furthermore, it could also be due, at least partly, to rapid expansion of the protein surface seen in FRET studies [47]. The time evolution of fluorescence intensity (Fig. 10.4a) shows that the fast phase associated with unfolding is essentially complete in ~10 s and hence the population of U_C is expected to be maximum around this time. The rotational dynamics of Trp53

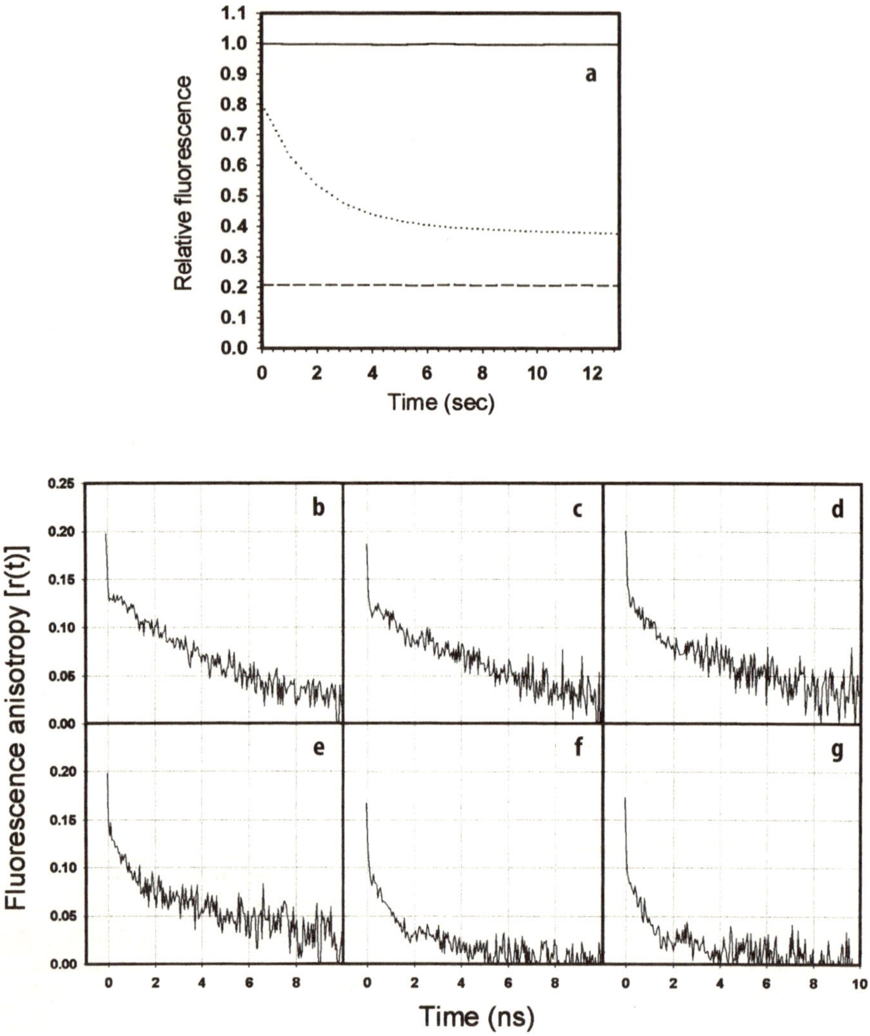

Fig. 10.4a–g. Time evolution of integrated fluorescence intensity (**a**) and time-resolved fluorescence anisotropy in the native form (**b**) and during unfolding of single tryptophan (Trp53) mutant of barstar in 5.2 M urea (**c–g**). Excitation and emission wavelengths were 295 and 340 nm, respectively. Trace **a** was recorded by using a SPEX T-format spectrofluorimeter. The horizontal continuous line (*top*) and the broken line (*bottom*) in **a** correspond to the signal levels at zero time and at the end of the unfolding process, respectively. Traces **b–g** were recorded by using the setup described in Fig. 3. Trace **b** corresponds to the native protein and 500-ms time slices were taken after 0.0 s (**c**), 4.0 s (**d**), 8.5 s (**e**), 45.5 s (**f**), and 228.5 s (**g**) during unfolding. Parameters recovered from analysis of such traces are given in Table 2

Table 10.2. Time evolution of rotational dynamics of Trp53 in barstar during unfolding

Time[a] (s)	Rotational correlation times (ns)[b], (amplitude)		Initial anisotropy, r_0
	φ_1 (β_1)	φ_2 (β_2)	
0.0	4.80 (1.00)	–	0.160
0.5	6.50 (0.83)	0.62 (0.17)	0.157
1.0	6.40 (0.74)	0.64 (0.26)	0.160
1.5	7.20 (0.70)	0.63 (0.30)	0.165
2.0	6.90 (0.69)	0.51 (0.31)	0.166
3.0	6.90 (0.56)	0.66 (0.44)	0.167
4.5	7.00 (0.61)	0.63 (0.39)	0.164
9.0	7.00 (0.50)	0.55 (0.50)	0.170
10.0	7.40 (0.53)	0.77 (0.47)	0.164
46.0	2.10 (0.39)	0.29 (0.61)	0.169
107.0	2.10 (0.31)	0.34 (0.69)	0.169
229.0	2.50 (0.26)	0.40 (0.74)	0.170
1534.0	2.70 (0.31)	0.31 (0.69)	0.170

[a] Data collected for 500 ms preceding the time mentioned, after solvent change from 0 to 5.2 M urea.

[b] Errors associated with φ_1, φ_2, β_1, and β_2 were ~0.3 ns, ~0.1 ns, ~0.05, and ~0.05, respectively.

monitored around 10 s showed two correlation times, viz, ~7 and ~0.8 ns. These two φ values can be assigned to the small (~25%) fraction of the native protein and the major fraction of the unfolded protein, respectively (Fig. 10.4a), when we take into account the relative amplitudes of the two φ values and the relative quantum yields of the folded and unfolded forms. Thus, these results indicate that the U_C form mentioned above is largely unstructured and is similar to U_T.

10.3 Time-Resolved Fluorescence Resonance Energy Transfer (tr-FRET) in Protein Folding

As mentioned earlier, the fluorescence lifetime of tryptophan is too sensitive to environment and this results in the inability to interpret changes in it in a unique way. However, when energy transfer from tryptophan to an intramolecular acceptor occurs, in a predominant manner by FRET [48], the fluorescence lifetime of the tryptophan becomes a sensitive monitor of intramolecular distances. Furthermore, in complex situations, such as those encountered even in small proteins, intramolecular distances could have distributions due to heterogeneity of structure. Such a distribution of intramolecular distances can be obtained when the fluorescence intensity decay profiles are analyzed to generate a distribution of lifetimes. Although there are a number of ways by which distribution of lifetimes can be obtained [10, 49–51], the maximum entropy method (MEM) provides the most unbiased distributions [9, 15–17]. Information on distance distribution is quite useful as it provides a quantitative estimate of the level of molecular level heterogeneity, which is a hallmark parameter in the energy landscape model of protein folding [42, 44, 45]. The first use of this (tr-FRET) method to study protein folding was demonstrated in the equilibrium denaturation of barstar [9]. Subsequently, the technique was applied to the folding kinetics of cytochrome c [40] and barstar [52]. The results from these experiments are described below. The only alternative to tr-FRET in quantifying the level of heterogeneity is single-molecule FRET studies [53, 54]. It should be noted that the information (i.e., population heterogeneity) content from both types of measurements are similar to each other. TCSPC-based fluorescence decay curves are in fact generated by watching a single photon, and hence a single molecule, at a time. Furthermore, typical TCSPC curves are generated from total photon counts in the range of several tens of millions. In contrast, single-molecule FRET observations [53, 54] rely on only a few hundreds of events to construct fluorescence intensity histograms.

10.3.1
tr-FRET Shows Incremental Unfolding of Barstar

The nonfluorescent acceptor group thionitrobenzoate (TNB) is a very efficient quencher of tryptophan fluorescence by FRET [55]. When TNB is covalently linked to Cys-82 of barstar it quenches the fluorescence of Trp53 dramatically (~95%) by FRET [9]. The quenching gets substantially relieved when the protein is denatured [9]. When the intensity decay curves were analyzed by MEM, the predominant single sharp peak at ~5 ns seen in the unlabeled N state of the protein (see above) shifted to a single peak at ~0.25 ns on labeling Cys-82 with TNB [9]. The distance between Trp53 and Cys82 (and also several other Cys side chains) estimated from this data by using Forster's theory [48] matches the NMR- or crystallography-derived distances. This is an important observation because it validates the assumptions, such as orientation factors and refractive index, used in Forster's equation while estimating the distance.

Titration of the TNB-labeled protein with denaturants such as urea or Gdn-HCl caused two effects: the peak position (~0.3 ns) shifted to a higher value and the width of the peak increased gradually with an increase in the concentration of denaturants until about the midpoint of unfolding transitions, monitored by steady-state spectroscopic probes such as fluorescence intensity or CD. A further increase in the concentration of denaturants beyond the midpoint of the transition resulted in a collapse of the peak corresponding to the N state and consolidation of the peak due to U-like forms in the MEM-generated lifetime distributions [9]. The most striking observation is the unique pattern of lifetime distribution seen in the middle region of the transition curve, demonstrating clearly the presence of intermediates. Steady-state probes such as fluorescence intensity (of the unlabeled protein) or CD failed to reveal the presence of intermediates during denaturant titration [9].

The following picture (Fig. 10.5) emerges from the observations mentioned above: induced by increasing concentration of denaturant, the protein in the N state becomes swollen and flexible in a continuous manner. This is reflected in the observed increase in both the intramolecular distance between Trp53 and Cys82 (seen as an increase in the lifetime of the TNB-labeled N state) and the

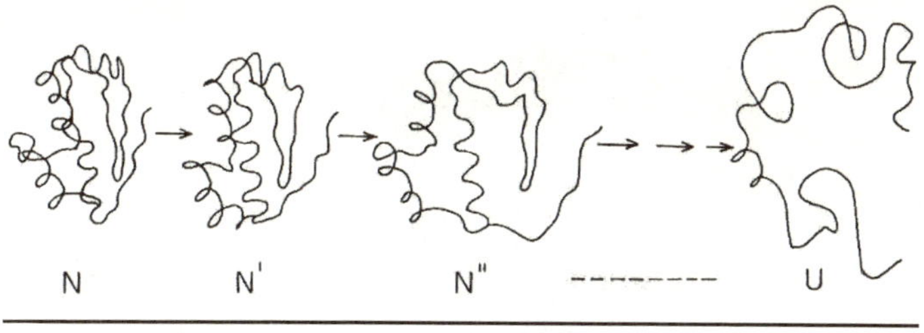

Fig. 10.5. Continuum of native-like states (N, N′, N″...) during equilibrium unfolding of barstar inferred from tr-FRET–MEM analysis (ref. [9]). The N-like states are separated from the unfolded (U) state by an energy barrier

distance distribution (reflected in the width of the lifetime distribution). These N-like forms of the protein resemble molten globule intermediates [32, 56] in their ~45% increase in volume and in being truly molten. However, passage of these N-like forms to the unfolded protein requires the protein to surmount an energy barrier similar to a first-order phase transition between the N and U states in a two-state model. Thus we see that transition between the N and U forms is *not* a two-step process even for a small single-domain protein. Commonly used techniques such as fluorescence intensity or CD are unable to differentiate between either the N-like or U-like forms and hence they report the process as a two-state process. In contrast, tr-FRET is able to identify the various N-like or U-like forms and report their population distribution. Such a new view of continuous change in structure has been seen only in simulations of off-lattice models of β-hairpin fragments of immunoglobulin-binding protein [57]. Molecular dynamic simulations [58–60] and energy landscape models predict that folding cannot be a two-state process when looked upon at the single molecule level. It is pleasing to note that the heterogeneity predicted by energy landscape models can be experimentally demonstrated at the macroscopic level and quantitated by the tr-FRET method outlined above.

10.3.2 Evolution of Population Heterogeneity During Folding of Barstar: Demonstration of "Folding Funnel"

As mentioned in the previous section, when FRET occurs between Trp53 and the TNB group covalently attached to a cysteine side chain located at various positions in the protein barstar, the lifetime distribution of Trp53 can be translated to provide intramolecular distances and distance distributions. This methodology offers the unique possibility of monitoring the time evolution of various intramolecular distances during the folding process. The ability of the tr-FRET technique to differentiate and quantitate various subpopulations within an ensemble (see Sect. 10.3.1) has been exploited to follow the structural composition of the intermediate ensemble I_N [52]. Furthermore, the dependence of the level of structural heterogeneity of I_N (see Sect. 10.2.4) on its stability, as well as the temporal evolution of heterogeneity during the transformation of I_N to N, could be followed.

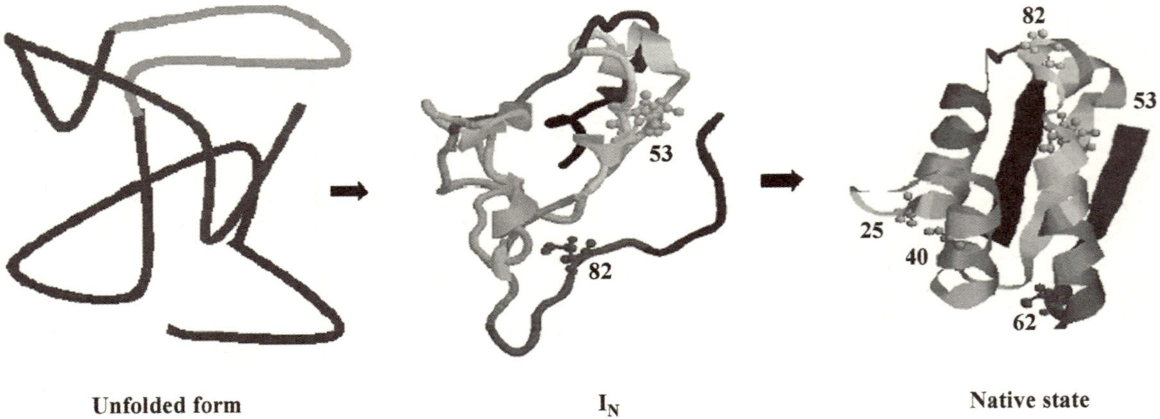

Fig. 10.6. Model for the folding intermediate I_N during folding of barstar. The I_N ensemble has an extended C-terminal region (containing cys82) but with remaining regions compact. In the N structure (generated by RASMOL using 1bta.pdb) are shown the single tryptophan at position 53 and the single cysteine side chains (25, 40, 62, and 82) used as acceptor sites

Four different single cysteine-containing mutants of barstar (cys25, cys40, cys62, and cys82) were studied [52] with the strategy mentioned above. Thus, four intramolecular distances could be monitored by this method. Folding was initiated by changing the solvent from denaturing (8 M urea) to nondenaturing conditions such as 0.5–2.0 M urea [27]. Fluorescence intensity decay profiles collected at ~1 min of refolding correspond predominantly to the I_N ensemble. Variation in the refolding conditions (0.5–2 M urea) offers variation in the stability of the I_N ensemble [35, 36]. It was found that in marginally stabilizing but still native-like conditions (say, 1.4 M urea) the I_N ensemble consists of populations with one or more intramolecular distances unfolded-like (Fig. 10.6). In strongly stabilizing conditions (such as 0.5 M urea), all four distances are essentially native-like in the entire population [52]. Thus, the heterogeneous population of the intermediate ensemble transforms into a more homogeneous one as the stability of the system is enhanced. A similar narrowing down of the structural heterogeneity is also seen during the kinetics of transformation of I_N to N [52]. Thus, these results from our tr-FRET studies provide the first direct experimental evidence in support of a funnel-shaped energy landscape for folding.

10.4 Conclusions and Outlook

Through the examples given in this chapter we have demonstrated the power of time-domain fluorescence techniques in bringing about deeper insights into the complex world of protein dynamics and protein folding dynamics. It was shown that the fluorescence depolarization kinetics of strategically located tryptophan side chains could be interpreted unambiguously, by and large, to visualize motional dynamics. Similarly, the fluorescence decay kinetics of tryptophan could be used as a handle for structure and dynamics when the dominant mechanism of deexcitation could be identified and engineered, such as FRET to appropriate groups at specific locations in the protein. Furthermore, the ability to quantitate the level of structural heterogeneity through the use of MEM-based lifetime distributions was demonstrated. Structural heterogeneity in complex systems is being recognized as a rule rather than exception. Hence we expect that the procedures elaborated here could be applied effectively in many complex systems apart from proteins.

Acknowledgements
We thank Drs. R. Swaminathan, G. S. Lakshmikanth, K. Sridevi, and Bhadresh Rami for their collaboration in the research work described here. We are grateful to Prof. N. Periasamy (TIFR, Mumbai) for providing us with the software used in the analysis of time-domain fluorescence data and advice in using them efficiently.

References

1. Fenimore PW, Frauenfelder H, McMohan BH, Parak FG (2002) Proc Natl Acad Sci USA 99:16047–16051
2. Hammes-Schiffer S (2002) Biochemistry 41:13335–13343
3. Lakowicz JR (1999) Principles of fluorescence spectroscopy. Plenum, New York; Brand L, Johnson ML (eds) (1997) Meth Enzymol 278

4. Lakowicz JR (ed) Topics in fluorescence spectroscopy : vol 1 Techniques (1991); vol 2 Principles (1991); vol 3 Biochemical applications (1992); vol 4 Probe design and chemical sensing (1994); vol 5 Nonlinear and two photon-induced fluorescence (1997); vol 6 Protein fluorescence (2000). Plenum, New York
5. Valeur B (2002) Molecular fluorescence: principles and applications. Wiley-VCH, Weinheim
6. Krishnamoorthy G (2003) Indian J Biophys Biochem 40:147–159
7. Ishima R, Torchia DA (2000) Nat Struct Biol 7:740–743
8. Krishnamoorthy G, Ira (2001) J Fluoresc 11:247–253
9. Lakshmikanth GS, Sridevi K, Krishnamoorthy G, Udgaonkar JB (2001) Nat Struct Biol 8:799–804
10. Klostemeier D, Millar DP (2002) Biopolymers 61:159–179; Eis PS, Kusba J, Johnson ML, Lakowicz JR (1993) J Fluoresc 3:23
11. Beechem JM, Brand L (1985) Annu Rev Biochem 54:43–71
12. Bajzer Z, Prendergast FG (1993) Biophys J 65:2313–2323
13. Ababou A, Bombarda E (2001) Protein Sci 10:2102–2113
14. Ross JBA, Wyssbrod HR, Porter A, Schwartz GP, Michaels CA, Laws WR (1992) Biochemistry 31:1585–1594
15. Swaminathan R, Krishnamoorthy G, Periasamy N (1994) Biophys J 67:2013–2023
16. Brochon JC (1994) Meth Enzymol 240:262–311
17. Swaminathan R, Periasamy N (1996) Proc Ind Acad Sci Chem Sci 108:39–49
18. Tanaka F, Mataga N (1992) In: Mataga N, Okada T, Masuhara H (eds) Dynamics and mechanisms of photoinduced electron transfer and related phenomena. Elsevier, New York, pp 501–512
19. Swaminathan R, Nath U, Udgaonkar JB, Periasamy N, Krishnamoorthy G (1996) Biochemistry 35:9150–9157
20. Rami BR, Krishnamoorthy G, Udgaonkar JB (2003) Biochemistry 42:7986–8000
21. Lakshmikanth GS, Krishnamoorthy G (1999) Biophys J 77:1100–1106
22. Pal S, Peon J, Zewail AH (2002) Proc Natl Acad Sci USA 99:1763–1768
23. Mandal D, Datta A, Pal SK, Bhattacharyya K (1998) J Phys Chem B 102:9070
24. Agashe VR, Shastry MC, Udgaonkar J (1995) Nature 377:754–757
25. Nolting B, Golbik R, Niera JL, Soler-Gonzalez AS, Schriebe G, Fersht AR (1997) Proc Natl Acad Sci USA 94:826–830
26. Rami BR, Udgaonkar JB (2002) Biochemistry 41:1710–1716
27. Sridevi K, Juneja J, Bhuyan A, Krishnamoorthy G, Udgaonkar JB (2000) J Mol Biol 302:479–495
28. Daggett V, Fersht AR (2003) Trends Biochem Sci 28:18–25
29. Baldwin RL (1989) Trends Biochem Sci 14:291–294
30. Khurana R, Udgaonkar JB (1994) Biochemistry 33:106–115
31. Swaminathan R, Periasamy N, Udgaonkar JB, Krishnamoorthy G (1994) J Phys Chem 98:9270–9278
32. Arai K, Kuwajima K (2000) Adv Protein Chem 53:209–282
33. Lakshmikanth GS (2001) PhD Thesis, University of Mumbai
34. Plaxo KW, Guijarro JN, Morton CJ, Pitkeathly M, Campbell ID, Dobson CM (1998) Biochemistry 37:2529–2537
35. Shastry MC, Aghashe VR, Udgaonkar JB (1994) Protein Sci 3:1409–1417
36. Shastry MC, Udgaonkar JB (1995) J Mol Biol 247:1013–1027
37. Beechem JM (1997) Meth Enzymol 278:24–49
38. Weikl TR, Dill KA (2003) J Mol Biol 329:585–598
39. O'Connor DV, Philips D (1984) Time-correlated single photon counting. Academic, London
40. Lyubovitsky JG, Gray HB, Winkler JR (2002) J Am Chem Soc 124:5481–5485
41. Bhuyan AK, Udgaonkar JB (1999) Biochemistry 38:9158–9168
42. Chan HS, Dill KA (1998) Proteins 30:2–33
43. Baldwin RL (1995) J Biomol NMR 5:103–109
44. Bryngelson JD, Onuchic JN, Socci ND, Wolynes PG (1995) Proteins 21:167–195
45. Wolynes PG, Onuchic JN, Thirumalai D (1995) Science 267:1619–1620

46. Pradeep L, Udgaonkar JB (2002) J Mol Biol 324:331–347
47. Sridevi K, Udgaonkar JB (2003) Biochemistry 42:1551–1563
48. Förster T (1967) In: Florkin M, Statz EH (eds) Comprehensive biochemistry, vol 22. Elsevier, New York, pp 61–77
49. Alcala JR, Gratton E, Prendergast FG (1987) Biophys J 51:597–604
50. Navon A, Ittah V, Landsman P, Scheraga HA, Hass E (2001) Biochemistry 40:105–118
51. Sotomayor CP, Aguilar LF, Cuevas FJ, Helms MK, Jameson DM (2000) Biochemistry 39:10928–10935
52. Sridevi K, Lakshmikanth GS, Krishnamoorthy G, Udgaonkar JB (unpublished observations)
53. Deniz AA, Laurence TA, Beligere GS, Dahan M, Martin AB, Chemka DS, Dawson PE, Shultz PG, Weiss S (2000) Proc Natl Acad Sci USA 97:5179–5184
54. Schiler B, Lipman EA, Eaton WA (2002) Nature 419:743–747
55. Wu P, Brand L (1994) Biochemistry 33:10457–10462
56. Ptitsyn OB (1992) In: Crieghton TE (ed) Protein folding. Freeman, New York, pp 243–300
57. Klimov DK, Thirumalai D (2000) Proc Natl Acad Sci USA 97:2544–2549
58. Day R, Bennion BJ, Ham S, Daggett V (2002) J Mol Biol 322:189–203
59. Pande VS (2003) Proc Natl Acad Sci USA 100:3555–3556
60. Karplus M, McCammon JA (2002) Nat Struct Biol 9:646–651

11 Time-Resolved Fluorescence and Two-Photon FCS Investigation of the Interaction of HIV-1 Nucleocapsid Protein with Hairpin Loop Oligonucleotides

J. Azoulay, S. Bernacchi, H. Beltz, J.-P. Clamme, E. Piemont, E. Schaub, D. Ficheux, B. Roques, J.-L. Darlix and Y. Mély

Keywords: Fluorescence correlation spectroscopy; Nucleocapsid protein ncp7; Fluorescence resonance energy transfer; Hairpin loop oligonucleotides

Abbreviations

FCS	Fluorescence correlation spectroscopy
Fl	5(and 6)-Carboxyfluorescein
FRET	Fluorescence resonance energy transfer
NC	Nucleocapsid protein of HIV-1
TMR	5(and 6)-Carboxytetramethylrhodamine
TPE	Two-photon excitation

11.1 Introduction

Fluorescence correlation spectroscopy (FCS) is a versatile technique for in vitro and in vivo investigations of biomolecular interactions that was introduced more than 30 years ago [1]. FCS is based on the analysis of the statistical fluctuations in the fluorescence intensity of a small illumination sample volume. This analysis leads to quantitative information about the processes that provoke these fluctuations. Basically, FCS is designed to measure the diffusion constants and concentrations of fluorescently labeled molecules [2–6]. Depending on the investigated system, additional parameters concerning photophysical reactions [2, 7–14], rotational dynamics [15, 16], protein self-association [17–22], binding equilibria [23–30] and kinetics [23, 31, 32], or conformational dynamics of nucleic acids [33–36] can be extracted. Using one-photon excitation, the sample volume is defined both by a focused laser beam and a confocal pinhole. In this respect, though the laser beam is focused, the excited volume is much larger than the sample volume. Therefore side effects like out-of-focus photobleaching are observed. This drawback is largely minimized by using two-photon excitation (TPE). Indeed, two-photon absorption is proportional to the square of illumination intensity, and thus for a tightly focused beam, the excitation of fluorescent molecules declines rapidly with the axial distance (roughly as z^{-4}) from the focal plane. This inherent excitation confinement provides 3D spatial resolution and a nearly 3D Gaussian observation volume. As a consequence, the background fluorescence level and out-of-focus photodamage are markedly reduced in comparison with one-photon excitation. Moreover, TPE also provides a large spectral separation between excitation and emission wavelengths, allowing background due to Raman and Rayleigh scattering of water to be largely excluded.

Time-resolved fluorescence measurements are also widely used for studying biological molecules and their interactions. This is because time-resolved data provide exquisite sensitive information about the environment of the fluorophores, the conformation of the biomolecules, interchromophore distances, and dynamical and photophysical processes [37–40]. While most investigations on proteins are based on the intrinsic fluorescence of Trp residues, nucleic acids are mainly monitored by using extrinsic fluorophores coupled at the 5′ and 3′ ends.

In this context, the aim of this paper is to review the potency of the combined use of time-resolved fluorescence and two-photon FCS techniques to characterize the interaction of NCp7, the nucleocapsid protein of HIV-1, with DNA hairpin loops [34, 35, 41]. NCp7 is a small basic protein, characterized by two CCHC zinc-finger motifs, that plays critical roles in HIV-1 structure and replication. The activities of NCp7 in the viral life cycle strongly rely on its nucleic acid chaperone properties [42–45], which allow NCp7 to lower the energy barrier for the breaking and the reformation of base pairs. These chaperone properties are notably important during the reverse transcription process, which consists in a complex series of reactions leading to the synthesis of a linear, double-stranded DNA copy of the viral RNA genome. Among these reactions, the first strand transfer is of particular interest. During this step, minus-strand strong-stop DNA [(–)ssDNA] is translocated to the 3′ end of the viral RNA genome, in a reaction mediated by base-pairing of the repeat (R) sequences at the 3′ ends of the RNA and DNA reactants. In the absence of NCp7, this reaction does not proceed efficiently [46–48] since both the TAR RNA sequence in the viral RNA genome and its complementary sequence, cTAR DNA, in (–)ssDNA (Fig. 11.1) are thought to fold into a stable stem loop structure [49–51].

In contrast, addition of NCp7 has been shown to largely stimulate minus-strand transfer by increasing the rate and extent of annealing [48, 52–55] and by blocking nonspecific self-primed reverse transcription [46, 47, 51, 53]. The initial step of the activation of minus-strand transfer is probably the destabilization of nucleic acid secondary and tertiary structures induced by the binding of NCp7 [56–58]. To further understand the molecular mechanism of this destabilization, we combined time-resolved fluorescence intensity and FCS measurements using a cTAR DNA sequence labeled at its 5′and 3′ extremities by a couple of fluorophores frequently used in fluorescence resonance energy transfer (FRET) [59, 60] and an NC(12–55) peptide (Fig. 11.1c) that contains the zinc-finger motifs but lacks the ability to aggregate the oligonucleotides [61].

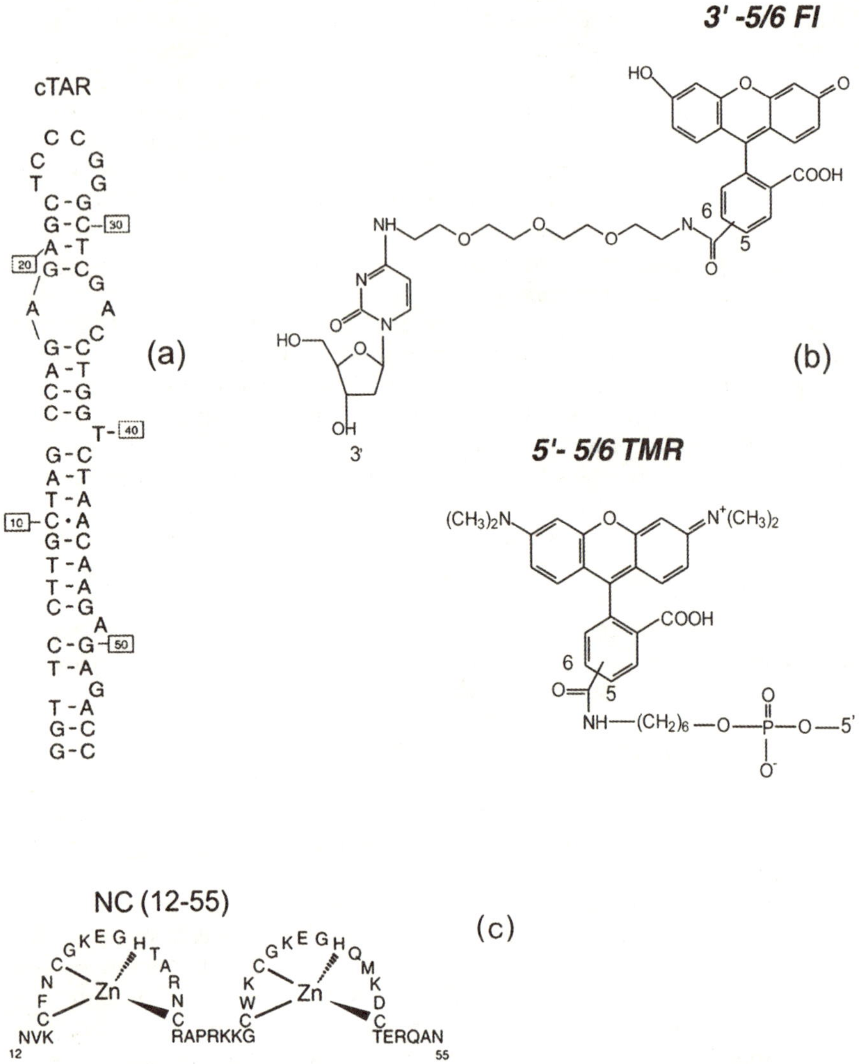

Fig. 11.1. **a** cTAR and **c** NC(12–55) sequences and **b** chemical structures of the dyes used in this study. The selected cTAR DNA sequence is the cDNA copy of the TAR RNA sequence from the MAL strain. The secondary structures of cTAR derivatives were predicted from that of TAR [49] and the mfold program [95] (http://www.bioinfo.rpi.edu/applications/mfold/old/dna/form1.cgi). Figure reproduced from [35] with permission of Elsevier

11.2 Materials and Methods

11.2.1 Materials

NC(12–55) peptide was synthesized as described previously [62]. The purity of the peptide was greater than 98%. The peptide was stored lyophilized in its zinc-bound form and an extinction coefficient of 5,700 M^{-1} cm^{-1} at 280 nm was used to determine its concentration.

Doubly and singly labeled DNA oligonucleotides were synthesized on a 0.2 μmole scale by IBA GmbH Nucleic Acids Product Supply (Göttingen, Germany). The 5′ terminus of the oligonucleotides was labeled by 5(and 6)-carboxytetramethyl-rhodamine (TMR) via an amino-linker with a six-carbon spacer arm. The 3′ terminus of the doubly labeled oligonucleotide was labeled with 5(and 6)-carboxy-fluorescein (Fl) using a special solid support with the dye already attached. Oligonucleotides were purified by the manufacturer by reversed-phase HPLC and polyacrylamide gel electrophoresis. The purity of the labeled oligonucleotides was greater than 93%. Experiments were performed in 25 mM Tris-HCl, pH 7.5, 30 mM NaCl, and 0.2 mM $MgCl_2$. An extinction coefficient of 521,900 M^{-1} cm^{-1} at 260 nm was used to calculate the concentration of cTAR. The absorption spectra were recorded on a Cary 400 spectrophotometer.

11.2.2 Steady-State and Time-Resolved Fluorescence Measurements

Fluorescence emission spectra were recorded on an SLM 48000 spectrofluorometer equipped with a thermostated cell compartment. Time-resolved fluorescence measurements were performed with a time-correlated single-photon counting technique using the stable excitation pulses provided by a pulse-picked frequency-tripled Ti–sapphire laser (Tsunami, Spectra Physics) pumped by a Millenia X laser (Spectra Physics). The temperature was maintained at 20 °C. The excitation pulses were at 470 nm, with a repetition rate of 4 MHz. The emission was collected through a polarizer set at the magic angle and a 4-nm band-pass monochromator (Jobin-Yvon H10) at 520 nm. The single-photon events were detected with a microchannel plate Hamamatsu R3809U photomultiplier coupled to a Phillips 6954 pulse preamplifier and recorded on a multichannel analyzer (Ortec 7100) calibrated at 25.5 ps/channel. The instrumental response function was recorded with a polished aluminum reflector, and its full width at half maximum height was 40 ps. Time-resolved data analysis was performed using the maximum entropy method and the Pulse5 software [63, 64]. For the analysis of the fluorescence decay, a distribution of 200 equally spaced lifetime values on a logarithmic scale between 0.01 and 20 ns was used. In all cases, the χ^2 values were close to 1.0, and the weighted residuals as well as the autocorrelation of the residuals were randomly distributed around zero, indicating an optimal fit.

11.2.3
FCS Setup

FCS measurements were performed on a two-photon platform (Fig. 11.2), as previously described [34, 65]. TPE was provided by a mode-locked Tsunami Ti:sapphire laser pumped by a Millenia V solid-state laser (Spectra Physics). Pulses of about 100 fs were produced at a wavelength of 850 nm. FCS measurements were performed on an Olympus IX70 inverted microscope equipped with a Märzhäuser motorized microscope stage. After a beam expander, parallel infrared laser light was focused into the sample by a water-immersion Olympus 60× objective (NA=1.2). The back aperture of the objective was slightly overfilled, creating a diffraction-limited focal spot of 0.6 μm diameter, which could be approximated by a Gaussian–Gaussian intensity profile. The measurements were carried out in eight-well Lab-Tek II chambers, using a 400 μl volume per well. The focal spot was set about 20 μm above the coverslip. The resulting fluorescence was collected through the same objective and directed by a dichroic mirror (COWL 750 nm; Coherent) to the lateral output of the microscope. After rejection of residual IR light by a BG 39 (Schott) filter and rejection of photons below 550 nm by a Coherent OG550 glass filter, the emitted photons were focused by a 200-mm achromatic lens on the 64-μm core of a multimode optical fiber coupled to an

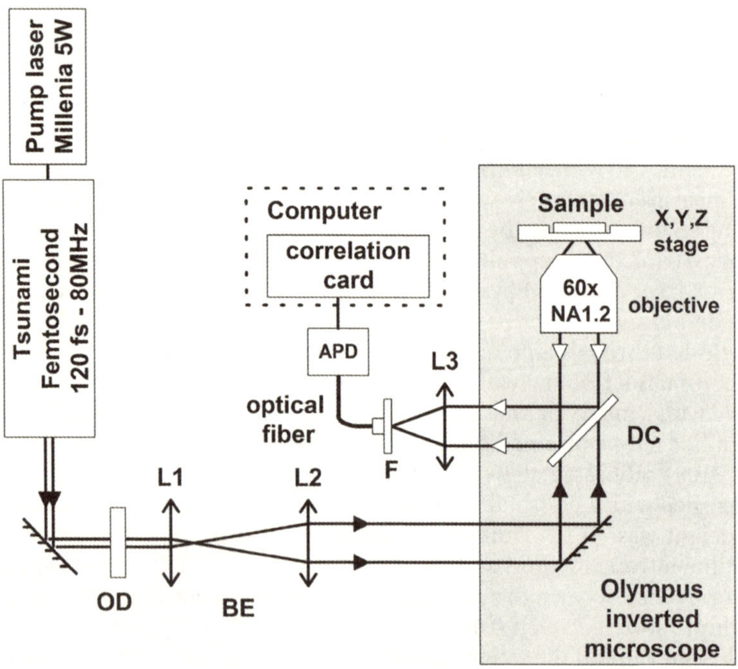

Fig. 11.2. Diagram of the experimental two-photon excitation FCS setup: *OD* optical density filter; *BE* beam expander made of two convergent lenses *L1*/*L2*; *DC* dichroic beamsplitter; *F* BG39 filter; *APD* avalanche photodiode

avalanche photodiode (EG&G SPCM-200 FC). The detector signal was correlated online by an ALV-5000E correlator (ALV, Germany). At this step, the normalized autocorrelation function, $G(\tau)$ was calculated from the fluorescence fluctuations, $\delta F(t)$, by:

$$G(\tau) = \langle \delta F(t)\, \delta F(t+\tau) \rangle / \langle F(t) \rangle^2 \qquad (11.1)$$

where $\delta F(t)$ is the difference between the fluorescence signal, $F(t)$, at a given time and the mean fluorescence signal, $\langle F(t) \rangle$. Moreover, $\delta F(t+\tau)$ designates the fluorescence fluctuations at a later time. Typical data recording times were 10×60 s and the optical collection efficiency was estimated to be about 1%.

An appropriate illumination intensity was selected by investigating the dependence of the apparent diffusion time, τ_{da}, of carboxytetramethylrhodamine (TMR) in buffer on the illumination power at the sample. τ_{da} was nearly constant up to 5 mW and then decreased at higher power. From the τ_{da} plateau value at low power, which represents the diffusion time in the absence of photobleaching, a lateral radius, ω_0, of 0.3 µm and an *s* value of 3 to 4 (using $D_{TMR}=2.8\times10^{-6}$ cm^2 s^{-1}) were determined for the sample volume. As a compromise between photobleaching and a good signal-to-noise ratio, we selected a power of 5–6 mW. At this power, the photon counting rate per molecule is around 4–5 kHz.

11.3 Results and Discussion

11.3.1 Time-Resolved Fluorescence Measurements

To interpret the data with the doubly labeled TMR-5′-cTAR-3′-Fl derivative, it was first necessary to analyze the data with the corresponding singly labeled cTAR-3′-Fl derivative. In the absence of NC(12–55), the fluorescence decay of cTAR-3′-Fl (Table 11.1) is characterized by two lifetimes, τ_2 and τ_3, which may be ascribed to two conformational states differing in the relative location of the dye with respect to the oligonucleotide [66–68]. In these conditions, the relative amplitudes, α_2 and α_3, may correspond to the fractional populations associated with these conformational states. The long-lived lifetime represents 85% and is typical of solvent-exposed Fl. Interestingly, addition of NC(12–55) induces only limited changes in both the lifetimes and associated amplitudes of cTAR-3′-Fl, suggesting that the peptide does not dramatically alter the Fl environment.

The decay of TMR-5′-cTAR-3′-Fl is more complex, being characterized by three lifetimes. Both the appearance of a short-lived component (0.12 ns) and the significant decrease of both intermediate- and long-lived lifetimes as compared to cTAR-3′-Fl are consistent with a FRET mechanism between Fl and TMR in the three conformational states associated with these lifetimes. Interestingly, each of these lifetimes is associated with a rather narrow distribution suggesting that only discrete degrees of stem opening are observed. Since the ratio of 3 between the mean lifetimes, $\langle \tau \rangle$, of cTAR-3′-Fl and TMR-5′-cTAR-3′-Fl is much lower than the ratio, R_m, of 15 between their steady-state fluorescence intensities,

Table 11.1. Time-resolved fluorescence parameters of singly and doubly labeled cTAR derivatives[a]

	t (°C)	r	α_0	τ_1 (ns)	α_1	τ_2 (ns)	α_2	τ_3 (ns)	α_3	$\langle\tau\rangle$ (ns)	R_m	K_d
cTAR-3′-Fl	20	–	–	–	–	2.03±0.10	0.15±0.01	4.28±0.02	0.85±0.01	3.94±0.02	–	–
	20	5	–	–	–	1.76±0.05	0.19±0.01	4.44±0.01	0.81±0.01	3.93±0.01	–	–
	60	–	–	–	–	2.00±0.15	0.23±0.04	4.00±0.05	0.77±0.03	3.54±0.01	–	–
TMR-5′-cTAR-3′-Fl	20	–		0.12±0.01	0.56±0.04	1.24±0.05	0.17±0.03	3.85±0.06	0.27±0.03	1.32±0.09	15	0.25
			0.80±0.03		0.12		0.03		0.05			
	20	5		0.18±0.03	0.30±0.01	1.27±0.02	0.45±0.04	3.69±0.08	0.25±0.01	1.55±0.04	4.71	1.2
			0.46±0.05		0.17		0.24		0.13			
	60	–	–	–	–	1.84±0.13	0.27±0.04	3.47±0.05	0.73±0.04	3.03±0.01		

[a] The fluorescence lifetimes, τ_1 to τ_3, the relative amplitudes, α_1 to α_3, and the mean lifetime, $\langle\tau\rangle$, are expressed as means ± standard error of the mean for three experiments. r designates the ratio of nucleotides to NC. The equilibrium constant, K_d, was calculated using: $K_d = [\text{open}]/[\text{close}] = (1-\alpha_0)/\alpha_0$. The relative amplitude, α_0, of the dark species is calculated by Eq. (11.2). Table reproduced from [41] with permission of Elsevier.

this strongly suggests the existence of dark species with very short (<20 ps) or null lifetime and a relative amplitude of 80%. These species may correspond to oligonucleotides where the terminal segment is closed and thus the Fl and TMR dyes are close together. The relative amplitude, α_0, associated with these dark species can be calculated by:

$$\alpha_0 = 1 - \frac{<\tau>_{\mathrm{Fl}}}{<\tau>_{\mathrm{m}} \times R_{\mathrm{m}}} \tag{11.2}$$

where $<\tau>_{\mathrm{Fl}}$ and $<\tau>_{\mathrm{m}}$ correspond to the measured mean lifetimes of cTAR-3′-Fl and TMR-5′-cTAR-3′-Fl, respectively. As a consequence, the relative amplitudes of the various lifetimes have been recalculated according to: $\alpha_{\mathrm{ic}}=\alpha_{\mathrm{i}}\,(1-\alpha_0)$, where α_i corresponds to the measured amplitude of the i-th lifetime (Table 11.1). Melting of TMR-5′-cTAR-3′-Fl (T_{m}=47 °C) by heating induces a disappearance of the dark species and the species associated with the short-lived component, strongly suggesting that these latter may correspond to non- or partly melted species. In addition, since the two remaining lifetimes are shorter than the corresponding ones in cTAR-3′-Fl at 60 °C, FRET is still observed in the fully melted species.

To get further information on the physical origin of these dark species, the absorption spectrum of TMR-5′-cTAR-3′-Fl was compared to that of an equimolar mixture of TMR-5′-cTAR and cTAR-3′-Fl (Fig. 11.3). The spectrum of the doubly labeled derivative differed from that of the mixture by a significant blue shift and absorbance increase of the Fl peak as well as by a red shift and absorbance decrease of the TMR peak [69]. Interestingly, heating TMR-5′-cTAR-3′-Fl above the melting temperature restores an absorption spectrum close to that of the mixture (data not shown), indicating that the integrity of the stem is required for the absorbance changes. In fact, the dramatic absorbance changes of TMR-5′-cTAR-3′-Fl as compared to the mixture of TMR-5′-cTAR and cTAR-3′-Fl are in line with those reported previously using different couples of dyes, namely (TMR, DABCYL) and (Rh6G, DABCYL) [70]. These changes in the absorption spectra are clearly incompatible with a Förster-type FRET mechanism, since the weak dipolar coupling that governs this mechanism does not induce any change in the

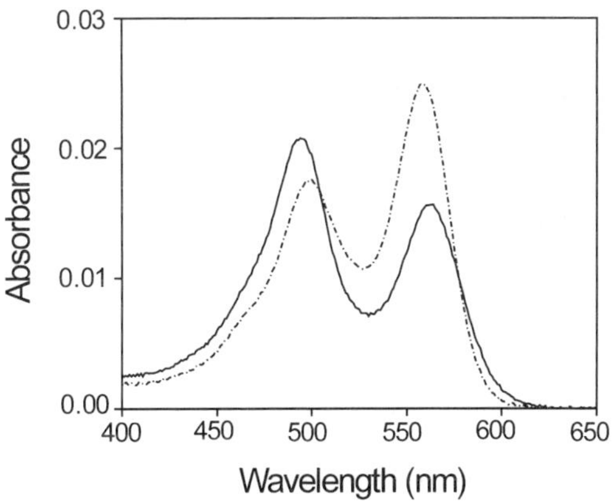

Fig. 11.3. Absorption spectra of the singly and doubly labeled cTAR derivatives. The spectra of TMR-5′-cTAR-3′-Fl (solid line) and an equimolecular mixture of TMR-5′-cTAR and cTAR-3′-Fl (dashed line) were performed at pH 7.5 and 20 °C

absorption spectrum [71]. In contrast, these spectral changes indicate that in the closed form of the stem, the fluorophore and the quencher form a ground-state intramolecular heterodimer. This allows a strong coupling between the transition dipoles of the two dyes, which causes the delocalization of excitation over the two dyes [71–73]. It follows that the spectral properties can no more be ascribed to the individual dyes but correspond to a unique optical signature of the heterodimer. Since this coupling strongly depends on the exact geometry of and distance between the dyes, the associated spectral changes could be used to investigate short-range (<20 Å) modifications of the stem structure. Moreover, when the heterodimer dissociates, FRET becomes applicable and its associated fluorescence changes could be used to characterize the interchromophore distances of partly melted species. Since FRET requires that the interchromophore distance is within a factor of 2 of the Förster critical distance (R_0=60 Å), this implies that interchromophore distances between 30 and 120 Å are measurable.

Two favorable features allow us to infer the interchromophore distances in this system. The first one is the large prevalence of the long-lived lifetime (representing more than 80%) for the fluorescence of Fl in the absence of acceptor (Table 11.1). Therefore the interchromophore distances may be calculated by comparing Fl lifetimes in TMR-5′-cTAR-3′-Fl to this long-lived component. The second favorable feature is the binding of both Fl and TMR dyes to cTAR through flexible linkers (Fig. 11.1) that probably allow both dyes to sample many orientations during the transfer period. Accordingly, as in the doubly labeled $tRNA^{Lys,3}$ derivative [74], a value of 2/3 may be reasonably assumed for the orientation factor κ^2 in the calculation of the Förster distance, R_0. Using the above assumptions, the energy transfer efficiencies, E, were calculated by:

$$E = 1 - \frac{\tau_i^{Fl,TMR}}{\tau_3^{Fl}} \tag{11.3}$$

where τ_3^{Fl} is the long-lived lifetime of cTAR-3′-Fl and $\tau_i^{Fl,TMR}$ is the i-th fluorescence lifetime of TMR-5′-cTAR-3′-Fl. The calculated E values are 0.97, 0.71, and 0.10, respectively, for the three fluorescent species of TMR-5′-cTAR-3′-Fl. In a next step, these values were used together with the R_0 value of the (Fl, TMR) couple to calculate the interchromophore distances, R, by using:

$$R = R_0 \left(\frac{1}{E} - 1 \right)^{\frac{1}{6}} \tag{11.4}$$

Accordingly, interchromophore distances of 34, 52, and 86 Å, respectively, were calculated for the three conformational states associated with the τ_1 to τ_3 lifetimes of TMR-5′-cTAR-3′-Fl. These distances could be compared to the theoretical distances calculated by assuming that an increasing number of segments at the oligonucleotide terminus are melted (Table 11.2). These theoretical distances were calculated using the wormlike chain (WLC) model [75] by considering the opened

Table 11.2. Calculated interchromophore distances for the melting of an increasing number of double-stranded segments in TMR-5′-cTAR-3′-Fl

n_{seg}^{a}	n_{nt}^{a}	$<R^2>^{1/2}$ (Å)[a]	R_{min}–R_{max} (Å)[a]
1	9	26	6–46
2	14	34	14–54
3	31	52	32–72
4	43	61	41–81
5	55	70	50–90

[a] n_{seg} designates the number of melted double-stranded segments starting from the terminal segment. This melting leads to the formation of two single-stranded segments that may be approximated as a continuous single strand (if we include the terminal base pair of the non-melted domain) with n_{nt} nucleotides. The average interchromophore distance, $<R^2>^{1/2}$, was calculated from Eq. (11.5). The minimum and maximum interchromophore distances, R_{min} and R_{max}, were calculated by assuming a linker length of 10 Å. The relevance of these distances is assessed by the good agreement between the 70 Å distance calculated for a fully melted oligonucleotide with the 80 Å distance calculated from the mean lifetimes of Fl-5′-cTAR-3′-TMR and Fl-5′-cTAR at 60 °C (Table 11.1) using. $E = 1 - \frac{<\tau>^{Fl,TMR}}{<\tau>^{Fl}}$ and Eq. (11.4). Table reproduced from [41] with permission of Elsevier.

fraction of the stem as a continuous single strand. By using this model, the mean square end-to-end distance, $<R^2>$, is given by:

$$<R^2> = 2PL\left[1 - \frac{P}{L}\left(1 - e^{-L/P}\right)\right] \qquad (11.5)$$

The contour length L is calculated assuming an internucleotide distance of 0.6 nm [76]. A persistence length, P, of 0.75 nm is used for the single strand [77]. Moreover, an additional 10 Å distance was considered to take into account the length of each spacer [78] and calculate the minimal distance, R_{min}, and the maximal distance, R_{max}, between the dyes.

If we assume that the flexible linkers adopt a large number of orientations with respect to the oligonucleotide ends, the average interchromophore distances may be close to the $<R^2>^{1/2}$ values and thus, the 0.12 ns lifetime may correspond to species where the terminal or the two terminal segments are melted. This conclusion was further confirmed by comparison with a cTAR mutant where the terminal bulge has been deleted [35]. The cooperative melting of the terminal or the two terminal double-stranded segments is consistent with the cooperative opening of DNA hairpin loops that has been shown to result in a two-state equilibrium between closed and open species [79]. Similarly, a cooperative opening until the fourth base pair from the end of a duplex has been shown to occur as a consequence of fraying [80–82]. Moreover, from comparison with several cTAR mutants, the 1.24 ns lifetime was inferred to correspond to species where the stem has been melted up to the ^{10}C–A^{44} mismatch in the central double-stranded segment or the central T^{40} bulge. Finally, the 3.85 ns component may correspond to fully melted species.

From our data, it thus appears that thermal fluctuations generate an equilibrium between closed and melted species at the stem terminus, consistent with a mechanism of fraying. This equilibrium is governed by an equilibrium constant, K_d=0.25 (Table 11.1), in excellent agreement with the K_d value for the fraying of the terminal G–C pair of a duplex [80] or the melting of the short stem of various DNA hairpin loops [79]. Moreover, since the fluorescent species with the shortest lifetime are the most populated (Table 11.1), it results that the melting of the terminal segment(s) is favored.

The major feature observed with the addition of NC(12–55) at a ratio, r, of nucleotide to peptide of 5 (which is expected to lead to a large coating of the oligonucleotide by the peptide) is the dramatic decrease of the relative amplitude associated with the dark species that drops to 46%. The dark species are essentially converted into species with intermediate- and long-lived lifetimes identical to those observed in the absence of peptide. This suggests that NC(12–55) does not generate new species but only shifts the equilibrium toward the open species already observed in the absence of peptide by increasing the level of fraying (K_d increases up to 1.2) of the stem terminus (Table 11.1). The closed species are mainly converted into the species with a 52 Å interchromophore distance, suggesting that the peptide favors the cooperative melting of the lower half of the cTAR stem.

The NC-induced destabilization was found to depend inversely on oligonucleotide stability and thus was more efficient for cTAR DNA than for the more stable TAR RNA sequence [41]. Moreover, by using cTAR mutants where either or both bulges at positions 49 and 52 were replaced by base pairs, it could be inferred that the destabilizing activity of NC(12–55) strongly relies on the two terminal bulges which cooperatively destabilize the cTAR secondary structure [35]. In addition, since cTAR DNA bulges or mismatches appear well conserved among all HIV-1 strains, a strong evolutionary relationship between the stability of the cTAR hairpin and NC destabilizing activity was deduced.

11.3.2 Fluorescence Correlation Spectroscopy

In order to determine the kinetics of fraying of cTAR derivatives and the effect of NC(12–55) on these kinetics, FCS with TPE was performed on labeled cTAR derivatives. Since in TMR-5′-cTAR-3′-Fl, the conformational fluctuations between the open fluorescent and closed nonfluorescent states are thought to contribute to the fluorescence fluctuations in the sample volume, we first performed FCS on the singly labeled TMR-5′-cTAR derivative (Fig. 11.4a) where fluctuations may only result from the translational diffusion of the labeled molecules in and out of the excitation volume.

By analyzing the fluorescence fluctuations through the autocorrelation function, $G(\tau)$, the apparent diffusion time, τ_{da}, as well as the mean number, N, of diffusing molecules in the sample volume could be recovered by:

$$G(\tau) = \frac{1}{N}\left(1+\frac{\tau}{\tau_{da}}\right)^{-1}\left(1+\frac{1}{s^2}\frac{\tau}{\tau_{da}}\right)^{-1/2}\left(1+\left(\frac{f_t}{1-f_t}\right)\exp\left(-\tau/\tau_t\right)\right) \qquad (11.6)$$

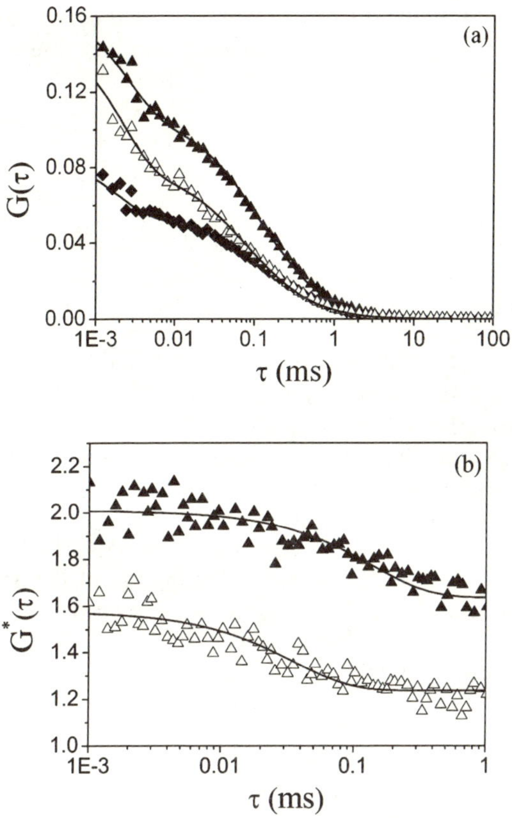

Fig. 11.4a, b. Dynamics of cTAR fraying. **a** Autocorrelation curves of TMR-5′-cTAR (◆) and TMR-5′-cTAR-3′-Fl in the absence (▲) or in the presence (△) of NC(12–55) added at r=5. *Solid lines* correspond to fits of the experimental points with Eq. (11.6) and the parameters of Table 3. **b** Ratio $G^*(\tau)$ between the autocorrelation curves of TMR-5′-cTAR-3′-Fl and TMR-5′-cTAR. The $G^*(\tau)$ ratios were obtained either in the absence (▲) or in the presence (△) of NC(12–55). The *solid line* is a three-parameter exponential fit to the data (see text) with the parameters given in Table 11.3. Figure reproduced from [35] with permission of Elsevier

where s designates the ratio between the axial and lateral radii of the sample volume. In addition to the diffusion process the TMR-5′-cTAR derivative, like most labeled molecules, undergoes photodynamic reactions involving the triplet state. The timescale (about 2 μs) of these reactions is much faster than the diffusion time and these reactions could thus be easily distinguished from the diffusion contribution in $G(\tau)$. The parameters associated with these fast reactions are f_t, the mean fraction of fluorophores in their triplet state and τ_t, the triplet state lifetime.

From the value of the apparent diffusion time, τ_{da}, deduced by fitting the autocorrelation data of TMR-5′-cTAR (Fig. 11.4a) with Eq. (11.6), a diffusion coefficient, D_{exp}=0.60 (±0.04)×10^{-6} cm^2 s^{-1}, was inferred by using free TMR in buffer as a reference. This D_{exp} value could be compared to the theoretical one, D_{th}, calculated by assuming that cTAR may be modeled as a rodlike double-stranded DNA of 27 bp:

$$D_{th} = \frac{k_b T}{3\eta\pi} \frac{1}{L} \left(\ln \frac{L}{l} + \gamma \right) \tag{11.7}$$

where k_b is the Boltzmann constant, T is the absolute temperature, and η is the viscosity of the solution. γ designates the end-correction parameter and is about 0.39 [83]. Assuming a rise per base of 3.4 Å, the length, L, of the rodlike DNA was calculated to be about 92 Å. Finally, the hydrodynamic diameter, l, for a number of DNA fragments with a length similar to cTAR was reported to be from 20.5 to 28 Å [84]. Using these values, we found that $0.74\times10^{-6} < D_{th} < 0.88\times10^{-6}$ cm^2 s^{-1}. The slower motion of cTAR as compared with the theory may be explained by deviations of cTAR from the rodlike model due to the bulges and loops (Fig. 11.1) as well as to the mechanism of fraying.

Figure 11.4a represents the autocorrelation curves of singly and doubly labeled cTAR sequences at the same concentration (100 nM). The average number of fluorescent molecules (which is equal to the inverse of the amplitude of $G(\tau)$ when $\tau \to 0$) for TMR-5′-cTAR-3′-Fl appears strongly decreased as compared to TMR-5′-cTAR, in line with the presence of a large fraction of dark species. As a consequence, the fraying may be treated as a two-state mechanism with a transition between an open fluorescent and a closed dark state. Using Eq. 11.6, we found a τ_{da} value that is about 27% less than that of TMR-5′-cTAR. Since the only fluorescent species are the partially melted forms of TMR-5′-cTAR-3′-Fl, which are expected to diffuse more slowly than the closed species, such a decrease seems to be paradoxical. However, this paradox is easily solved if fraying induces additional fluorescence fluctuations with kinetic rate constants similar to or faster than the diffusion rate constant. It follows that the autocorrelation function of TMR-5′-cTAR-3′-Fl may be written as:

$$G(\tau) = G_D(\tau) \times \left[1 + \left(\frac{f_t}{1 - f_t - f_c} \right) \exp(-\tau/\tau_t) + \left(\frac{f_c}{1 - f_c - f_t} \right) \exp(-\tau/\tau_r) \right] \tag{11.8}$$

where $G_D(\tau)$ is the autocorrelation function resulting from diffusion and f_c is the fraction of molecules in the closed form. The chemical reaction time, τ_r, corresponds to $\tau_r^{-1} = k_{op} + k_{cl}$, where k_{op} and k_{cl} are respectively the opening and closing rate constants of the secondary structure of cTAR involved in the fraying mechanism. The τ_r parameter could be extracted from the ratio, $G^*(\tau)$, of the autocorrelation function of TMR-5′-cTAR-3′-Fl to that of TMR-5′-cTAR [79]. Since τ_r is expected to be much larger than τ_t, this latter may be neglected and thus, $G^*(\tau)$ becomes a simple monoexponential function fitted by $G^*(\tau) = A + B\exp(-t/\tau_r)$. In keeping with our two-state model, an excellent fit of the experimental $G^*(\tau)$ function (Fig. 11.4b) was obtained with the above three-parameter exponential, using $\tau_r = 160(\pm40)$ µs. The effective opening, k_{op}, and closing, k_{cl}, rate constants could then be deduced by:

Table 11.3. Kinetics of fraying of cTAR derivatives

nt/NC	τ_r (µs)	k_{op} (s^{-1})	k_{cl} (s^{-1})
–	160±40	1,250±150	5,000±600
5	<30	>18,000	>15,400

The chemical rate constants, τ_r, were deduced from the fits of the data in Fig. 11.4b to $G^*(\tau)=A+B\exp(-t/\tau_r)$. The opening rate constant, k_{op}, and the closing rate constant, k_{cl}, were deduced from Eq. (11.9). The results are expressed as means ± standard error of the mean for at least two independent experiments. Table reproduced from [35] with permission of Elsevier.

$$k_{op} = \tau_r^{-1} \frac{K_d}{1 + K_d} \qquad k_{cl} = \tau_r^{-1} \frac{1}{1 + K_d} \tag{11.9}$$

Using K_d=0.25 (Table 11.1), values of 1,250 and 5,000 s^{-1} were calculated for k_{op} and k_{cl}, respectively (Table 11.3).

Addition of NC(12–55) (at an NC to nt molar ratio of 1:5) decreased the diffusion coefficient, D_{exp}, to a value of 0.46(±0.03)×10^{-6} cm^2 s^{-1}. This change of D_{exp} was expected, since both the increase of the molecular weight due to the binding of the peptide and the deviation from a rodlike structure associated with the peptide-induced partial melting of cTAR (Table 11.1) are thought to decrease the D_{exp} value. Moreover, in agreement with the time-resolved fluorescence measurements, addition of NC(12–55) to TMR-5′-cTAR-3′-Fl significantly increased the effective number of fluorescent species (Fig. 11.4a). The $G^*(\tau)$ ratio of the autocorrelation function of TMR-5′-cTAR-3′-Fl to that of TMR-5′-cTAR in the presence of NC(12–55) is reported in Fig. 11.4b and appears to decay much faster than in the absence of NC(12–55). According to the limitations of the technique, only an upper limit of 30 µs could be deduced for the reaction time τ_r. Nevertheless, it is likely that τ_r may be close to this upper limit. Calculation of the kinetic rate constants from τ_r indicated that NC(12–55) increases k_{op} by about one order of magnitude and k_{cl} by at least a factor of 3. This major effect of NC on k_{op} and thus on the opening frequency clearly strengthens the hypothesis that NC lowers the energy barrier for breakage of base pairs [44, 85, 86]. Moreover, the increase in k_{cl} may result from the neutralization of DNA phosphate charges by the positively charged NC, which may decrease the repulsion between the single-stranded segments.

11.4 Conclusion

Based on our data, a tentative mechanism may be proposed for the destabilization of the lower part of the cTAR stem by NC (Fig. 11.5). First, due to fraying, the 3- to 5-terminal nucleotides on the 5′ end and the 4 to 7 terminal nucleotides of the 3′ end of cTAR DNA are transiently present as single-stranded segments (step 1). Due to the preferential affinity of NC for single-stranded versus double-

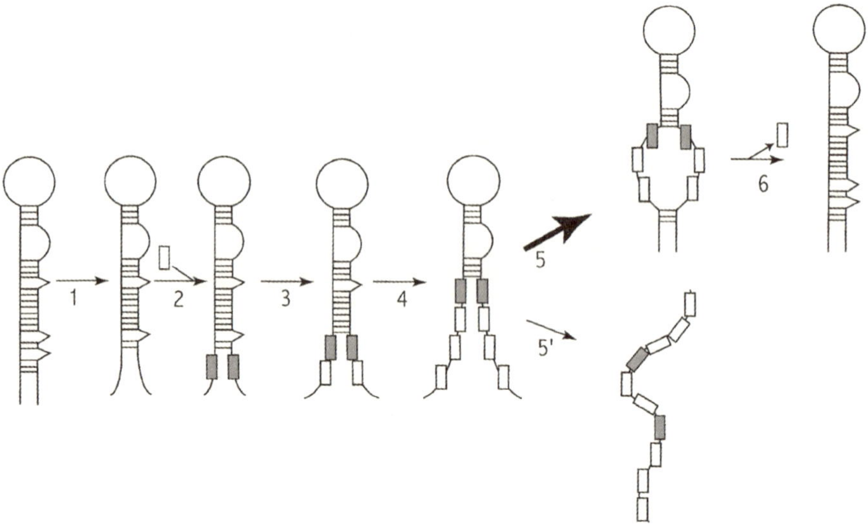

Fig. 11.5. Proposed model for cTAR destabilization by NC. Peptides are represented by *rectangles*. The *shaded rectangles* correspond to the first peptides that bind to cTAR DNA. Steps 1 to 6 are described in the text. Figure reproduced from [35] with permission of Elsevier

stranded sequences [87–89], this may constitute initial binding sites for NC (step 2). From these sites, NC may then destabilize the neighboring base pairs, allowing the sliding of the peptide [90] and cooperative binding of additional peptides [91] (step 3). Due to the limited stability of the cTAR stem, the melting may then easily propagate up to the ^{10}C–A^{44} mismatch or the T^{40} bulge (step 4). At this level, it is likely that the stability of the remaining double-stranded segments may be high enough to prevent the full melting of cTAR. As a consequence, there may be a competition between melting extension (step 5′) and closure of the opened sequences (steps 5 and 6). Since closure probably occurs when the single-stranded segments are still partially covered by NC (step 5), the resulting neutralization of DNA phosphate charges by the positively charged NC may decrease the repulsive interactions between the single-stranded segments [58, 61] and explain the NC-induced increase in the k_{cl} value. By symmetry to the mechanism of stem opening, its closure is expected to be strongly cooperative and thus rapidly induce NC release (step 6). According to the limited population of fully melted species observed by time-resolved fluorescence (Table 1), it is believed that full melting is a minor pathway.

One important remaining question is to determine the degree of cTAR melting required to initiate the hybridization with TAR RNA. Nevertheless, since cTAR is expected to form a loop–loop kissing complex with TAR [92, 93], it may be speculated that the increase of the frequency of the open state may increase the probability of initiating the nucleation of the extended duplex. However, this increase is not the sole favoring event since the NCp7-induced increase of k_{op} is limited to one order of magnitude while the increase of the hybridization kinetics between cTAR and TAR is about three orders of magnitude [94]. It is likely that the neutralization by NC positive charges of the negative phosphate groups of the oligonucleotides [58,61] reduces the electrostatic repulsion between the complementary sequences and thus further favors their interaction.

Acknowledgements
This work was supported by grants from the Agence Nationale de Recherches sur le SIDA (ANRS). HB was a fellow from the Ministère de la Recherche et des Technologies. JA was a fellow from Sidaction. SB was a fellow from ANRS and Sidaction. JPC was a fellow from the Ligue contre le Cancer région Alsace.

References

1. Magde D, Elson EL, Webb WW (1972) Phys Rev Lett 29:705
2. Thompson NL, Lieto AM, Allen NW (2002) Curr Opin Struct Biol 12:634
3. Thompson NL (1991) Fluorescence correlation spectroscopy. In: Lakowicz JR (ed) Topics in fluorescence spectroscopy. Plenum, New York, p 337
4. Elson EL, Magde D (1974) Biopolymers 13:1
5. Magde D, Elson EL, Webb WW (1974) Biopolymers 13:29
6. Hess ST, Huang S, Heikal AA, Webb WW (2002) Biochemistry 41:697
7. Haupts U, Maiti S, Schwille P, Webb WW (1998) Proc Natl Acad Sci USA 95:13573
8. Widengren J, Mets G, Rigler R (1999) Chem Phys 250:171
9. Widengren J, Schwille P (2000) J Phys Chem A 104:6416
10. Schwille P, Kummer S, Heikal AA, Moerner WE, Webb WW (2000) Proc Natl Acad Sci USA 97:151
11. Jung G, Bräuchle C, Zumbusch A (2001) J Chem Phys 114:3149
12. Widengren J, Mets U, Rigler R (1995) J Phys Chem 99:13368
13. Widengren J, Rigler R, Mets U (1994) J Fluoresc 4:255
14. Kask P, Piksarv P, Mets U (1985) Eur Biophys J 12:163
15. Ehrenberg M, Rigler R (1974) Chem Phys 4:390
16. Aragon SR, Pecora R (1976) J Chem Phys 64:1791
17. Bieschke J, Giese A, Schulz-Schaeffer W, Zerr I, Poser S, Eigen M, Kretzschmar H (2000) Proc Natl Acad Sci USA 97:5468
18. Hinterdorfer P, Gruber HJ, Striessnig J, Glossmann H, Schindler H (1997) Biochemistry 36:4497
19. Pitschke M, Prior R, Haupt M, Riesner D (1998) Nat Med 4:832
20. Tjernberg LO, Pramanik A, Bjorling S, Thyberg P, Thyberg J, Nordstedt C, Berndt KD, Terenius L, Rigler R (1999) Chem Biol 6:53
21. Post K, Pitschke M, Schafer O, Wille H, Appel TR, Kirsch D, Mehlhorn I, Serban H, Prusiner SB, Riesner D (1998) Biol Chem 379:1307
22. Jansen K, Schafer O, Birkmann E, Post K, Serban H, Prusiner SB, Riesner D (2001) Biol Chem 382:683
23. Pramanik A, Rigler R (2000) FCS—Analysis of ligand–receptor interactions in living cells. In: Rigler R, Elson ES (eds) Fluorescence correlation spectroscopy: theory and application. Springer, Berlin Heidelberg New York, p 101
24. Matayoshi E, Swift K (2000) Application of FCS to protein–ligand interactions: comparison with fluorescence polarization. In: Rigler R, Elson ES (eds) Fluorescence correlation spectroscopy: theory and applications. Springer, Berlin Heidelberg New York, p 84
25. Van Craenenbroeck E, Vercammen J, Matthys G, Beirlant J, Marot C, Hoebeke J, Strobbe R, Engelborghs Y (2001) Biol Chem 382:355
26. Chen Y, Muller JD, Tetin SY, Tyner JD, Gratton E (2000) Biophys J 79:1074
27. Schuler J, Frank J, Trier U, Schafer-Korting M, Saenger W (1999) Biochemistry 38:8402

28. Lagerkvist AC, Foldes-Papp Z, Persson MA, Rigler R (2001) Protein Sci 10:1522
29. Hasler K, Panke O, Junge W (1999) Biochemistry 38:13759
30. Wohland T, Friedrich K, Hovius R, Vogel H (1999) Biochemistry 38:8671
31. Neumann T, Kirschstein SO, Camacho Gomez JA, Kittler L, Unger E (2001) Biol Chem 382:387
32. Bismuto E, Gratton E, Lamb DC (2001) Biophys J 81:3510
33. Wallace MI, Ying L, Balasubramanian S, Klenerman D (2000) J Phys Chem B 104:11551
34. Azoulay J, Clamme JP, Darlix JL, Roques BP, Mely Y (2003) J Mol Biol 326:691
35. Beltz H, Azoulay J, Bernacchi S, Clamme JP, Ficheux D, Roques B, Darlix JL, Mely Y (2003) J Mol Biol 328:95
36. Bonnet G, Krichevsky O, Libchaber A (1998) Proc Natl Acad Sci USA 95:8602
37. Lackowicz JR (1999) Principles of fluorescence spectroscopy, 2nd edn. Plenum, New York
38. Beechem JM, Brand L (1985) Annu Rev Biochem 54:43
39. Engelborghs Y (2001) Spectrochim Acta A 57:2255
40. Klostermeier D, Millar DP (2001) Biochemistry 40:11211
41. Bernacchi S, Stoylov S, Piemont E, Ficheux D, Roques BP, Darlix JL, Mely Y (2002) J Mol Biol 317:385
42. Herschlag D (1995) J Biol Chem 270:20871
43. Herschlag D, Khosla M, Tsuchihashi Z, Karpel RL (1994) EMBO J 13:2913
44. Rein A, Henderson LE, Levin JG (1998) Trends Biochem Sci 23:297
45. Tsuchihashi Z, Brown PO (1994) J Virol 68:5863
46. Guo J, Henderson LE, Bess J, Kane B, Levin JG (1997) J Virol 71:5178
47. Lapadat-Tapolsky M, Gabus C, Rau M, Darlix JL (1997) J Mol Biol 268:250
48. Lapadat-Tapolsky M, Pernelle C, Borie C, Darlix JL (1995) Nucleic Acids Res 23:2434
49. Baudin F, Marquet R, Isel C, Darlix JL, Ehresmann B, Ehresmann C (1993) J Mol Biol 229:382
50. Jeeninga RE, Huthoff HT, Gultyaev AP, Berkhout B (1998) Nucleic Acids Res 26:5472
51. Driscoll MD, Hughes SH (2000) J Virol 74:8785
52. Guo J, Wu T, Bess J, Henderson LE, Levin JG (1998) J Virol 72:6716
53. Guo J, Wu T, Anderson J, Kane BF, Johnson DG, Gorelick RJ, Henderson LE, Levin JG (2000) J Virol 74:8980
54. Davis WR, Gabbara S, Hupe D, Peliska JA (1998) Biochemistry 37:14213
55. You JC, McHenry CS (1994) J Biol Chem 269:31491
56. Johnson PE, Turner RB, Wu ZR, Hairston L, Guo J, Levin JG, Summers MF (2000) Biochemistry 39:9084
57. Hargittai MR, Mangla AT, Gorelick RJ, Musier-Forsyth K (2001) J Mol Biol 312:985
58. Williams MC, Rouzina I, Wenner JR, Gorelick RJ, Musier-Forsyth K, Bloomfield VA (2001) Proc Natl Acad Sci USA 98:6121
59. Tyagi S, Kramer FR (1996) Nat Biotechnol 14:303
60. Bonnet G, Tyagi S, Libchaber A, Kramer FR (1999) Proc Natl Acad Sci USA 96:6171
61. Stoylov SP, Vuilleumier C, Stoylova E, De Rocquigny H, Roques BP, Gerard D, Mely Y (1997) Biopolymers 41:301
62. De Rocquigny H, Ficheux D, Gabus C, Fournie-Zaluski MC, Darlix JL, Roques BP (1991) Biochem Biophys Res Commun 180:1010
63. Livesey AK, Brochon JC (1987) Biophys J 52:693
64. Brochon JC (1994) Meth Enzymol 240:262
65. Clamme JP, Azoulay J, Mély Y (2003) Biophys J 84:1960
66. Edman L, Mets U, Rigler R (1996) Proc Natl Acad Sci USA 93:6710
67. Widengren J, Dapprich J, Rigler R (1997) Chem Phys 216:417
68. Vamosi G, Gohlke C, Clegg RM (1996) Biophys J 71:972
69. Bernacchi S, Piemont E, Potier N, Dorsselaer A, Mely Y (2003) Biophys J 84:643
70. Bernacchi S, Mely Y (2001) Nucleic Acids Res 29:E62
71. Kasha M (1963) Radiat Res 20:55
72. Kasha M (1991) Basic Life Sci 58:231
73. Scholes DS, Ghiggino KP (1994) J Phys Chem 98:4580

74. Chan B, Weidemaier K, Yip WT, Barbara PF, Musier-Forsyth K (1999) Proc Natl Acad Sci USA 96:459
75. Rivetti C, Walker C, Bustamante C (1998) J Mol Biol 280:41
76. Ying L, Wallace MI, Balasubramanian S (2000) J Phys Chem B 104:5171
77. Smith C, Lee SW, Wong E, Gallardo H, Page K, Gaspar O, Lebkowski J, Gilboa E (1996) Antiviral Res 32:99
78. Wennmalm S, Edman L, Rigler R (1997) Proc Natl Acad Sci USA 94:10641
79. Bonnet G, Krichevsky O, Libchaber A (1998) Proc Natl Acad Sci USA 95:8602
80. Nonin S, Leroy JL, Gueron M (1995) Biochemistry 34:10652
81. Leroy JL, Charretier E, Kochoyan M, Gueron M (1988) Biochemistry 27:8894
82. Cain RJ, Glick GD (1997) Nucleic Acids Res 25:836
83. Kovacic RT, van Holde KE (1977) Biochemistry 16:1490
84. Tirado MM, Garcia de la Torre J (1979) J Chem Phys 71:2581
85. Herschlag D, Khosla M, Tsuchihashi Z, Karpel RL (1994) EMBO J 13:2913
86. Urbaneja MA, Wu M, Casas-Finet JR, Karpel RL (2002) J Mol Biol 318:749
87. Sykora KW, Moelling K (1981) J Gen Virol 55:379
88. Vuilleumier C, Bombarda E, Morellet N, Gerard D, Roques BP, Mely Y (1999) Biochemistry 38:16816
89. Urbaneja MA, Wu M, Casas-Finet JR, Karpel RL (2002) J. Mol Biol 318:749
90. Tisne C, Roques BP, Dardel F (2001) J Mol Biol 306:443
91. Mely Y, de Rocquigny H, Sorinas-Jimeno M, Keith G, Roques BP, Marquet R, Gerard D (1995) J Biol Chem 270:1650
92. Berkhout B, Vastenhouw NL, Klasens BI, Huthoff H (2001) RNA 7:1097
93. Chang KY, Tinoco I Jr (1997) J Mol Biol 269:52
94. You JC, McHenry CS (1994) J Biol Chem 269:31491
95. SantaLucia J Jr (1998) Proc Natl Acad Sci USA 95:1460

Part 4

Application of Fluorescence Spectroscopy to DNA and Drug Delivery

12 Fluorescence Techniques in Non-Viral Gene Therapy

N. Adjimatera, A. P. Neal and I. S. Blagbrough

Keywords: Non-viral gene therapy; Fluorescent lipopolyamines; (in vivo) FCS, DNA condensation, Fluorescence microscopy, FRET, Polyamines, Spermine

Abbreviations

1,4-DHP	1,4-Dihydropyridine
DOPE	Dioleoylphosphatidylethanolamine
EGFP	(Enhanced) green fluorescent protein
EthBr	Ethidium bromide
FACS	Fluorescence-activated cell sorting
FCS	Fluorescence correlation spectroscopy
FFS	Fluorescence fluctuation spectroscopy
FLIM	Fluorescence lifetime microscopy
FRET	Fluorescence resonance energy transfer
GFP	Green fluorescent protein
HTAB	Hexadecyltrimethylammonium bromide
LS	Light scattering
NBD-PE	*N*-4-Nitrobenzo-2-oxa-1,3-diazole phosphatidylethanolamine
N/P	Ammonium/phosphate charge ratio
NPC	Nuclear pore complex
N-Rh-PE	1,2-Dioleoyl-*sn*-glycero-3-phosphatidylethanolamine-*N*-lissamine rhodamine B sulphonyl
NVGT	Non-viral gene therapy
ODN	Oligonucleotide
PEI	Polyethylenimine
PNA	Peptide nucleic acid
Ru-BD	$[Ru(bpy)_2(dppz)]^{2+}$ (bpy=2,2′-bipyridine, dppz=dipyrido[3,2-*a*:2′,3′-*c*]phenazine)

12.1 Introduction to Non-Viral Gene Therapy and its Development

Gene therapy has been defined as “the opportunity for the treatment of genetic disorders in adults and children by genetic modification of human body cells” (UK Health Minister’s Advisory Committee, 1995) [1]. There have been 636 clinical gene therapy studies run worldwide, mostly in the USA (79.4%) and in the UK (6.8%) [2]. Recent experiments show that gene therapy is not exclusive to genetic diseases such as cystic fibrosis, cancer, melanoma, or severe recombined immunodeficiencies, but it is also possible for other types of diseases including viral infections [1, 2]. These new opportunities make gene therapy research challenging in a highly competitive area.

The delivery of a DNA medicine into target cells is currently (mainly) achieved by two means: viral vectors and non-viral vectors. Though viral vectors are found to be highly efficient in transfection, the severe immune response/toxicity from the viral genome and the limited DNA packing size (the "payload") are major therapeutic problems. As a result, non-viral gene therapy (NVGT) vectors have been synthesized as an alternative safe system with possible control over their DNA payload and their manufacture, suitable for clinical use [1, 3–10]. According to Felgner et al.'s NVGT nomenclature guidelines, the two major synthetic gene delivery systems are: lipoplex (cationic lipids-nucleic acid complex), e.g. lipopolyamines, cationic liposomes; and polyplex (cationic polymer-nucleic acid complex), e.g. polyethylenimine (PEI), polylysine [11]. Though initially the efficiency of NVGT vectors was less than that of viral vectors [1, 3–10, 12, 13], the performance gap has been rapidly closed [14].

Gene therapy is a complex pro-drug strategy [15]. It is currently believed that the delivered plasmid DNA must dissociate, at some point, from the lipoplex for gene expression, transcription and translation to the desired therapeutic protein. There are intracellular barriers to these processes [16–18]. Polyamines and lipopolyamines have recently been developed from a natural DNA condensing agent, spermine, for applications in NVGT (Fig. 12.1). The mechanisms involved and the rate-limiting steps are still not well understood [3].

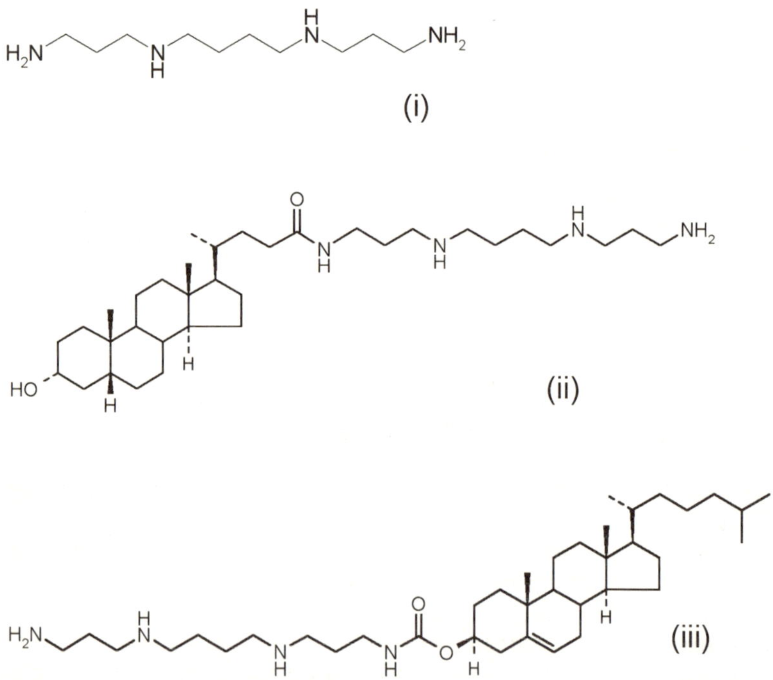

Fig. 12.1. Spermine (i) and two lipopolyamines based on lithocholic acid amide (ii) and cholesteryl carbamate (iii)

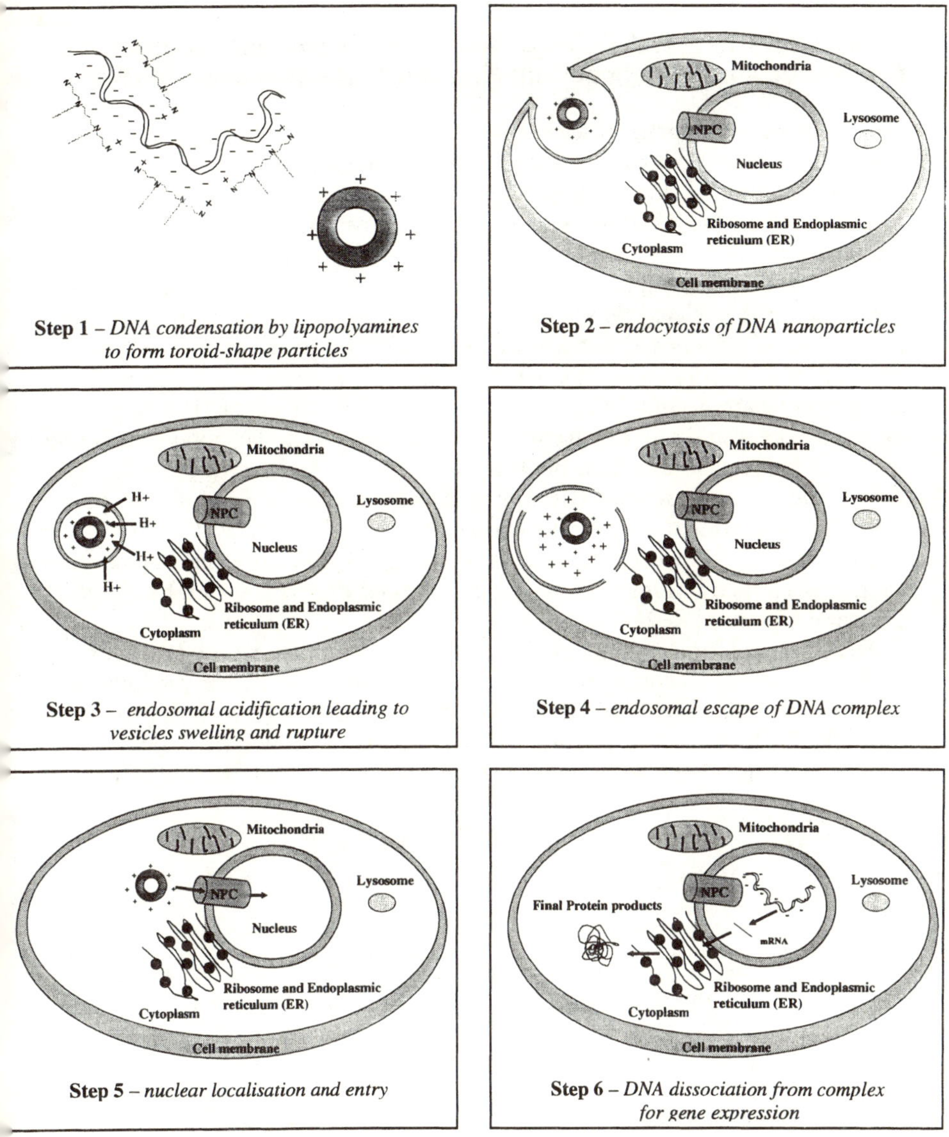

Fig. 12.2. The key steps in the mechanism of non-viral gene therapy

In the first step in the mechanism of NVGT (Fig. 12.2), the phosphate anions of DNA are neutralized by lipopolyamines (ammonium ions) to form condensed nanoparticles [3, 19, 20]. DNA complexes then enter cells by endocytosis. Endosomal escape, involving proton pumping, releases the delivered complexes to the cytoplasm. Nuclear entry, a crucial step for gene expression in NVGT, is achieved by nuclear localization and then transport through the nuclear pore complex (NPC) [3, 17].

12.2 Using Fluorescence Techniques to Determine the Efficiency of DNA Condensing Agents: an Important First Step in the Mechanism of NVGT

The development of gene-based medicines as a new class of pharmaceuticals is clearly important. DNA condensation by polyamine conjugates is a key first step, and fluorimetric, high-throughput assays have been developed to aid in the characterization of NVGT delivery systems. Negatively charged (due to the phosphate groups) DNA was bent by the addition of NVGT delivery agents carrying positive charges. This change in conformation (bending) presumably starts at certain specific nucleic acid sequences, as studied by Hud and co-workers [21], and results in condensed nanometre-sized particles in the range 10–180 nm, typically 50–150 nm in outer diameter. Nanoparticles, toroidal in shape, were observed in polyamine-induced DNA condensation [19, 22]. This DNA compaction facilitates the stability in extracellular compartments, cellular uptake and other intracellular processes such as nuclear entry [17, 19]. Additionally, the aggregation issue of DNA particles is important to the bioavailability of therapeutic DNA. The adjustment of complex charge and ionic strength is critical to their colloidal stability. However, the presence of an excess of positive charges may lead to undesired interactions with negatively charged extracellular molecules such as serum albumin [23, 24].

DNA intercalating agents have been widely used to determine DNA conformational changes, including the characterization of NVGT delivery systems, e.g. ethidium bromide (EthBr) (Fig. 12.3), YOYO-1, acridine orange and PicoGreen [25]. The EthBr assay, first introduced by LePecq and Paoletti in 1967, has been mostly used by NVGT researchers to assess the DNA condensation efficiency

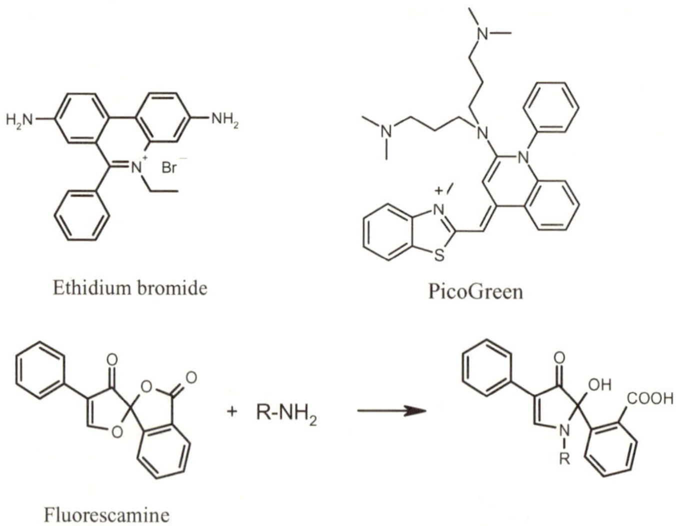

Fig. 12.3. Ethidium bromide, PicoGreen, fluorescamine, and the reaction of fluorescamine with a primary amine

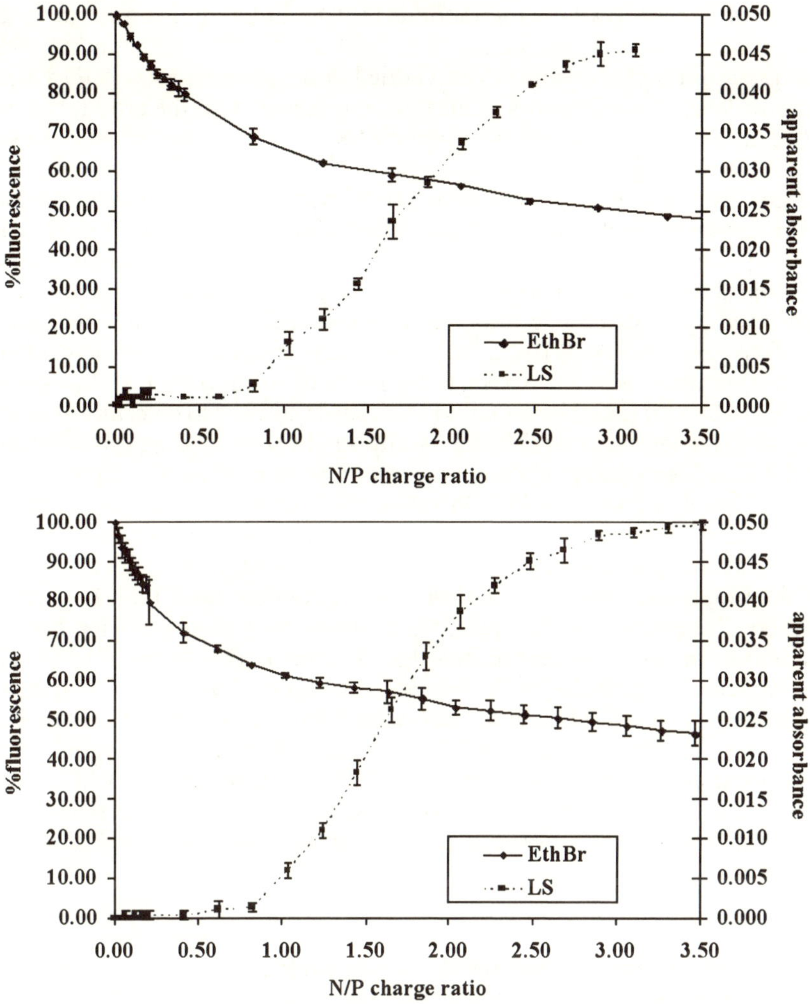

Fig. 12.4. EthBr and LS assay of pEGFP (*above*) and pGL3 (*below*) by spermine

of vector systems. EthBr, intercalated in stacks of DNA base pairs, fluoresces at 600 nm by direct excitation at 546 nm, or more efficiently through energy transfer from DNA base by excitation at 260 nm. On DNA condensation at increasing ammonium/phosphate (N/P) charge ratio [11], a decrease in EthBr fluorescence intensity was measured [26, 27]. This improved methods reported by Geall and Blagbrough [26] offers a rapid and sensitive analysis of lipoplex formation.

Fluorescamine, a non-fluorescent molecule, is also useful in studying NVGT [28, 29]. Fluorescamine easily reacts with primary amine functional groups of polyamines, forming a fluorescent molecule (Fig. 12.3). On salt formation between DNA phosphates and polyammonium ions, the reactivity to fluorescamine of these primary amine groups is eliminated. This observation allowed a study of DNA-polyamine interactions [30], including polyamine-mediated DNA condensation. Thus, the reaction between free (i.e. unbound to DNA) primary amines on polyamines and added fluorescamine was used to determine the level of condensation. There is fluorescence from the product formed between fluorescamine and these amine functional groups, observed using λ_{ex}=392 nm and λ_{em}=480 nm.

PicoGreen, an (expensive commercial) cyanine dye, becomes intensely fluorescent with high quantum yield and molar extinction coefficient when binding to as little as 25 pg/ml dsDNA [31, 32]. The binding between chitosan (i.e. copolymers of glucosamine and *N*-acetyl glucosamine) and DNA was confirmed by PicoGreen, a fluorescent nucleic acid stain which only reacts with free (uncomplexed) DNA. Free DNA was extracted by gradient centrifugation and mixed with PicoGreen to form a fluorescent complex (λ_{ex}=480 nm and λ_{em}=520 nm) [28]. The unbound DNA level was used to determine the condensation performance of chitosan. Additionally, DNA loading can be studied by the PicoGreen assay of DNA after the digestion of nanocomplexes with chitosanase and lysozyme [33]. PicoGreen was compared to a large number of common nucleic acid stains based on EthBr, acridine orange, YOYO-1, etc. Fluorescence spectroscopy was compared to agarose gel electrophoresis for the biophysical characterization of DNA condensation [25]. In addition to these fluorescence techniques, the formation of nanoparticles by DNA condensation can be observed in a light scattering (LS) assay [34, 35]. This is complementary to the DNA condensation assay (Fig. 12.4); UV apparent absorption is measured at λ>300 nm (where there is no absorption by DNA). Precipitation after aggregation follows a plateauing of the LS signal, and it does not increase the absorption above 300 nm. However, the DNA concentration used in this assay was in tenfold excess compared to the EthBr assay, given the low sensitivity of this experiment and the lack of a fluorescence indicator.

12.3 Conjugation of Lipopolyamines to Fluorophores: Probes Derived from DNA Delivery Agents

Fluorescent lipopolyamines were synthesized by Fmoc (fluoren-9-ylmethoxycarbonyl) chemistry in order to label a specific position with a chosen fluorophore [3, 36, 37]. Fluorescence techniques, such as fluorescence correlation spectroscopy (FCS) [38–40], fluorescence resonance energy transfer (FRET) [41, 42] and confocal fluorescence microscopy [43], have recently been applied in NVGT. Thus, by introducing these probes to the DNA to be delivered during the condensation process, NVGT events can be followed by a range of available fluorescence techniques either spectroscopically or microscopically. Our designed fluorescent lipopolyamines (Fig. 12.5) are important tools for studying the intracellular fate of DNA nanoparticles (Fig. 12.6) [3, 36, 37].

Byk, Scherman and co-workers [44–46] have designed and synthesized polyamine–hydrocarbon lipid conjugates. A rhodamine derivative of a lipopolyamine (RPR 121653) was also synthesized and studied as an NVGT probe

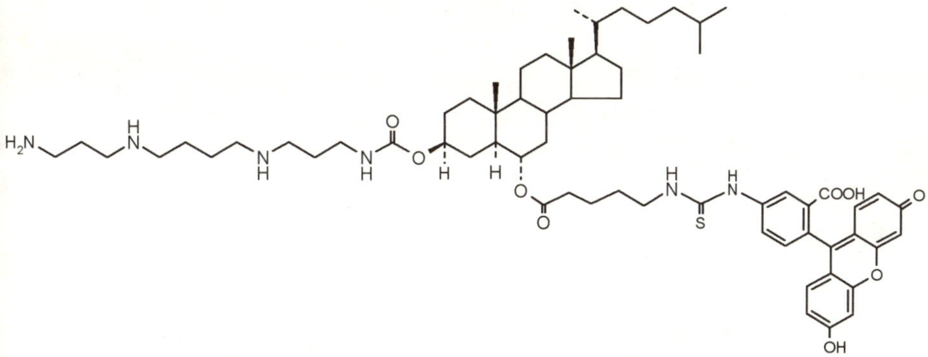

Fig. 12.5. An example of our novel fluorescent lipopolyamine conjugates, carrying a DNA condensing moiety (from spermine), a lipophilic steroid (a *trans*-AB-oxygenated cholestane), and coupled to a fluorescent tag

(Fig. 12.7) [44]. Structure modification was also carried out by introducing a disulphide bridge (a reduction-sensitive functional group) at different positions in the backbone of the lipids. The disulphide is incorporated in order to afford another escape mechanism for the DNA from the lipopolyamine carrier, taking advantage of the cytosolic cellular reducing medium, such as the plasma membrane or cytoplasmic reductases. Early reduction led to undesirable DNA release, i.e. total disruption of the particles in the early phase of delivery yielded total loss of transfection, but some positions for the disulphide bridge afforded increased transfection efficiency [45, 46].

Fluorescent derivatives of a small lipid molecule such as *N*-Rh-PE (1,2-dioleoyl-*sn*-glycero-3-phosphatidylethanolamine-*N*-lissamine rhodamine B sulphonyl) and NBD-PE (*N*-4-nitrobenzo-2-oxa-1,3-diazole phosphatidylethanolamine) are commercially available (Avanti Polar Lipids, AL, USA). These molecules (Fig. 12.7), incorporated into cationic liposomes in NVGT, enable the ability to track fluorescently the progress of transfection from liposomes (*vide infra*) [42, 47]. Fluorescent labelling of polyplex NVGT carriers has also recently been reported, with Oregon Green-PEI [43] and Texas Red-chitosan (*vide infra*) [48].

Ideal fluorophores are primarily chosen for their photostability profile. Naylor et al. [49] showed in interesting experiments that the cellular uptake and metabolism of fluorescent fatty acid analogues were different. The polar fluorophore has poor cellular uptake, and anthracene or pyrene fluorophores were regarded as the most extensively incorporated into cellular lipids. From these findings, it is expected that the molecular hydrophobicity of fluorophores may play an important role in fluorescent lipid (fatty acid)-based NVGT vector design.

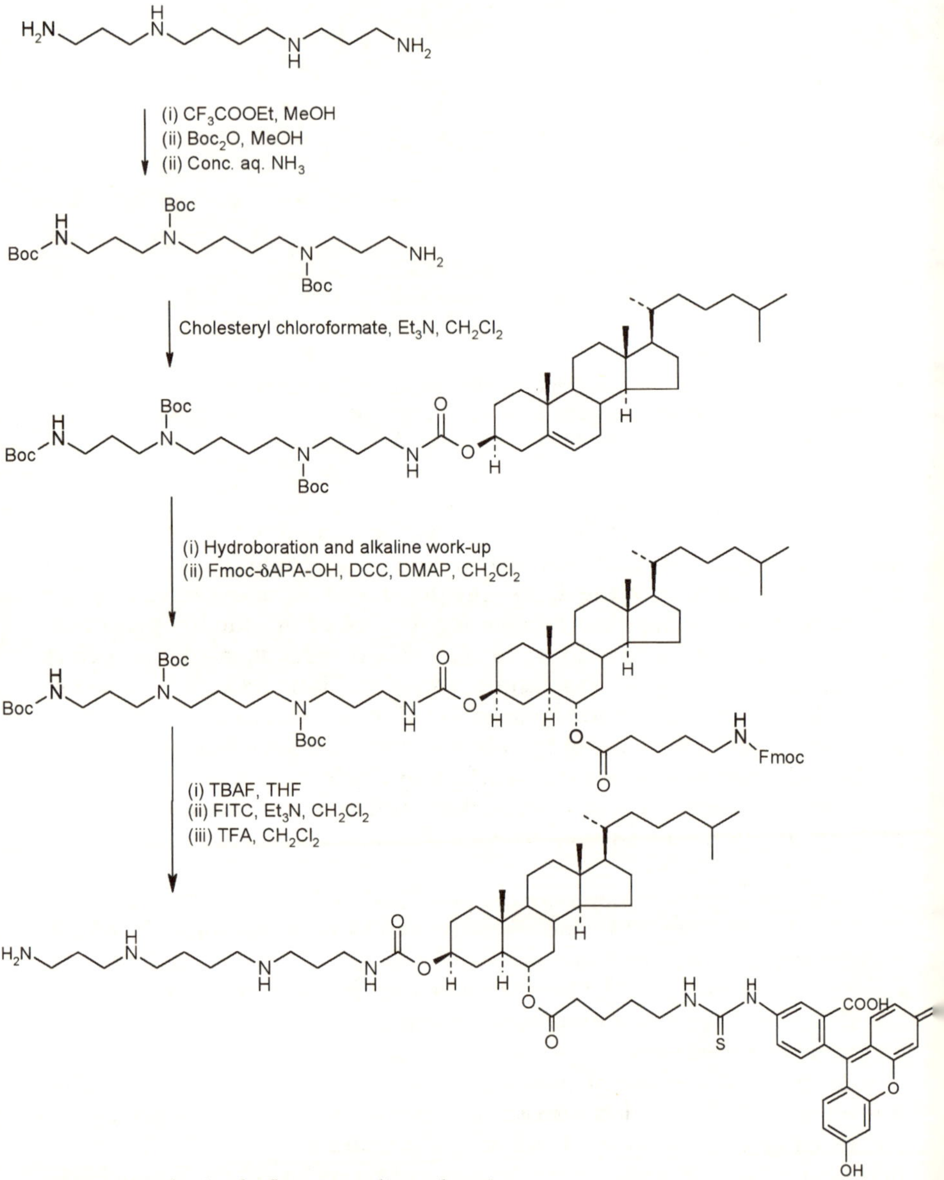

Fig. 12.6. Synthesis of a fluorescent lipopolyamine

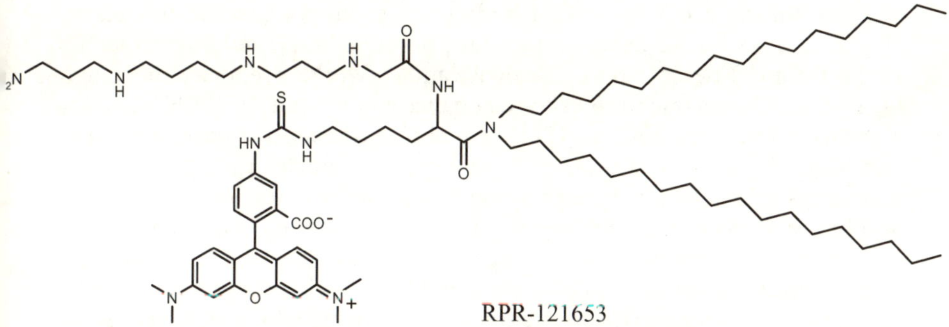

RPR-121653
(Byk and co-workers[44])

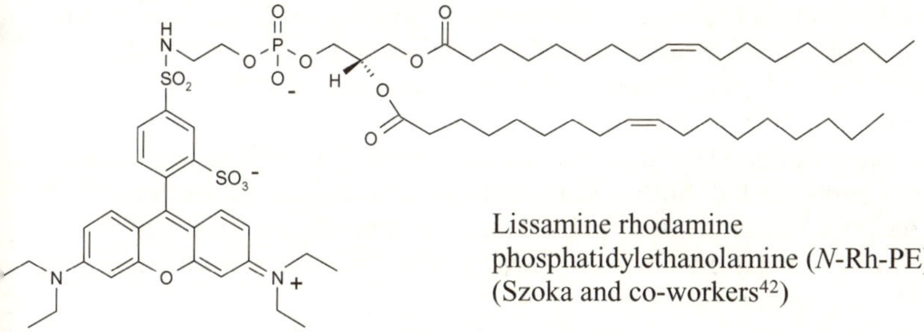

Lissamine rhodamine phosphatidylethanolamine (*N*-Rh-PE)
(Szoka and co-workers[42])

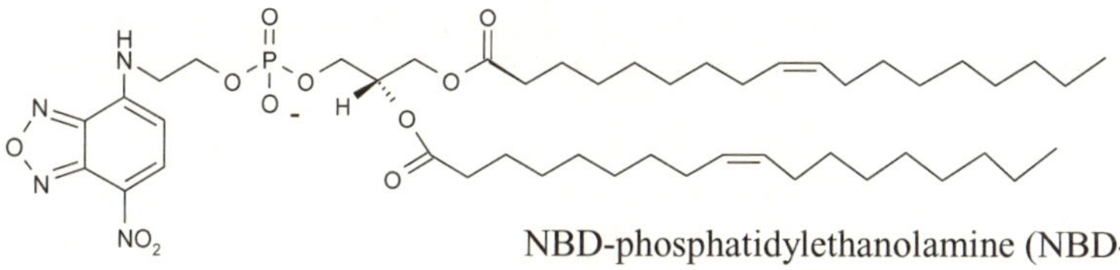

NBD-phosphatidylethanolamine (NBD-PE)
(Szoka and co-workers[42])

Fig. 12.7. Recently reported fluorescent probes

12.4 Preparation of Fluorescent Macromolecules

DNA can be labelled with fluorescent molecules by chemical techniques such as photocrosslinking (e.g. ethidium monoazide [50–52], *p*-azido-tetrafluoro-benzyl-lissamine [53], or dinitrophenyl [54]). Aryl azides can be photoactivated with UV light to generate highly reactive aryl nitrenes which bind to the aromatic bases of DNA. Covalent labelling without photoactivation is also possible, such as the Label IT kit (Mirus Ltd.) [41, 55]. Fluorescent DNA has also been used to study NVGT barriers, such as cell entry [52, 56] and nuclear entry [52, 57].

Some fluorophores for DNA labelling are sequence specific, for example Hoechst 33258 for AT-bases and minor groove [58–61], and 7-aminoactinomycin D for GC-bases [58–61]. Lakowicz and co-workers reported the imaging of cell nuclei based on fluorescence resonance energy transfer (FRET) and fluorescence lifetime microscopy (FLIM) to obtain unbiased maps of spatial distribution of the AT- and GC-rich DNA regions in nuclei. In FLIM, the image contrast is insensitive to concentration, but sensitive to the local environment and interactions of fluorophores such as FRET, allowing examination of the proximity between donors and acceptors. This selective labelling may lead to a better understanding of the starting point for DNA bending, the first step for condensation to form nanoparticles [58–61]. This labelling strategy can also be used together with DNA intercalating dyes, such as propidium iodide, to form a FRET pair [62]. Better fluorophores are being continually developed to increase sensitivity and photostability. Recent studies have demonstrated that silver metallic particles can increase the quantum yield and decrease the lifetimes of nearby fluorophores, including double-stranded DNA oligomers labelled with Cy3 or Cy5 in close proximity to silver particles [63]. The metal-ligand complex, $[Ru(bpy)_2(dppz)]^{2+}$ (bpy=2,2′-bipyridine, dppz=dipyrido-[3,2-*a*:2′,3′-*c*]phenazine) (Ru-BD), was utilized to study nucleic acid dynamics. This Ru-BD complex has both a long lifetime and a molecular light switch property on DNA binding, due to protection of its dppz ligand from water. The slow rotational correlation times appeared to be consistent with the bending motions of the plasmids [64, 65].

In addition to using fluorescent plasmids as a probing strategy for NVGT, protein markers have been employed to follow the transfection outcome, i.e. gene expression. Such protein markers include: (enhanced) green fluorescent protein (EGFP), which fluoresces naturally; β-galactosidase, which can turnover a pro-fluorescent substrate, e.g. one based on umbelliferone; and luci-

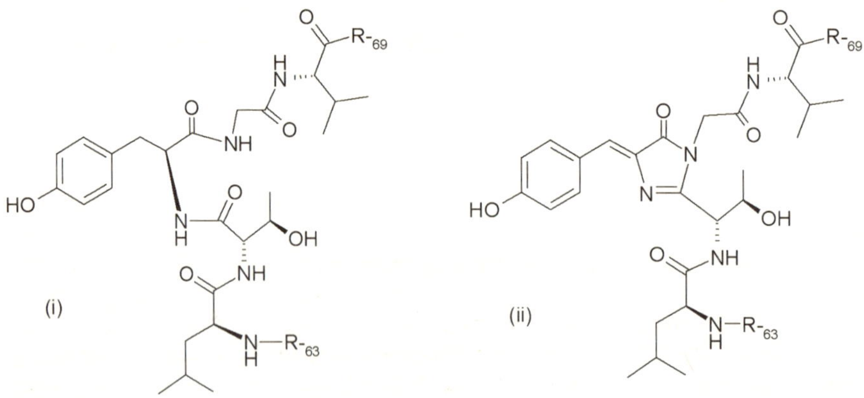

Fig. 12.8. The fluorophore in EGFP. (i) EGFP amino acid sequence L(64)<u>TYG</u>V(68) shown. (ii) Post-translational cyclization of amino acids 65–67 forming hydroxybenzylidene-imidazolidinone, an EGFP fluorophore with λ_{ex}=488 nm (red-shifted from wild GFP protein) and λ_{em}=507 nm

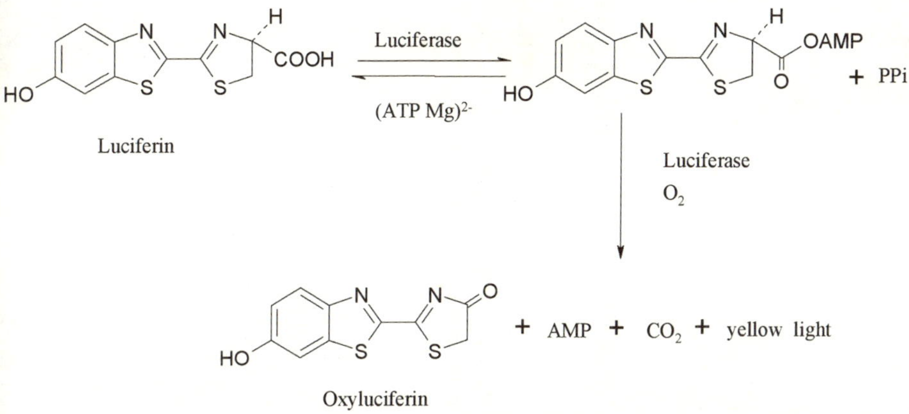

Fig. 12.9. The biochemical reaction of luciferin oxidation by luciferase to oxyluciferin

ferase, which drives luciferin oxidation and generates light. The chemistry that underpins the use of these reporter systems is highlighted below. By using pEGFP [66] as delivered DNA, the EGFP chromophore (a substituted 4-hydroxybenzylidene imidazolidinone Fig. 12.8) was detected in successfully transfected cells by FACS (fluorescence-activated cell sorting). Luminescence involves chemical reactions producing structurally different products which emit light. This bioluminescence method has also been used to quantify NVGT efficiency [67]. Firefly luciferase (*Photinus pyralis*) [67], which is widely used for gene expression, catalyses luciferin oxidation (Fig. 12.9) generating oxyluciferin and detectable yellow light.

12.5 Lipopolyamines and Cationic Lipids Used in Transfection

LipoGen (Invivogen) (Fig. 12.10) is a lipospermine with two oleoyl groups at N^2 and N^3 of spermine and therefore only two positively charged (primary) amines. It was prepared as a non-liposomal formulation. The lipophilic modification aims to facilitate the transfection process (e.g. potentially through enhanced DNA condensation, cell entry, endosomal escape). This lipophilic modification of the spermine structure resulted in a more efficient pEGFP cDNA condensation (Fig. 12.11) (15% residual fluorescence in the EthBr assay, at N/P charge ratio 2.5) compared to tetracationic spermine (50% at N/P charge ratio 3.0). The in vivo transfection of pEGFP (2 μg/well) with LipoGen was carried out using FEK4 [68] (2.5×10^4 cells/well at 50% confluence), incubated for 4 h, then the transfection was stopped by removal of the DNA complex and replacement with foetal calf serum-containing media; FACS analysis was performed 48 h post-transfection. Lipofectin (DOTMA/DOPE=1:1 w/w, Invitrogen) (Fig. 12.10), the commonly used transfection liposomal reagent containing cationic lipid (with one positive charge) and helper lipid (DOPE), was also used in this experiment for a comparison. The fluorescent cell counts observed in all N/P ratios were higher in LipoGen-

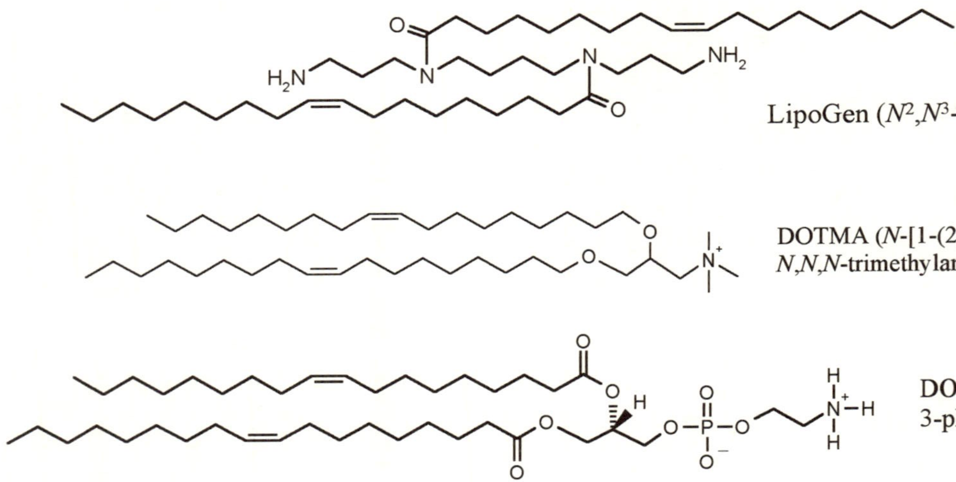

LipoGen (*N*[2],*N*[3]-Dioleoylspermine)

DOTMA (*N*-[1-(2,3-Dioleyloxy)propyl]-*N*,*N*,*N*-trimethylammonium chloride)

DOPE (1,2-Dioleoyl-*sn*-glycero-3-phosphoethanolamine)

Fig. 12.10. LipoGen and Lipofectin

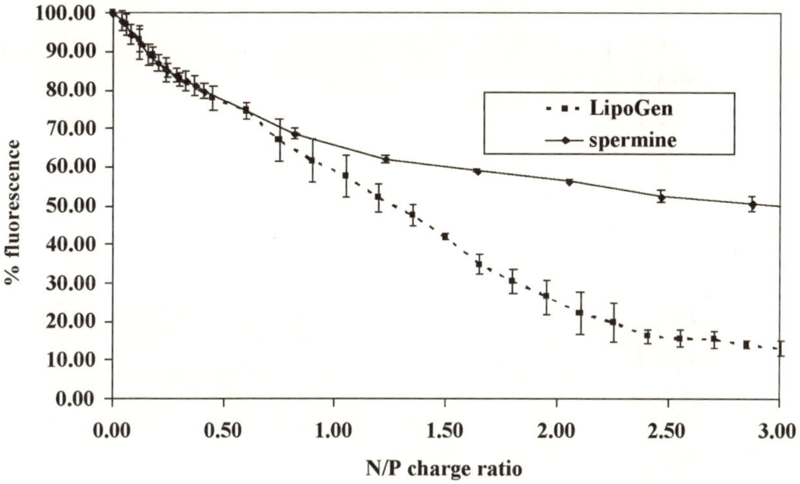

Fig. 12.11. pEGFP-EthBr fluorescence quenching by LipoGen and spermine

mediated transfection. Thus, the lipid moiety in lipospermine is playing an important role in in vitro transfection. The optimal charge ratio for FEK4 transfection with pEGFP-LipoGen complex is around 2.5, which corresponds to the optimal DNA condensation N/P ratio (from the EthBr assay). Though increased condensation leads to higher transfection efficacy (Fig. 12.12), the high N/P ratio (more than 5.0) results in less efficient gene delivery overall. A similar relationship between N/P charge ratio and transfection efficiency was also observed in Lipofectin-mediated transfection.

A highly fluorescent preparation of plasmid DNA was generated by hybridizing a fluorescently labelled peptide nucleic acid (PNA) to the plasmid and using this to study the biodistribution of conformationally and functionally intact plasmid DNA in living cells after cationic lipid-mediated transfection. This method enables the mechanism of plasmid delivery and nuclear import by synthetic gene delivery systems to be elucidated [69]. Using a fluorescent plasmid expressing GFP enabled simultaneous co-localization of both plasmid and expressed protein in living cells and in real time. GFP was shown to be expressed in cells containing detectable nuclear fluorescent plasmid; though conjugated, it still underwent transcription and translation [69].

Double-charged 1,4-dihydropyridine (1,4-DHP) amphiphiles condense ethidium monoazide-crosslinked fluorescent DNA and efficiently transfect cells in vitro. Confocal laser fluorescence microscopy was used to investigate the intracellular distribution of these nanoparticles. The biophysical properties, as a function of structure–activity relationships, determine the intracellular kinetics and transfection efficiency. Not one property among high cellular uptake, membrane destabilizing activity or buffering capacity alone is sufficient to achieve high transfection yields. Overall, there is a complex interplay of various factors that determine intracellular kinetics and, consequently, transfection [70, 71].

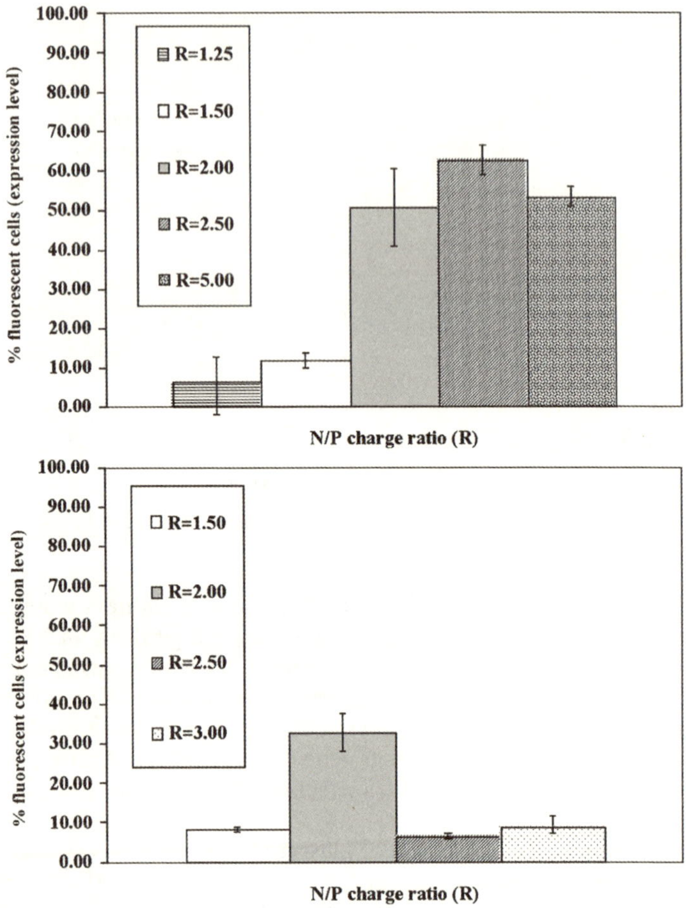

Fig. 12.12. FEK4 transfection of pEGFP complexes by LipoGen (*above*) and Lipofectin (*below*)

Cationic liposomes are useful to transfer genes into eukaryotic cells in vitro and in vivo. However, liposomes with good transfection efficiency are often cytotoxic, and often require serum-free conditions for optimal activity. A new cationic liposome formulation efficiently delivered a plasmid DNA for GFP into more than 80% of the cultured human cell hybrids derived from HeLa cells and normal fibroblasts. FACS analysis revealed that the efficiency of the GFP gene expression was 40–50% in a tumour-suppressed cell hybrid, while it was greatly reduced in the tumorigenic counterpart. The enhanced GFP expression in tumour-suppressed cell hybrids was quantitatively well correlated with a prolonged presence of the plasmid DNA, which had been labelled with the fluorescent probe ethidium monoazide, within the cells. The stability of the plasmid DNA inside the cell is a crucial step in this liposome-mediated gene expression. The mechanisms by which cationic liposomes mediate gene transfer into eukaryotic cells are being studied [72].

Polylysine-molossin is a 31 amino acid synthetic peptide that acts as a DNA vector in vitro for cell lines and for the cornea. It incorporates the 15 amino acid integrin-binding domain of the venom of the American pit viper, *Crotalus molossus* molossus, as the targeting moiety and a chain of 16 lysines as the DNA-binding moiety. Binding and tissue distribution of the vector-DNA complexes were followed using fluorescein-labelled DNA. Long exposure times (2–3 h) to the transfection medium were essential for substantial gene transfer. Although exposure to chloroquine for 8–10 h after uptake of vector/DNA complexes was essential for optimal gene transfer, exposure of complexes to even 1% serum before transfection markedly inhibited gene transfer. Careful attention to several parameters of little importance in vitro needs to be paid for optimal in vivo application of DNA vector systems [73].

12.6 Association and Dissociation Studies of DNA Complexes Through Fluorescence Correlation Spectroscopy (FCS)

FCS is a technique where fluctuations in the detected fluorescence from small molecules (such as DNA intercalating probes) are used to study the dynamic processes on the molecular scale, including DNA conformational changes resulting from polycation-mediated condensation [74]. This technique, first introduced by Magde et al. in 1972, to measure the EthBr diffusion and binding to double-stranded DNA, has been undergoing major technical improvements following the implementation of confocal microscopy in 1993 [74]. The count rate, diffusion time and particle numbers observed by FCS at the single molecule level and their correlations can be used to differentiate the nature of polycation-DNA [38–40, 75, 76] and oligonucleotide condensation [77]. Kral et al. reported the DNA complex association mediated by spermine. The pHbetaAPr-1-neo (10 kbp; contour length 3.4 μm) was labelled with EthBr and propidium iodide, and then titrated with spermine to form condensed DNA particles. The diffusion time, count rate and particle number decreased when the charge ratio (N/P) was increased, suggesting the dissociation of dyes from condensed DNA [38, 40]. The correlation plots of these FCS parameters versus condensing agent concentration can be a source of additional information about the nature of cationic compound–DNA interactions. In similar experiments, hexadecyltrimethylammonium bromide (HTAB)-mediated DNA condensation showed different correlation plots from spermine, and possible differences in DNA conformation were deduced [39].

FCS with two-photon excitation was used to characterize the complexes formed by rhodamine-labelled 25-kDa PEI or DNA plasmid molecules by Clamme et al. [75, 78]. FCS results revealed that fluorescent PEI in the complex solution at the N/P ratio used in transfection was 86% in a free PEI form leading to cell toxicity. As an application, FCS was also used to monitor the purification of PEI/DNA complexes by ultrafiltration as well as the heparin-induced dissociation of the complexes. Purification of the complexes is therefore important in lowering possible toxicity from uncomplexed polycations along NVGT vectors [75].

12.7
DNA Complexes and Their Intracellular Trafficking: Monitoring by Fluorescence (Förster) Resonance Energy Transfer (FRET)

FRET is gaining in importance as a technique for studying the mechanism and barriers in NVGT. A change in conformation on condensing DNA leads to a change in the distance between two fluorophores, which can be demonstrated through the random double labelling of DNA for FRET [41] or double labelling of the DNA and carriers [79, 80]. The stability in physiological media (in the presence of 20% serum) of polyplex- and lipoplex–non-viral gene delivery vectors, e.g. Lipofectamine, poly-L-lysine and poly(ethylene glycol)-poly(L-lysine) block copolymer, has been evaluated by detecting the conformational change in plasmid DNA labelled simultaneously with fluorescein (energy donor) and X-rhodamine (energy acceptor) and examining the condensation by FRET [41, 81]. FRET was also used to estimate the distance of closest approach of supercoiled plasmid DNA to the lipid bilayer of cationic liposomes. The structure of the negatively charged complexes is consistent with DNA extending from the surface of the particles, whereas those possessing an excess of positive charge were multi-lamellar aggregates with the DNA effectively condensed between lipid bilayers. Complexes between these two states consisted of weighted fractions of these two species [79].

FRET has also been used to monitor interactions between Cy3-labelled plasmid DNA and NBD-labelled cationic liposomes. Significantly, the time allowed for complex formation affected the in vitro luciferase transfection efficiencies of DOPE-based lipoplexes. Lipoplexes prepared with a 1 h incubation had much higher transfection efficiencies than samples with 1 min or 5 h incubations. The molar charge ratio of DOTAP to negatively charged phosphate in the DNA also affected the interaction between liposomes and plasmid DNA, and interactions stabilized more rapidly at higher positive charge ratios. Lipoplexes formulated with DOPE were more resistant to high ionic strength than complexes formulated with cholesterol [80].

Prior to cell entry of DNA nanoparticles, binding of DNA complexes with extracellular serum, which results in lower NVGT bioavailability, can also be studied by FRET [23]. Clear implications for clinical intravenous lipofection come from FRET and EthBr intercalation studies of intravenous gene delivery using cationic lipid vectors to achieve systemic gene expression in the (mouse) lung. DOPE vectors stayed poorly in the lung and were barely active in transfecting cells. However, cholesterol-containing vectors had a rapid aggregation, a slow disintegration, and were highly efficient in transfecting cells in vivo [23]. Endosome escape is also a key barrier to study microscopically. Fluorescent lipids in the bilayer membranes, as a FRET probe complementary pair to the fluorescent DNA carriers, were used in membrane fusion/leakage studies [82, 83]. FRET interaction of fluorescent lipopolyamine-DNA complexes with NLS can be studied to gain insights into nuclear localization, nuclear entry processes and the timing of DNA release from the lipoplex [42].

12.8 Fluorescence Microscopy in NVGT

The intracellular processes in NVGT can be imaged by scanning and confocal fluorescence microscopy [42, 47, 48, 84]. Szoka and co-workers were pre-eminent in this research area. They proposed a mechanism for fluorescein-labelled oligonucleotide (ODN) release from cationic rhodamine-labelled liposome complexes, showing that the fluorescent lipid remained in the cytoplasm. ODN displacement from the complex was studied by FRET. They proposed that the complex, after internalization by endocytosis, induces flip-flop of anionic lipids from the cytoplasmic facing monolayer. Anionic lipids laterally diffuse into the complex and form a charged neutralized ion pair with the cationic lipids. This leads to displacement of the ODN from the cationic lipid and its release into the cytoplasm (Fig. 12.13) [42].

A lung inflation-fixation protocol to examine the distribution and gene transfer efficiency of fluorescently tagged lipoplexes using fluorescence confocal microscopy within thick lung tissue sections was used to investigate the observation that intravenous (i.v.) administration of lipoplex was superior to intratracheal (i.t.) administration for gene transfer in the murine lung. A fluorescent ODN was used as a marker for cytoplasmic release of nucleic acids. Not unexpectedly, toxicity was associated with high local concentrations of

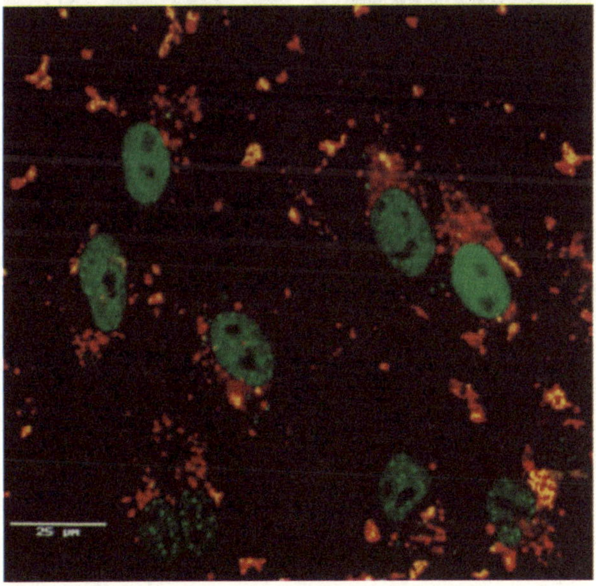

Fig. 12.13. Intracellular distribution of fluorescent lipids and fluorescent oligonucleotide (F-ODN). F-ODN were associated with DOTAP/N-Rh-PE at a 10:1 charge ratio. CV-1 cells were incubated in serum-free medium with complexes for 3 h at 37 °C and imaged using a confocal microscope. (Zelphati O, Szoka FC Jr, Proc Natl Acad Sci USA, 1996, 93:11493–11498; Copyright (1996) National Academy of Sciences, USA)

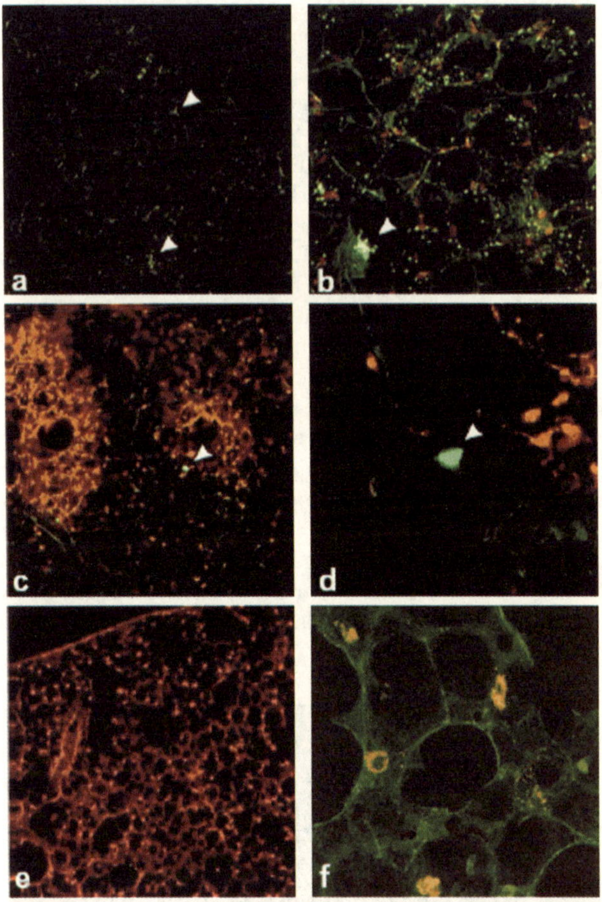

Fig. 12.14a–f. Simultaneous fluorescent oligonucleotide (*red*) localization and GFP (*green*) expression in the mouse lung after i.v. lipoplex administration (**a**, **b**), i.t. lipoplex administration (**a**, **c**, **e** at ×100) and high magnification (**b** at ×400, **d** and **f** at ×600). GFP expression shown by *arrowheads*. In panels **a** and **b**, a fluorescein-PE marker was also used and can be distinguished as green punctuates, distinct from GFP expression (*arrowheads*). The green reticular fluorescence observed in panels **b** and **f** is due to autofluorescence that is visualized at the gain used to image the section. No fluorescent lectin was used in these studies [47]. (Used with permission of Nature)

cationic lipoplexes. The ratio of GFP-expressing cells to fluorescent nuclei indicated that capillary endothelial cells were more efficient in gene expression per delivery event than were pulmonary epithelial cells. Thus, the greater gene expression efficiency of i.v.-administered lipoplexes was due not only to the initial distribution, but also to the greater efficiency of the vascular endothelial cells to appropriately traffic and express the foreign gene (Fig. 12.14) [47].

The seminal work carried out by Godbey, Wu and Mikos, reported in 1999, on PEI stands as a major contribution in this research area of NVGT [43].

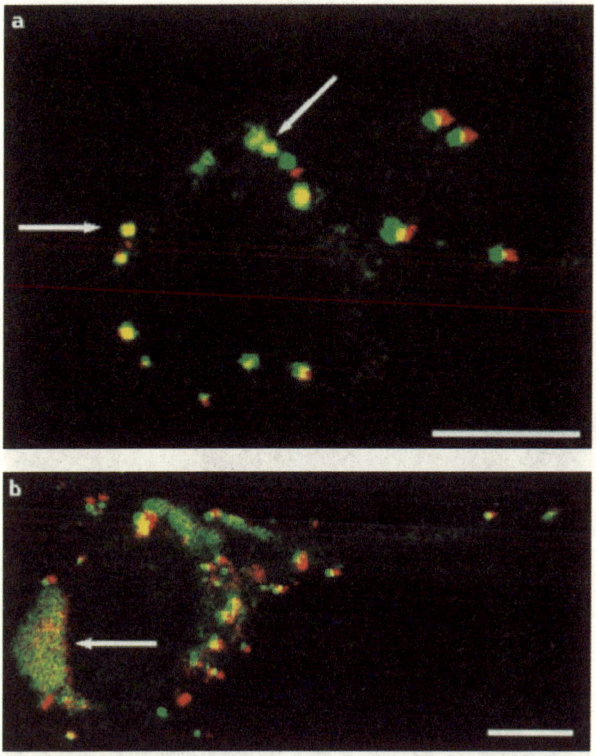

Fig. 12.15 a, b. Tracking of double-labelled PEI/DNA complexes. The fluorescence patterns for single-labelled complexes are also seen for double-labelled complexes. **a** At 2 h post-transfection, visible complexes appear as clumps on the cell's exterior, as indicated by *arrows*. **b** At 3 h post-transfection, both surface aggregation and endosomes are visible. The *arrow* indicates endosomal formation. **c, d** s. p. 220

Their rigorous proof of mechanism, using fluorescent labelling and confocal microscopy, followed PEI/DNA complexes from endocytosis to gene expression. Significantly, the cationic polymer PEI (with or without DNA) underwent nuclear localization (Fig. 12.15) [43].

Chitosan, a polymer of glucosamine and *N*-acetylglucosamine, is useful as a non-viral vector for gene delivery. Although there are several reports of chitosan in gene delivery, studies of the effects on transfection and the chitosan-specific transfection mechanism are still few. Sato and co-workers have recently studied the transfection mechanism of plasmid/chitosan complexes as well as the relationship between transfection activity and cell uptake using fluorescein isothiocyanate-labelled plasmid and Texas Red-labelled chitosan. Factors that increased transfection activity and cell uptake included: molecular mass of chitosan (40 or 84 kDa), stoichiometry of the chitosan nitrogen to DNA phosphate (N/P ratio) in the complex was 5, and the transfection medium contained 10% serum at pH 7.0. For details of the transfec-

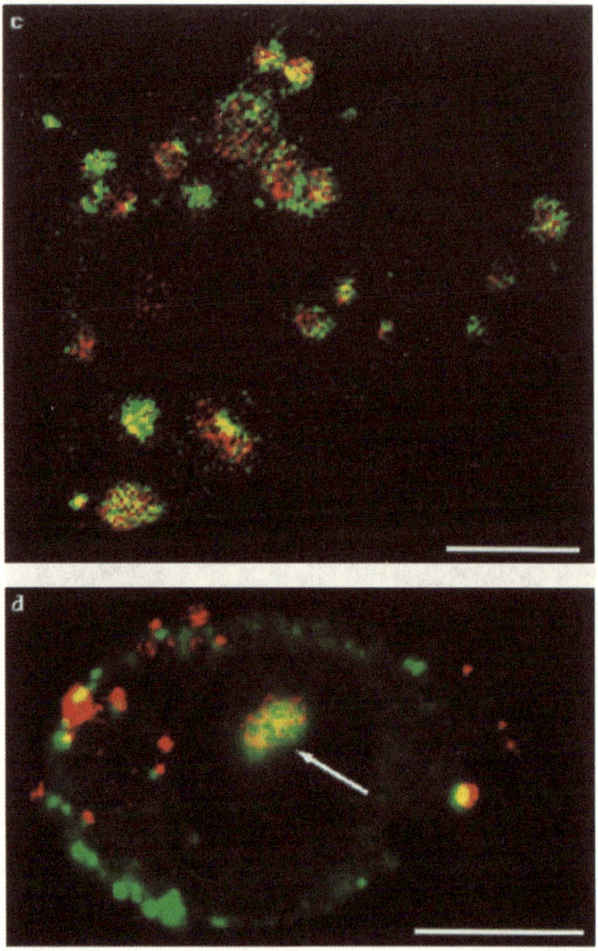

Fig. 12.15 c,d. **c** At 4 h post-transfection, endosomes containing both PEI and DNA are visible throughout the cell cytoplasm. **d** At 4.5 h post-transfection, fluorescent structures containing both PEI and DNA inside the cell nucleus are present, as indicated by the *arrow*. (*Bar*=10 μm). (Godbey WT, Wu KK, Mikos AG, Proc Natl Acad Sci USA, 1999, 96:5177–5181; Copyright (1999) National Academy of Sciences, USA)

tion mechanism, they found that plasmid/chitosan complexes condense to form large aggregates (5–8 μm) which absorb to the cell surface. Plasmid/chitosan complexes are endocytosed and then released from endosomes due to osmotic swelling of endosome, in addition to possible swelling of plasmid/chitosan complex, causing the endosome to rupture. These complexes were observed to accumulate in the nucleus using confocal laser scanning microscopy (Fig. 12.16) [48].

Polyethylenimine (PEI) is one of the most efficient polymeric non-viral vectors for gene therapy, but one suffering from the serious limitation of

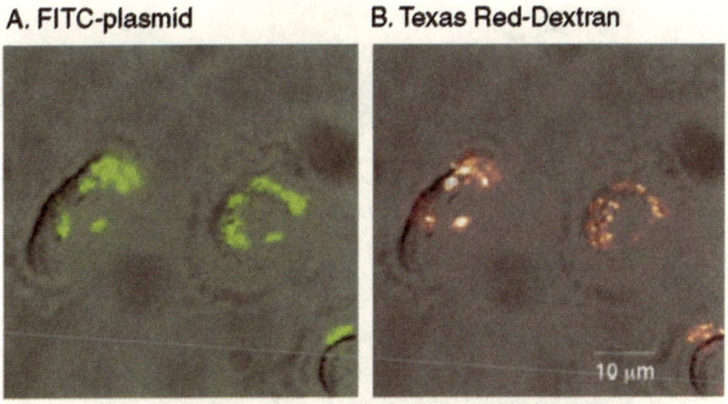

Fig. 12.16 A, B. Intracellular distribution of FITC-plasmid/chitosan complexes (**A**, *green*) and Texas Red-dextran (**B**, *red*) were observed by a confocal fluorescence microscope at 1 h post-incubation in SOJ cells. The confocal images show overlaid images of the fluorescent probe and the phase contrast. Molecular mass of chitosan was 40 kDa. (*Bar*=10 μm) [48]. (Used with permission of Elsevier)

significant cytotoxicity when used in large amounts for transfection. What is the role of endocytosis in the transfection of synchronized L929 fibroblasts by PEI/DNA complexes? Employing a combination of confocal microscopy and FACS, using the endocytosis marker FM4-64 and PEI/DNA complexes labelled either with the DNA intercalator YOYO-1 or with fluorescein covalently linked to PEI, Mely and co-workers showed that nanoparticles were typically taken up within 10 min in endosomes that did not exceed 200 nm in diameter. The location then became perinuclear and fusion between late endosomes was shown to occur. In L929 cells, escape of the complexes from the endosomes is a major barrier in transfection [84]. Comparison with the intracellular trafficking of the same complexes in EA.hy 926 cells (Godbey, Wu and Mikos [43]) revealed that endocytosis of PEI/DNA complexes is strongly cell-dependent (Fig. 12.17) [84].

Mely and co-workers [85] have recently reported their continuing studies on PEI. Fluorescence probes such as Nile red, cSNARF-1 and cyanine dye $DiSC_2(3)$, coupled with the technique of picosecond time-resolved fluorescence microscopy, were used to show that cytoplasmic pH increased by 0.1–0.4 units when cells were treated with PEI. Mely, Behr and co-workers [86] have also developed and studied a new cysteine detergent, ornithinyl-cysteinyl-tetradecylamide (C-14-CO), able to convert itself, via oxidative dimerization, into a cationic cystine lipid. Using fluorescence techniques, they characterized the structures of plasmid DNA lipoplexes [86].

12.8.1 New Emerging Fluorescence Techniques to Explore in NVGT Research

The non-invasive technique of FCS, with its high spatial resolution of less than 0.5 μm, extracts information about molecular dynamics from the tiny fluctuations that can be observed in the emission of small ensembles of fluorescent

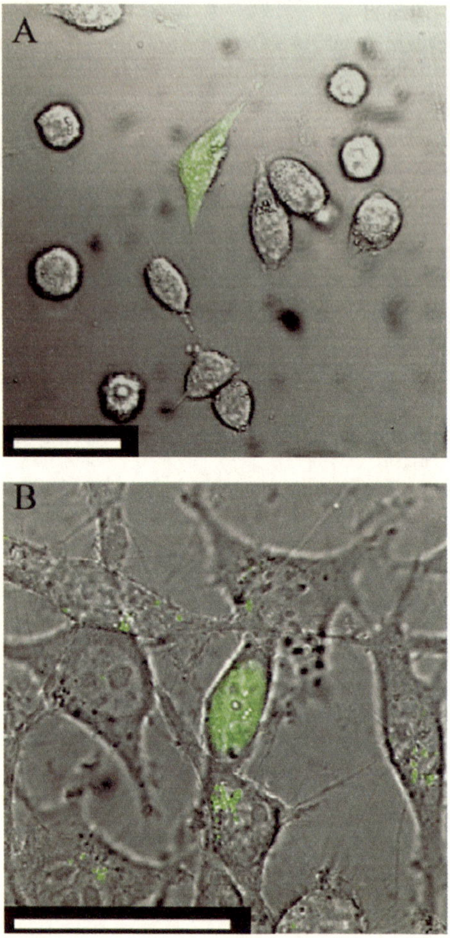

Fig. 12.17 A, B. Transfection of L929 fibroblasts with a GFP-coding plasmid complexed with PEI. The cells were incubated with a GFP-coding plasmid complexed with either unlabelled (**A**) or FITC-labelled PEI (**B**). At 24 h post-transfection, a positive transfection is noted by the expression of the diffuse fluorescence of the green fluorescent protein (**A**) that could be easily differentiated from the fluorescence of the FITC-labelled complexes (**B**). (*Bar*= 50 μm) [84]. (Used with permission of Elsevier)

molecules in thermodynamic equilibrium. More non-invasive fluorescence techniques were developed and this allows NVGT research to be performed with intact cells directly, such as in vivo FCS [87, 88]. FCS is becoming increasingly popular as a technique to add chemical and biophysical information, e.g. particle mobility, local concentrations, rate constants for association and dissociation processes, enzyme kinetics and molecular interactions, to live cell images obtained by microscopy or other techniques. Recent examples of both auto- and cross-correlation applications further demonstrate the potential of FCS for cell biology [87]. Thus, FCS is a versatile technique, particularly attractive for affording quantitative assessment of interactions and dynamics of small molecular quantities in biologically relevant systems [88]. Additionally, the combination of fluorescence techniques is a newly emerging strategy to obtain more information for a better understanding of the complex biological events in NVGT, e.g. the FCS measurement of FRET between the donor (Alexa488) and acceptor (Cy5) fluorophores, and in particular studying fluctuations generated by *trans-cis* isomerization of the acceptor dye, which can be conveniently measured by FCS [89]. The dissociation events of rhodamine green-labelled oligonucleotide (ODN 20-mer) and DNA from rhodamine green- or Cy5 red-labelled cationic polymer carriers (e.g. PLL) can also be studied by dual-colour fluorescence fluctuation spectroscopy (FFS), as recently reported by Lucas et al. [90]. In their studies, both the ODN and the cationic polymers were fluorescently labelled, and the results were compared with data obtained from single-colour FFS in which only the ODN or the cationic polymers were fluorescently marked [90].

Single-molecule fluorescence techniques have been increasingly used in biomolecular studies [91, 92]. The kinetics of spermidine-mediated linear DNA condensation were monitored by optical tweezers and fluorescence imaging at the single molecule level. Two steps, i.e. medium flow speed-dependent lag period and collapse, were observed in the DNA condensation process. The observed lag time suggests that loop formation at the end of the DNA may be a prerequisite for DNA condensation [93].

12.9 Conclusions

Fluorescence techniques can potentially play important roles in all areas of NVGT research. EthBr and other DNA intercalating dyes (as a simple, rapid analytical screen) contribute to the discovery of novel DNA condensing agents. The NVGT efficiency can be evaluated by using plasmid carrying fluorescent protein and luminescence-associated enzymes. Fluorescent lipopolyamines were designed and synthesized, enabling the intracellular tracking of DNA complexes to reveal (both spectroscopically and microscopically) the key steps and barriers in gene delivery. Fluorescence microscopy (e.g. GFP monitoring) has been intensively employed in NVGT imaging. Since the introduction of confocal microscopy, FCS has undergone major developments and has been used to monitor biological events including transfection. Moreover, fluorescent labelling of DNA can be used together with lipopolyamine probes for studies of fluorophore interactions, e.g. by FRET for studies of the (dis-)association of DNA and its cationic carrier, NLS or other biomolecules or intracellular organelles. Newly developed fluorescence techniques, e.g. in vivo FCS and FRET by FCS, are certainly valuable tools for ongoing research towards the goal of efficient NVGT. Designed small molecule probes will play their part in this.

Acknowledgements
We acknowledge the financial support of the EPSRC (a studentship to APN) and an ORS award from Universities U.K. (partial support of studentship fees to NA). We are grateful to Dr C. Pourzand (University of Bath) for cell biology skills and useful discussions and to Prof. R.M. Tyrrell (University of Bath) for the FEK4 cell line. In particular, we are grateful for the generous agreement to reproduce the colour plates (Figs. 12.13–12.17) provided by Profs. Mely, Mikos, Sato and Szoka, and their respective publishers.

References

1. Brooks G (2002) Gene therapy: the use of DNA as a drug. Pharmaceutical Press, London
2. Wiley (2003)http://www.wiley.com/legacy/wileychi/genmed/clinical/
3. Blagbrough IS, Geall AJ, Neal AP (2003) Polyamines and novel polyamine conjugates interact with DNA in ways that can be exploited in non-viral gene therapy. Biochem Soc Trans 31:397–406
4. Kumar VV, Singh RS, Chaudhuri A (2003) Cationic transfection lipids in gene therapy: successes, set-backs, challenges and promises. Curr Med Chem 10:1297–1306
5. Schmidt-Wolf GD, Schmidt-Wolf IGH (2003) Non-viral and hybrid vectors in human gene therapy: an update. Trends Mol Med 9:67–72
6. Liu F, Huang L (2002) Development of non-viral vectors for systemic gene delivery. J Control Release 78:259–266
7. Godbey WT, Mikos AG (2001) Recent progress in gene delivery using non-viral transfer complexes. J Control Release 72:115–125
8. Li S, Huang L (2000) Non-viral gene therapy: promises and challenges. Gene Ther 7:31–34
9. Lollo CP, Banaszczyk MG, Chiou HC (2000) Obstacles and advances in non-viral gene delivery. Curr Opin Mol Ther 2:136–142
10. MacLachlan I, Cullis P, Graham RW (1999) Progress towards a synthetic virus for systemic gene therapy. Curr Opin Mol Ther 1:252–259
11. Felgner PL, Barenholz Y, Behr JP, Cheng SH, Cullis P, Huang L, Jessee JA, Seymour L, Szoka F, Thierry AR, Wagner E, Wu G (1997) Nomenclature for synthetic gene delivery systems. Hum Gene Ther 8:511–512
12. El Aneed A (2004) An overview of current delivery systems in cancer gene therapy. J Control Release 94:1–14
13. Roux D, Cheneviera P, Pott T, Navailles L, Regev O, Monval OM (2004) Conception and realization of a non-cationic non-viral DNA vector. Curr Med Chem 11:169–177
14. Russell SJ (2003) Rise of the nanomachines. Nat Biotechnol 21:872–873
15. Kamiya H, Akita H, Harashima H (2003) Pharmacokinetic and pharmacodynamic considerations in gene therapy. Drug Discov Today 8:990–996
16. Ruponen M, Honkakoski P, Ronkko S, Pelkonen J, Tammi M, Urtti A (2003) Extracellular and intracellular barriers in non-viral gene delivery. J Control Release 93:213–217
17. Wiethoff CM, Middaugh CR (2003) Barriers to nonviral gene delivery. J Pharm Sci 92: 203–217
18. Ferrari S, Geddes DM, Alton EWFW (2002) Barriers to and new approaches for gene therapy and gene delivery in cystic fibrosis. Adv Drug Deliv Rev 54:1373–1393
19. Vijayanathan V, Thomas T, Thomas TJ (2002) DNA nanoparticles and development of DNA delivery vehicles for gene therapy. Biochemistry 41:14085–14094
20. Vijayanathan V, Thomas T, Shirahata A, Thomas TJ (2001) DNA condensation by polyamines: a laser light scattering study of structural effects. Biochemistry 40:13644–13651
21. Hud NV, Plavec J (2003) A unified model for the origin of DNA sequence-directed curvature. Biopolymers 69:144–159
22. Golan R, Pietrasanta LI, Hsieh W, Hansma HG (1999) DNA toroids: stages in condensation. Biochemistry 38:14069–14076
23. Li S, Tseng WC, Stolz DB, Wu SP, Watkins SC, Huang L (1999) Dynamic changes in the characteristics of cationic lipidic vectors after exposure to mouse serum: implications for intravenous lipofection. Gene Ther 6:585–594

24. Oupicky D, Konak C, Dash PR, Seymour LW, Ulbrich K (1999) Effect of albumin and polyanion on the structure of DNA complexes with polycation containing hydrophilic nonionic block. Bioconjug Chem 10:764–772
25. Welz C, Fahr A (2001) Spectroscopic methods for characterization of nonviral gene delivery systems from a pharmaceutical point of view. Appl Spectrosc Rev 36:333–397
26. Geall AJ, Blagbrough IS (2000) Rapid and sensitive ethidium bromide fluorescence quenching assay of polyamine conjugate–DNA interactions for the analysis of lipoplex formation in gene therapy. J Pharm Biomed Anal 22:849–859
27. Gershon H, Ghirlando R, Guttman SB, Minsky A (1993) Mode of formation and structural features of DNA-cationic liposome complexes used for transfection. Biochemistry 32: 7143–7151
28. Corsi K, Chellat F, Yahia L, Fernandes JC (2003) Mesenchymal stem cells, MG63 and HEK293 transfection using chitosan-DNA nanoparticles. Biomaterials 24:1255–1264
29. Read ML, Etrych T, Ulbrich K, Seymour LW (1999) Characterisation of the binding interaction between poly(L-lysine) and DNA using the fluorescamine assay in the preparation of non-viral gene delivery vectors. FEBS Lett 461:96–100
30. Bode J, Willmitzer L (1975) Application of fluorescamine to the study of protein-DNA interactions. Nucleic Acids Res 2:1951–1965
31. Cosa G, Focsaneanu KS, Mclean JRN, McNamee JP, Scaiano JC (2001) Photophysical properties of fluorescent DNA-dyes bound to single- and double-stranded DNA in aqueous buffered solution. Photochem Photobiol 73:585–599
32. Singer VL, Jones LJ, Yue ST, Haugland RP (1997) Characterization of PicoGreen reagent and development of a fluorescence-based solution assay for double-stranded DNA quantitation. Anal Biochem 249:228–238
33. Leong KW, Mao HQ, Truong L, Roy K, Walsh SM, August JT (1998) DNA-polycation nanospheres as non-viral gene delivery vehicles. J Control Release 53:183–193
34. Geall AJ, Taylor RJ, Earll ME, Eaton MAW, Blagbrough IS (2000) Synthesis of cholesteryl polyamine carbamates: pKa studies and condensation of calf thymus DNA. Bioconjug Chem 11:314–326
35. Wilson RW, Bloomfield VA (1979) Counterion-induced condensation of deoxyribonucleic acid. A light-scattering study. Biochemistry 18:2192–2196
36. Neal AP, Blagbrough IS (2001) Design and synthesis of fluorescent cholesterol and lithocholic acid polyamine conjugates. Abstr Pap Am Chem Soc 221:332–MEDI
37. Neal AP, Blagbrough IS (2001) Fluorescent steroidal lipopolyamine conjugates for monitoring gene delivery. Abstr Pap Am Chem Soc 221:352–MEDI
38. Kral T, Hof M, Langner M (2002) The effect of spermine on plasmid condensation and dye release observed by fluorescence correlation spectroscopy. Biol Chem 383:331–335
39. Kral T, Hof M, Jurkiewicz, P, Langner M (2002) Fluorescence correlation spectroscopy (FCS) as a tool to study DNA condensation with hexadecyltrimethylammonium bromide (HTAB). Cell Mol Biol Lett 7:203–211
40. Kral T, Langner M, Benes M, Baczynska D, Ugorski M, Hof M (2002) The application of fluorescence correlation spectroscopy in detecting DNA condensation. Biophys Chem 95:135–144
41. Itaka K, Harada A, Nakamura K, Kawaguchi H, Kataoka K (2002) Evaluation by fluorescence resonance energy transfer of the stability of nonviral gene delivery vectors under physiological conditions. Biomacromolecules 3:841–845
42. Zelphati O, Szoka FC Jr (1996) Mechanism of oligonucleotide release from cationic liposomes. Proc Natl Acad Sci USA 93:11493–11498
43. Godbey WT, Wu KK, Mikos AG (1999) Tracking the intracellular path of poly(ethylenimine)/DNA complexes for gene delivery. Proc Natl Acad Sci USA 96:5177–5181
44. Byk G, Dubertret C, Escriou V, Frederic M, Jaslin G, Rangara R, Pitard B, Crouzet J, Wils P, Schwartz B, Scherman D (1998) Synthesis, activity, and structure–activity relationship studies of novel cationic lipids for DNA transfer. J Med Chem 41:224–235
45. Byk G, Scherman D (2000) Genetic chemistry: tools for gene therapy coming from unexpected directions. Drug Dev Res 50:566–572

46. Byk G, Wetzer B, Frederic M, Dubertret C, Pitard B, Jaslin G, Scherman D (2000) Reduction-sensitive lipopolyamines as a novel nonviral gene delivery system for modulated release of DNA with improved transgene expression. J Med Chem 43:4377–4387
47. Uyechi LS, Gagne L, Thurston G, Szoka FC (2001) Mechanism of lipoplex gene delivery in mouse lung: binding and internalization of fluorescent lipid and DNA components. Gene Ther 8:828–836
48. Ishii T, Okahata Y, Sato T (2001) Mechanism of cell transfection with plasmid/chitosan complexes. Biochim Biophys Acta Biomembr 1514:51–64
49. Naylor BL, Picardo M, Homan R, Pownall HJ (1991) Effects of fluorophore structure and hydrophobicity on the uptake and metabolism of fluorescent lipid analogs. Chem Phys Lipids 58:111–119
50. Bolton PH, Kearns DR (1978) Spectroscopic properties of ethidium monoazide: a fluorescent photoaffinity label for nucleic acids. Nucleic Acids Res 5:4891–4903
51. Cantrell CE, Yielding KL, Pruitt KM (1979) Efficiency of photolytic binding of ethidium monoazide to nucleic acids and synthetic polynucleotides. Mol Pharmacol 15:322–330
52. Zabner J, Fasbender AJ, Moninger T, Poellinger KA, Welsh MJ (1995) Cellular and molecular barriers to gene transfer by a cationic lipid. J Biol Chem 270:18997–19007
53. Neves C, Byk G, Escriou V, Bussone F, Scherman D, Wils P (2000) Novel method for covalent fluorescent labeling of plasmid DNA that maintains structural integrity of the plasmid. Bioconjug Chem 11:51–55
54. Keller GH, Huang DP, Manak MM (1989) Labeling of DNA probes with a photoactivatable hapten. Anal Biochem 177:392–395
55. Slattum PS, Loomis AG, Machnik KJ, Watt MA, Duzeski JL, Budker VG, Wolff JA, Hagstrom JE (2003) Efficient in vitro and in vivo expression of covalently modified plasmid DNA. Mol Ther 8:255–263
56. Yoshinaga T, Yasuda K, Ogawa Y, Takakura Y (2002) Efficient uptake and rapid degradation of plasmid DNA by murine dendritic cells via a specific mechanism. Biochem Biophys Res Commun 299:389–394
57. Byrnes CK, Nass PH, Shim J, Duncan MD, Lacy B, Harmon JW (2002) Novel nuclear shuttle peptide to increase transfection efficiency in esophageal mucosal cells. J Gastrointest Surg 6:37–42
58. Murata S, Herman P, Lin HJ, Lakowicz JR (2000) Fluorescence lifetime imaging of nuclear DNA: effect of fluorescence resonance energy transfer. Cytometry 41:178–185
59. Murata S, Herman P, Mochizuki K, Nakazawa T, Kondo T, Nakamura N, Lakowicz JR, Katoh R (2003) Spatial distribution analysis of AT- and GC-rich regions in nuclei using corrected fluorescence resonance energy transfer. J Histochem Cytochem 51:951–958
60. Murata S, Herman P, Lakowicz JR (2001) Texture analysis of fluorescence lifetime images of AT- and GC-rich regions in nuclei. J Histochem Cytochem 49:1443–1451
61. Murata S, Herman P, Lakowicz JR (2001) Texture analysis of fluorescence lifetime images of nuclear DNA with effect of fluorescence resonance energy transfer. Cytometry 43:94–100
62. Murata S, Kusba J, Piszczek G, Gryczynski I, Lakowicz JR (2000) Donor fluorescence decay analysis for energy transfer in double-helical DNA with various acceptor concentrations. Biopolymers 57:306–315
63. Malicka J, Gryczynski I, Maliwal BP, Fang JY, Lakowicz JR (2003) Fluorescence spectral properties of cyanine dye labeled DNA near metallic silver particles. Biopolymers 72:96–104
64. Kang JS, Abugo OO, Lakowicz JR (2002) Dynamics of supercoiled and relaxed pTZ18U plasmids probed with a long-lifetime metal-ligand complex. J Biochem Mol Biol 35:389–394
65. Kang JS, Abugo OO, Lakowicz JR (2002) Dynamics of supercoiled and linear pTZ18U plasmids observed with a long-lifetime metal-ligand complex. Biopolymers 67:121–128
66. Cormack BP, Valdivia RH, Falkow S (1996) FACS-optimized mutants of the green fluorescent protein (GFP). Gene 173:33–38
67. Deluca M, McElory WD (1978) Purification and properties of firefly luciferase. Meth Enzymol 57:3–15

68. Vile GF, Tyrrell RM (1993) Oxidative stress resulting from ultraviolet-A irradiation of human skin fibroblasts leads to a heme oxygenase-dependent increase in ferritin. J Biol Chem 268:14678–14681
69. Zelphati O, Liang XW, Hobart P, Felgner PL (1999) Gene chemistry: functionally and conformationally intact fluorescent plasmid DNA. Hum Gene Ther 10:15–24
70. Hyvonen Z, Plotniece A, Reine I, Chekavichus B, Duburs G, Urtti A (2000) Novel cationic amphiphilic 1,4-dihydropyridine derivatives for DNA delivery. Biochim Biophys Acta 1509: 51–466
71. Hyvonen Z, Ruponen M, Ronkko S, Suhonen P, Urtti A (2002) Cellular and intracellular factors influencing gene transfection mediated by 1,4-dihydropyridine amphiphiles. Eur J Pharm Sci 15:449–460
72. Serikawa T, Suzuki N, Kikuchi H, Tanaka K, Kitagawa T (2000) A new cationic liposome for efficient gene delivery with serum into cultured human cells: a quantitative analysis using two independent fluorescent probes. Biochim Biophys Acta Biomembr 1467:419–430
73. Collins L, Sawyer GJ, Zhang XH, Gustafsson K, Fabre JW (2000) In vitro investigation of factors important for the delivery of an integrin-targeted nonviral DNA vector in organ transplantation. Transplantation 69:1168–1176
74. Krichevsky O, Bonnet G (2002) Fluorescence correlation spectroscopy: the technique and its applications. Rep Progr Phys 65:251–297
75. Clamme JP, Azoulay J, Mely Y (2003) Monitoring of the formation and dissociation of polyethylenimine/DNA complexes by two-photon fluorescence correlation spectroscopy. Biophys J 84:1960–1968
76. Van Rompaey E, Engelborghs Y, Sanders N, De Smedt SC, Demeester J (2001) Interactions between oligonucleotides and cationic polymers investigated by fluorescence correlation spectroscopy. Pharm Res 18:928–936
77. Jurkiewicz P, Okruszek A, Hof M, Langner M (2003) Associating oligonucleotides with positively charged liposomes. Cell Mol Biol Lett 8:77–84
78. Clamme JP, Krishnamoorthy G, Mely Y (2003) Intracellular dynamics of the gene delivery vehicle polyethylenimine during transfection: investigation by two-photon fluorescence correlation spectroscopy. Biochim Biophys Acta Biomembr 1617:52–61
79. Wiethoff CM, Gill ML, Koe GS, Koe JG, Middaugh CR (2002) The structural organization of cationic lipid-DNA complexes. J Biol Chem 277:44980–44987
80. Zhang Y, Garzon-Rodriguez W, Manning MC, Anchordoquy TJ (2003) The use of fluorescence resonance energy transfer to monitor dynamic changes of lipid-DNA interactions during lipoplex formation. Biochim Biophys Acta Biomembr 1614:182–192
81. Itaka K, Harada A, Yamasaki Y, Nakamura K, Kawaguchi H, Kataoka K (2004) In situ single cell observation by fluorescence resonance energy transfer reveals fast intra-cytoplasmic delivery and easy release of plasmid DNA complexed with linear polyethylenimine. J Gene Med 6:76–84
82. Pal R, Barenholz Y, Wagner RR (1988) Pyrene phospholipid as a biological fluorescent probe for studying fusion of virus membrane with liposomes. Biochemistry 27:30–36
83. Struck DK, Hoekstra D, Pagano RE (1981) Use of resonance energy transfer to monitor membrane fusion. Biochemistry 20:4093–4099
84. Remy-Kristensen A, Clamme JP, Vuilleumier C, Kuhry JG, Mely Y (2001) Role of endocytosis in the transfection of L929 fibroblasts by polyethylenimine/DNA complexes. Biochim Biophys Acta 1514:21–32
85. Ira A, Mely Y, Krishnamoorthy G (2003) DNA vector polyethylenimine affects cell pH and membrane potential: a time-resolved fluorescence microscopy study. J Fluoresc 13: 339–347
86. Lleres D, Dauty E, Behr JP, Mely Y, Duportail G (2001) DNA condensation by an oxidizable cationic detergent. Interactions with lipid vesicles. Chem Phys Lipids 111:59–71
87. Bacia K, Schwille P (2003) A dynamic view of cellular processes by in vivo fluorescence auto- and cross-correlation spectroscopy. Methods 29:74–85
88. Haustein E, Schwille P (2003) Ultrasensitive investigations of biological systems by fluorescence correlation spectroscopy. Methods 29:153–166

89. Widengren J, Schweinberger E, Berger S, Seidel CAM (2001) Two new concepts to measure fluorescence resonance energy transfer via fluorescence correlation spectroscopy: theory and experimental realizations. J Phys Chem A 105:6851–6866
90. Lucas B, Van Rompaey E, De Smedt SC, Demeester J, Van Oostveldt P (2002) Dual-color fluorescence fluctuation spectroscopy to study the complexation between poly-L-lysine and oligonucleotides. Macromolecules 35:8152–8160
91. Weiss S (1999) Fluorescence spectroscopy of single biomolecules. Science 283:1676–1683
92. Sako Y, Yanagida T (2003) Single-molecule visualization in cell biology. Nat Cell Biol SS1–SS5
93. Su TJ, Theofanidou E, Arlt J, Dryden DTF, Crain J (2004) Single molecule fluorescence imaging and its application to the study of DNA condensation. J Fluoresc 14:65–69

13 Fluorescence Applications in Targeted Drug Delivery

K. BRYL and M. LANGNER

Keywords: Targeted drug delivery systems; Aggregate formation; Fluorescence resonance energy transfer; Fluorescence correlation spectroscopy

Abbreviations

DPH	1,6-Diphenyl-1,3,5-hexatriene
FA	Fluorescence anisotropy
FCS	Fluorescence correlation spectroscopy
FIDA	Fluorescence intensity distribution analysis
FRAP	Fluorescence recovery after photobleaching
FRET	Fluorescence resonance energy transfer
NBD	*N*-(7-nitrobenz-2-oxa-1,3-diazol-4-yl)
PCH	Photon-counting histogram analysis

13.1 Introduction

The concept of targeted drug delivery systems is based on the assumption that different functions can be assigned to specific molecules which are then assembled together to form an aggregate. In such an approach drug distribution, accumulation, and release will depend solely on the properties of the carrier. The functional division between the active compound and carrier is a concept that has found its full manifestation within new strategies employed in, for example, gene therapy [1–4]. To ensure the desired aggregate behavior, a delicate balance between its stability and ability to integrate with the targeted cells needs to be achieved. There are numerous data regarding aggregate formation, their interaction with whole cells, and particular components [5–7]. Despite a large body of research, few general conclusions regarding aggregate production and how its properties determine its fate inside the body have been reached. Due to this, each active compound requires separate carrier design. In practice, this means the properties, efficacy, and all relevant pharmacological parameters of each compound need to be determined uniquely. Such a situation makes the development of new supramolecular formulations expensive and highly time consuming.

The aggregate formation process depends on a large number of parameters that depend on the aggregate components' properties individually, and these cannot be translated easily to properties of the ensemble as a whole. The resulting particulate structure should fulfill the requirements of a drug carrier, i.e., it has to possess the desired pharmacokinetic and pharmacodynamic parameters. Therefore, the development of targeted drug delivery systems, contrary to the process of drug discovery, is concerned with carrier properties rather than drug activity itself. To rationalize the research and development stage, relevant parameters need to be determined and selected, appropriate techniques employed, and their relevance to the prediction of drug carrier behavior in vivo assessed.

A variety of methods are used for drug discovery, ranging from computer simulation and physicochemical characterization to testing on cell cultures [8, 9]. All these sophisticated experimental methodologies have limited use in developing targeted drug delivery systems, since they focus mainly on the binding efficiency of the active compound with relevant receptors [10, 11].

The development process of drug-carrying aggregates should focus on the design and construction stages which ensure the desired biodistribution and effective delivery of the carried cargo to the targeted tissue. An efficient drug carrier should be characterized by a number of traits and functions encoded in aggregate composition, topology, and stability. These are:

- Ability to incorporate a broad spectrum of drugs
- Maintenance of drug activity
- Prevention of drug degradation
- Immunological neutrality
- Recognition of the target
- Generation of therapeutic drug concentrations
- Release of the drug at the target
- Biodegradability
- Biological inertness
- Feasible large-scale and cost-effective production

Developing supramolecular aggregates for delivering biologically active compounds requires new procedures to be implemented and innovative techniques to be applied, which will ensure the required technological precision. There is also a need for the statistical analysis of the resulting aggregate distribution on a single particle level. To achieve this, a number of aggregate parameters need to be selected and measured. These include, for example, aggregate size, surface electrostatic potential, the content of the active compound, topology (i.e., the spatial organization of all components), stability, and behavior when exposed to enzymatic hydrolysis [12–14]. Each of these parameters cannot be trivially measured, especially when the aggregate population distribution needs to be determined simultaneously.

The major issues in targeted drug delivery system development are:

Aggregate formation
- Formation process kinetics
- Topology of the intermediates
- Formation process energetics

Aggregate characterization
- Physicochemical properties
- Stability

- Topology
- Homogeneity

Aggregate fate in blood
- Aggregate component stability exposure to enzymatic hydrolysis
- Interaction with blood components
- The effect of dilution – dynamic instability

Aggregate interaction with targeted cells
- Aggregate association with the cellular plasma membrane
- Exchange of components between the aggregate and cell
- Internalization
- Disassembly
- Intracellular distribution

13.2 Fluorescence Techniques as Tools for the Development of Targeted Drug Delivery Systems

Fluorescence techniques are especially suitable for studying complex supramolecular assemblies. A large number of assays have already been developed in the course of studies on macromolecular and supramolecular assemblies. Experience gained during basic research on, for example, biological and model membranes is particularly valuable, as a number of issues regarding complex multiphase systems have already been addressed and successfully resolved [15–25]. In addition, fluorescence techniques offer a broad spectrum of methods and parameters which can be measured in a time- and cost-effective way. Even more importantly, they are suitable to be automated and adapted to high-throughput screening protocols [26–28].

13.2.1 The Supramolecular Aggregate Formation Process

The properly constructed aggregate should have parameters that satisfy the requirements of an efficient drug carrier. The desired chemical composition and the proper relative location of its components should be achieved. Obtaining the desired aggregate topology is a challenging experimental task, as it depends not only on component properties but also on the formation processes. Therefore, appropriate aggregate preparation conditions and well-controlled formation kinetics are necessary [4]. Aggregates are formed from various materials, including natural lipids, nucleic acids, proteins, and synthetically derived polymers [5–7, 12, 29, 30]. As aggregates must be formed from many different molecules, multilabeling approaches and complex experimental designs need to be used in order to obtain all the information necessary to determine the resulting structure and properties.

Among the techniques that directly probe the structure of macromolecular ensembles the most popular are:

- X-ray diffraction
- NMR spectroscopy
- Electron microscopy
- Scanning probe techniques

The last allow measurements with nanometer resolution. However, they require large quantities of purified components, are usually performed under nonphysiological conditions, are inefficient for statistical sample evaluation, and are seldom suitable for observing molecular reactions in real time and with a high-throughput setup. On the other hand, fluorescence techniques can provide a wide range of experimental approaches that enable structural parameters and their change in time to be measured. There are generally two fluorescence labeling strategies for supramolecular ensembles:

- Covalent attachment of fluorophores to aggregate components
- Noncovalent association of dyes with selected compounds or the aggregate as a whole

The aggregate formation process is driven predominantly by the hydrophobic effect. Hence, at some point during the aggregate formation process a multiphase system will appear, which indicates that the fluorescent label may be exposed to changing conditions and environments. Due to this, its location and fluorescence properties should be carefully planned and controlled. Certain probes allow specific properties to be measured selectively at various regions of the aggregate. The appearance of a hydrophobic phase and its organization can be monitored with hydrophobic dyes such as 1,6-diphenyl-1,3,5-hexatriene (DPH) or excimer-forming dyes [31, 32]. Their fluorescence properties depend on the organization of the hydrophobic phase, meaning that the polarization of DPH and amount of pyrene dye excimer vary with aggregate packing [33].

Complex interface regions can be characterized with respect to their relevant parameters, including surface pH [34], surface electrostatic potential [35], small molecule adsorption [36], and local conformational changes [37]. The existence of multiple phases and steric constraints may cause problems in the determination of fluorescent dye location. Regardless of how the probe is associated with the aggregate, its affinity toward various phases needs to be known in advance. For example, the amphiphilic *N*-(7-nitrobenz-2-oxa-1,3-diazol-4-yl) (NBD) fluorophore, covalently bound to a lipid molecule, will position itself in the interface region of the membrane regardless of which part of the lipid it is attached to [38, 39]. Fluorescent dyes frequently used in FCS measurements of the cell, i.e., rhodamine green, Alexa dyes, and far red Cy-5, may interact with various membrane components, leading to artifacts [40]. Newly developed dyes are intended to meet these strict criteria and have strong preferences toward the selected phase. For example, DY-630 and DY-635 are designed to be hydrophilic [41].

Assays developed for membrane fusion can be easily adopted to follow the kinetics of aggregate assembly formation. For example, dequenching [42, 43] and fluorescence resonance energy transfer (FRET) [44] are especially valuable. FRET can also be employed to measure the relative location of aggregate compounds [45], and when it is combined with information obtained by collisional, static, and chemical quenching, the aggregate topology can be well characterized [20,46,47]. Different types of information can be gathered using fluorescence anisotropy (FA), which is suited to detecting changes in molecule mobility. Molecular interactions usually result in a change in fluorescent probe rotational mobility, which can be translated into a change in FA. An additional advantage of the FA approach is that it is independent of total fluorescence intensity. Recently, microplate readers with (sub)nanomolar sensitivity have become commercially available for FA. Such instruments enable FA to be applied to high-throughput screenings [48].

When monitoring the time dependence of aggregate formation, a multi-labeling approach and appropriate data analysis are needed. The simultaneous measurement of different aggregate compounds, and their mutual interaction, mobility, and excited state lifetime, provide information on their relative position and progress of association. In addition, combining this classical fluorescence with an FCS approach may provide measurements of the sample homogeneity. The recently introduced burst-integrated fluorescence lifetime spectroscopy technique may prove to be useful for measuring macromolecule conformational changes induced by their association. This technique combines the advantages of selective time-resolved fluorescence and correlation spectroscopy [26,49].

In order to monitor aggregate formation, the incorporated fluorescent probe should not interfere with the ongoing processes. Because aggregation depends predominantly on the participating molecules' chemical structure, any additional modifications may change the experimental conditions. Therefore, the probe concentration regimes and its most neutral locations need to be determined in advance. We have shown, for example, that the association of propidium iodide with DNA alters its conformation and condensation induced by cationic compounds [50–52].

The inherent properties of aggregate suspensions are intensive light scattering and possible background fluorescence, making fluorescence intensity measurements difficult. Time-resolved fluorescence techniques are successful at eliminating these problems [26,27,53].

13.2.2
Selected Aggregate Parameters – Relevance and Measurement Techniques

In general, there are two factors that determine aggregate behavior in the organism: size and surface properties.

13.2.2.1
Aggregate Size

Aggregate size has been correlated with its ability to accumulate in tumor tissues via what is known as passive targeting [10]. Small liposomes can enter the tumor through ill-formed capillaries, whereas large ones remain in circulation and are subject to elimination processes [54]. Aggregate size is a result of the formation process and the dimensions of the encapsulated macromolecules. The development of a carrier for genetic information is a good illustration of this problem. There are still no reliable methods available for condensing nucleic acid to reasonably sized aggregates. In studies on nucleic acid condensation, various fluorescence techniques have been used [2, 6, 55]. When conformational changes are measured for macromolecules, there is an inherent difficulty associated with the system. Large molecules like DNA normally have a number of different configurations in a single preparation, i.e., topoisomers [56]. Consequently, the sample always contains a certain number of subpopulations, each of which is likely to behave differently [57]. There is also another problem that results from the existence of topoisomers, namely that sample labeling can alter conformer distribution and hence change the experimental conditions [50–52]. Some solutions to these problems are offered by the FCS technique, which indicates that the multipopulation system has been formed [50–52].

13.2.2.2
Capacity to Carry the Active Compound

In the case of liposomes, the capacity to carry an active compound is measured by the entrapped volume which depends on size, composition, and lamellarity. Most methods used to evaluate this parameter utilize the fluorescence exclusion technique, where the entrapped volume is determined by the amount of probe (calcein or carboxyfluorescein) retained in the sample after filtration [58].

13.2.2.3
Aggregate Stability

One of the major problems limiting the widespread use of supramolecular drug carriers is their stability, both chemical and physical. Chemical instability can result from enzymatic hydrolysis or component oxidation [54, 59]. Well-characterized fluorescence approaches can be adopted to estimate the progress of enzymatic or oxidation reactions, i.e., the extent of oxidation occurring in the hydrophobic phase can be evaluated by *cis*- and *trans*-parinaric acids [60], or by a BODIPY analog BP-C11 (4,4-difluoro-5-(4-phenyl-1,3-butadienyl)-4-bora-3a,4a-diaza-*s*-indacene-3-undecanoic acid). The latter changes its fluorescence spectral maximum upon oxidation from 591 to 520 nm [61]. The enzymatic hydrolysis of aggregate components can be followed by approaches similar to those already available with fluorescence assays [62, 63].

Physical instability can result from active compound leakage, aggregation, and/or aggregate fusion. Carboxyfluorescein is a model of a water-soluble drug and its dequenching upon dilution can be utilized to evaluate its retention in the carrier. Release of the model molecule can be quantified by an increase in fluorescence intensity (excitation/emission maxima 490/520 nm) [64]. The extent of drug leakage is very important not only for preparing "good quality" carriers, but also for evaluating the extent to which the encapsulated compound destabilizes the membrane. For example, Engelke et al. [65] have demonstrated that tamoxifen (a nonsteroidal antiestrogen drug used in the treatment and prevention of breast, liver, pancreas, and brain cancers [66]) induces the release of carboxyfluorescein entrapped in liposomes as a result of permanent bilayer disruption and transient hole formation. Supramolecular drug carriers may aggregate and/or fuse, hence altering their physiochemical parameters. FRET and quenching/dequenching techniques have been successfully applied to studying such phenomena [16, 67, 68].

13.2.2.4
Aggregate Surface Properties

The aggregate surface is the region where interactions with proteins take place, which essentially decide how the immunological system perceives the structure. There are a number of parameters relevant to these processes, including steric constraints imposed by surface-associated polymers, surface electrostatic potential, and the packing of molecules forming the surface [5, 7, 69–72].

13.2.2.4.1
Aggregate Surface Electrostatic Potential

One of the most important surface parameters is the surface electrostatic charge. It influences the mechanism and extent of aggregate-cell interaction and protein adsorption, and is frequently an immunostimulating factor [70, 73–75]. For example, aggregate surface electrostatic charge influences association with carcinoma cells [70], negative surface charge associated with phosphatidylserine (PS) induces thrombosis blood coagulation [76], and positive surface charge induces aggregate fusion with biological membranes [54]. In addition, surface charge affects formulation physical stability [77]. Surface electrostatic potential can be measured with electrophoretic methods; nonetheless, the results are difficult to interpret when surface topology is complicated, i.e., when polymers are grafted. Better methods are based on changing fluorescent dye properties, whose association depends on surface electrostatic potential [35], or on the quenching of surface-associated probes by charged ions [78, 79]. Another interesting approach is based on the correlation between probe emission maximum and the aggregate surface properties [80].

13.2.2.4.2
Aggregate components mobility

The mobility of molecules within the aggregate is frequently measured by membrane viscosity (or membrane fluidity, its reciprocal). Depending on the membrane viscosity, lipids and other aggregate-associated elements exhibit different vertical and lateral displacements as well as the capability to exchange between various aggregates or aggregate and cell membranes. These phenomena affect aggregate stability in vivo, and may become a tool for designing changes in aggregate properties as a function of time upon the dissociation and/or association of various compounds [72, 81]. The mobilities of aggregate-forming molecules are routinely measured by fluorescence techniques including FRET, fluorescence recovery after photobleaching (FRAP – in which the diffusivity of a fluorophore is measured) [82], fluorescence anisotropy (in which fluorophore mobility is related to local viscosity), and the application of environment-sensitive fluorescent probes. The last category includes those compounds whose emission wavelengths shift with the viscosity of the medium (for example, 2-dimethylamino-6-lauroylnaphthalene [83]), and whose fluorescence quantum yield depends on the viscosity (for example, 9-(dicyanovinyl)-julolidine and its analogs [84, 85]).

13.2.2.5
Aggregate Topology

Aggregate surface heterogeneity affects receptor aggregation which in turn may affect its cell association and internalization capabilities. Therefore, information concerning lateral aggregate surface component distribution is of functional significance. Fluorescent probes have been used to study domain organization on lipid bilayer surfaces for a long time [86]. For example, FRET is very sensitive to alterations in inter- and intramolecular distances, because its efficiency depends on the average distance between the donor and acceptor [87, 88]. Dyes whose spectra are sensitive to the surrounding physical state (prodan and laurdan) are also commonly used to characterize aggregate organization [89]. Lateral inhomogeneity may result from the existence of structural defects, compromising aggregate integrity, causing active compound release, and/or stimulating hydrolytic enzyme activity [46, 47, 63].

Another important component parameter is their relative location, especially with regard to their access to the surface. For example, the exposure of nucleic acid may result in its degradation and/or immunostimulation. There are no efficient fluorescence methods that provide details about aggregate topology. However, the quenching of dyes attached to macromolecules by quenchers located in the aggregate itself or in the aqueous phase may provide some data in analogy to the methodologies employed in membrane protein studies [25, 90]. A number of the dynamic quenching approaches, including quenching by small water-soluble molecules [20] or by halides and spin labels attached to aggregate compounds, may provide information on the relative position of macromolecules [25, 90]. Dynamic and static quenching data for multipopulations carry inherent interpretation difficulties. Chemical quenching is a solution to some of these problems. The fluorescent probe transforms into a nonfluorescent form when exposed to a water-soluble reagent (for instance, dithinite). Such methodology allows the amount of accessible dye and consequently labeled molecules to be determined quantitatively. In addition, this approach allows the leakage of the liposome content to be measured [91].

13.2.2.6
Homogeneity of Aggregate Preparation

Despite much effort, only a few aggregate-based products are on the market. The major reasons for this are inherent problems with aggregate production (i.e., their homogeneity is not certain) and the lack of precise sample evaluation methods. Bulk fluorescence techniques are volume-limited and prone to artifacts when performed under high adsorption or light scattering conditions. In addition, samples with nonuniform aggregate populations are difficult or impossible to analyze, even qualitatively. Applying FCS and techniques based on time-resolved approaches may overcome these limitations [92–94]. Different processes and molecules can be distinguished by their characteristic fluctuations. Using this approach, a wide range of parameters have already been measured, including ligand–receptor interactions (e.g., equilibrium constants, the stoichiometry of binding and reactant masses), aggregation, dimerization, chemical reaction rates, and macromolecule conformations. The advantages of FCS lie in short measurement times (milliseconds to minutes) and the small amount of material needed (nanomolar to micromolar concentrations and volumes as low as 1 μl). FCS has been already used to estimate aggregate destabilization and content release [95]. Variations of the method (dual-color fluorescence cross-correlation) enable the dynamics of two molecules to be followed simultaneously [96]. New methods have recently been derived from the original FCS concept: fluorescence intensity distribution analysis (FIDA) and photon-counting histogram analysis (PCH) [97, 98], which are both based on investigating equilibrium fluctuations and their statistical analysis.

13.2.3
Aggregate Intracellular Fate

Knowledge regarding aggregate disintegration and component distribution within the cell is of vital importance when the active compound is directed toward targets in the cytoplasm. This is the case when the genetic material is involved [6]. To study such processes, living cells have to be used as a platform for acquiring biological information that is relevant to modern drug discovery and drug delivery systems. This new perspective is cell-centric rather than focusing on gene or protein isolation. A cell-centric perspective must integrate into a single platform the reagents that report the states of molecular processes within cell, the automated detection and the analysis of these processes, and cellular knowledge [99]. A variety of labeling strategies have been tested and a large number of fluorescent probes developed in order to study various intracellular events. It is presently possible to measure intracellular ion concentration and pH [100]. Organelles can be selectively labeled and their functional parameters assessed [22]. A good example of such an approach is the studies carried out by Pagano's group. They monitored the distribution of labeled lipids within the cell [39, 60, 101], resulting in the development of a diagnostic protocol based on the quantification of cell fluorescent images [102, 103].

Ligand-receptor interaction type, membrane crossing, and intracellular compartmentalization measured in situ are subject to extensive research effort that is aimed at developing cellular tests for evaluating drug activity and ability to reach the target structures. For these purposes, various approaches and their combinations have been used, including TRF, anisotropy, FCS, and FRET in microscopic setups [18, 40, 94, 104–107]. A combination of fluorescence signals from labeled cell structures and tagged carrier components can provide valuable information on the fate of the delivered material and processes leading to its release. For example, the exchange of aggregate components with cellular structures and/or their internalization can be evaluated with well-known fusion assays [108, 109]. The double labeling of lipid–DNA aggregates is used for determining the intracellular distribution of each aggregate component independently [110], whereas fluorescence microscopy allows information on the second to hour timescale to be gathered. FCS and FRAP techniques allow local cytoplasmic properties encountered by an exogenous compound to be determined [111]. It is instructive to follow the path of each component when a multicomponent aggregate is internalized by the cell. Double labeling in FCS experiments has already been used to follow hylotoxin internalization by endocytosis [96]. It has even been shown that by employing the time-resolved technique, seven-color sample quantitative analysis is feasible by using appropriate data analysis methods [112].

However, nothing more important has been done for cellular research since the introduction of the green fluorescent protein (GFP). GFP of *Aequorea victoria* has been used as a tool to monitor and localize gene expression, proteins, and molecular interactions in living cells. By genetic engineering, GFP can be fused as a tag to a protein of interest. In addition, several mutants of GFP have been created with optimized expression, stability, and color (excitation and emission maximum ranging from 380 to 514 nm and from 440 to 527 nm, respectively) [8, 21, 113, 114].

A combination of GFP and fluorescent probe associated with the aggregate can be used to evaluate intracellular distribution with molecular resolution using FRET. GFP can be used to detect endogenous and exogenous compounds, as recently demonstrated by detecting the intracellular distribution of inositol [107]. Additional possibilities are available when the suppressor tRNA technology is applied. This technique enables the site-specific introduction of unnatural amino acids into proteins [11]. In protein research, the technique has been exploited by applying various tryptophan analogs with modified fluorescence properties ([25] and citations therein).

When fluorescence is used to monitor an intracellular event, a certain amount of caution is needed. Certain probes can spontaneously exchange between aggregate and various cellular structures, leading to erroneous conclusions [115]. Fluorescent probe sensitivity to various conditions also needs to be thoroughly evaluated, as cell compartments can differ substantially in their physicochemical parameters, which may affect the dye fluorescence properties [116].

The possibilities of cell labeling are supported by the rapid growth of imaging techniques and data acquisition. Recent advances have greatly enhanced the possibilities of monitoring intracellular events in real time. For example, information about the spatial distribution of fluorophore and its molecular environment is available through fluorescence lifetime imaging microscopy (FLIM), in which the lifetime of a fluorophore depends on its environment. Using this method, the oligomerization of labeled epidermal growth factor receptors has been investigated [117]. Nipkow microscopy and automated image analysis combined with multichannel data acquisition presently opens the door to rapid data collection in multisample screens and live cells [27].

13.3 Perspectives

Multidisciplinary approaches that combine informatics, chemistry, biophysics, and biology are needed in order to develop efficient and reliable procedures for designing and testing new supramolecular aggregates [118]. The most promising progress in increasing the predictability of drug discovery is created by new scientific disciplines:

- Functional genomics, which allows for the generation of new molecular targets and ensures the transfer of active compound to selected cellular compartments.
- Combinatorial chemistry, which generates the molecular diversity necessary to select appropriate active compounds for new targets and extends the available arsenal of components for constructing the carrier.
- High-throughput screening (HTS), which selects compounds and delivery vehicles efficiently [105].

Since the selection of aggregate components is a multiparameter problem and there are no theories for the prediction of aggregate properties and biodistribution, adopting combinatorial methods in the development process is natural. Aggregate structure is sensitive to even minute changes in component chemical makeup. Therefore, screening methods can explore the multidimensional aggregate conformational space, as they have done in the past for protein folding problems [69]. The main interests in developing new combinatorial assays are reducing screening cost and time and increasing the number of samples.

An interesting development in the field is the modular approach. It is based on the assumption that the aggregate can be formed from interchangeable elements which can then be combined at will into a desired form. This greatly simplifies the carrier construction process. Such an approach has been presented by Dr. Allen's group. The aggregate is functionally divided into three key elements: the active compound, the capsule, and the targeting element. An appropriate combination of these three elements in the aggregate allows a broad spectrum of satisfactory formulations to be constructed [119]. All these new concepts, combined with a growing efficiency of screening procedures in vitro and on cell models, will provide new possibilities for the development of efficient targeted drug delivery systems.

References

1. Anwer K, Meaney C, Kao G, Hussain N, Shelvin R, Earls RM, Leonard P, Quezada A, Rolland AP, Sullivan SM (2000) Cancer Gene Ther 7:1156
2. Byk G, Scherman D (2000) Drug Dev Res 50:566
3. Faneca H, Simoes S, Pedroso de Lima MC (2002) Biochim Biophys Acta 1567:23
4. Pedroso de Lima MC, Simoes S, Pires P, Faneca H, Duzgunes N (2001) Adv Drug Deliv Rev 47:277
5. Langner M, Kral T (1999) Pol J Pharmacol 51:211
6. Langner M (2000) Cell Mol Biol Lett 5:295
7. Langner M (2000) Pol J Pharmacol 52:3
8. Fernandes PB (2003) Curr Opin Chem Biol 2:597
9. White RE (2000) Annu Rev Pharmacol Toxicol 40:133
10. Iqbal SS, Mayo MW, Bruno JG, Bronk BV, Batt CA, Chambers JP (2000) Biosens Bioelectron 15:549
11. Hovius R, Vallotton P, Wohland T, Vogel H (2000) TiPS 21:266
12. Torchilin VP, Lukyanov AN (2003) Drug Discov Today 8:259
13. Hafez IM, Cullis PR (2001) Adv Drug Deliv Rev 47:139
14. Mills JK, Needham D (1999) Expert Opin Ther Patents 9:1499
15. Amanullah A, Hewitt CJ, Nienow AW, Lee C, Chartrain M, Buckland BC, Drew SW, Woodley JM (2002) Biotechnol Bioeng 80:239
16. Blumenthal R, Gallo SA, Viard M, Raviv Y, Puri A (2002) Chem Phys Lipids 116:39
17. Bobrovnik SA (2003) J Biochem Biophys Methods 55:71
18. Cohen-Kashi M, Moshkov S, Zurgil N, Deutsch M (2002) Biophys J 83:1395
19. Domanov YA, Gorbenko GP (2002) Biophys Chem 99:143
20. Geddes CD (2001) Meas Sci Technol 12:R53
21. Giuliano KA, Taylor DL (1998) Trends Biotechnol 16:135
22. Hahn K, Touchkine A (2002) Curr Opin Cell Biol 14:167
23. Hillisch A, Lorenz M, Diekmann S (2001) Curr Opin Struct Biol 11:201
24. Krishna MMG, Srivastava A, Periasamy N (2001) Biophys Chem 90:123
25. Ladokhin AS, Jayasinghe S, White SH (2000) Anal Biochem 285:235
26. Eggeling C, Fries JR, Brnd L, Gunter R, Seidel CAM (1998) Proc Natl Acad Sci USA 95:1556
27. Eggeling C, Band L, Jager DUS (2003) Drug Discov Today 8:632
28. Pope AJ, Haupts UM, Moore KJ (1999) Drug Discov Today 4:350
29. Kataoka K, Harada A, Nagasaki Y (2001) Adv Drug Deliv Rev 47:113
30. Torchilin VP (2001) J Control Release 73:137
31. Galla H-J, Hartmann W (1980) Chem Phys Lipids 27:199
32. Kaiser RD, London E (1998) Biochemistry 37:8180
33. Van der Meer BW (1988) In: Hilderson HJ (ed) Subcellular biochemistry, vol 13. Plenum, New York, p 1
34. Kraayenhof R, Sterk GJ, Sang HW (1993) Biochemistry 32:10057
35. Fromherz P (1989) Meth Enzymol 171:376
36. Langner M, Gabrielska J, Przestalski S (2000) Appl Organometal Chem 14:25
37. Pal R, Petri WA, Ben-Yashar V, Wagner RR, Barenholz Y (1985) Biochemistry 24:573
38. Chattopadhyay A (1990) Chem Phys Lipids 53:1
39. Hao M, Maxfield FR (2000) J Biol Chem 275:15279
40. Dittrich P, Malvezzi-Campeggi F, Jahnz M, Schwille P (2001) Biol Chem 382:491
41. Czerny P, Lehmann F, Wentzel M, Buschann V, Dietrich A, Mohr GJ (2001) Biol Chem 382:495
42. Stegmann T, Schoen P, Bron R, Wey J, Bartoldusy I, Ortiz A, Nieva JL, Wilschut J (1993) Biochemistry 32:11330
43. MacDonald RI (1990) J Biol Chem 265:13533
44. Selvin PR (1995) Meth Enzymol 246:300
45. Keller RCA, Silvius JR, de Kruijff B (1995) Biochim Biophys Res Commun 207:508
46. Langner M, Hui SW (1991) Chem Phys Lipids 60:127

47. Langner M, Hui SW (1993) Chem Phys Lipids 65:23
48. Hemmila I (1988) Scand J Clin Lab Invest 48:389
49. Maita S, Haups U, Webb WW (1997) Proc Natl Acad Sci USA 94:11753
50. Kral T, Langner M, Benes M, Baczynska D, Ugorski M, Hof M (2002) Biophys Chem 95:135
51. Kral T, Hof M, Langner M (2002) Biol Chem 383:331
52. Kral T, Hof M, Jurkiewicz P, Langner M (2002) Cell Mol Biol Lett 7:203
53. Rogers MV (1997) Drug Discov Today 2:156
54. Sharma A, Sharma US (1997) Int J Pharmaceutics 154:123
55. Miller AD (1998) Angew Chem Int Ed 37:1768
56. Langowski J, Kapp U, Klenin K, Vologosdkii A (1994) Biopolymers 34:639
57. Tobias I (1998) Biophys J 74:2545
58. Kawakami K, Nishihara Y, Hirano K (1999) Anal Biochem 269:139
59. Samuni AM, Lipman A, Barenholz Y (2000) Chem Phys Lipids 105
60. Maier O, Oberle V, Hoekstra D (2002) Chem Phys Lipids 116:3
61. Borst JW, Visser NV, Kouptsova O, Visser AJ (2000) Biochim Biophys Acta 1487:61
62. Henshaw JB, Olsen CA, Farnbach AR, Nielsen KH, Bell JD (1998) Biochemistry 37:10709
63. Davidsen J, Jorgensen K, Andersen TL, Mouritsen OG (2003) Biochim Biophys Acta 1609:95
64. Bryl K, Kedzierska S, Taylor A (2000) J Fluoresc 10:209
65. Engelke M, Bojarski P, Blos R, Diehl H (2001) Biophys Chem 90:157
66. Wiseman H (1994) TiPS 15:83
67. Bryl K, Kedzierska S, Laskowska M, Taylor A (2000) Eur J Biochem 267:794
68. Nunes-Correia I, Eulalio A, Nir S, Duzgunes N, Ramalho-Santos J, Pedroso de Lima MC (2002) Biochim Biophys Acta 1561:65
69. Drumond DC, Meyer O, Hong K, Kirpotin DB, Papahadjopoulos D (1999) Pharmacol Rev 51:691
70. Baczynska D, Widerak K, Ugorski M, Langner M (2001) Z Naturforsch 56c:872
71. Chiu GNC, Bally MB, Mayer LD (2002) Biochim Biophys Acta 1560:37
72. Vrhovnik K, Kristl J, Sentjurc M, Smid-Korbal J (1998) Pharm Res 15:523
73. Campbell RB, Fukumura D, Brown EB, Mazzola LM, Izumi Y, Jain RK, Torchilin VP, Munn LL (2002) Cancer Res 62:6831
74. Miller CR, Bondurant B, McLean SD, McGovern KA, O'Brien DF (1998) Biochemistry 37:12875
75. Nakanishi T, Kunisawa J, Hayashi A, Tsutsumi Y, Kubo K, Nakagawa S, Nakanishi M, Tanaka K, Mayumi T (1999) J Control Release 61:233
76. Chiu GNC, Bally MB, Mayer LD (2003) Biochim Biophys Acta 1613:115
77. Sharma A, Mayhew E, Straubinger RM (1993) Cancer Res 53:5877
78. Langner M, Cafiso D, Marcelja S, McLaughlin S (1990) Biophys J 57:335
79. Wlniski AP, Eisenberg M, Langner M, McLaughlin S (1988) Biochemistry 27:386
80. Singh Y, Gulyani A, Bhattacharya S (2003) FEBS Lett 541:132
81. Li WM, Xue L, Mayer LD, Bally MB (2001) Biochim Biophys Acta 1513:193
82. Axlrod D, Koppel DE, Schlessinger J, Elson R, Webb WW (1976) Biophys J 16:1055
83. Parasassi T, DiStefano M, Loiero M, Ravagnan G, Gratton E (1994) Biophys J 66:763
84. Kung CE, Reed JK (1986) Biochemistry 25:6114
85. Haidekker MA, Ling T, Anglo M, Stevens HY, Frangos JA, Theodorakis EA (2001) Chem Biol 8:123
86. Davenport L (1997) Meth Enzymol 278:487
87. Silvius JR (2003) Biophys J 85:1034
88. Pedersen S, Jorgensen K, Baekmark TR, Mouritsen OG (1996) Biophys J 71:554
89. Campbell RB, Balasubramanian SV, Straubinger RM (2001) Biochim Biophys Acta 1512:27
90. Ladokhin AS (1999) Anal Biochem 276:65
91. McIntyre JC, Sleight RG (1991) Biochem 30:11819
92. Thompson NL (1991) In: Lakowicz JR (ed) Topics in fluorescence spectroscopy techniques. Plenum, New York, p 337
93. Medina MA, Schwille P (2002) Bioessays 24:758

94. Brock R, Jovin TM (1998) Cell Mol Biol 44:847
95. Pramanik A, Thyberg P, Rigler R (2000) Chem Phys Lipids 104:35
96. Heinze KG, Koltermann A, Schwille P (2000) Proc Natl Acad Sci USA 97:10377
97. Kask P, Palo K, Ullmann D, Gall K (1999) Proc Natl Acad Sci USA 96:13756
98. Chen Y, Muller JD, So PT, Gratton E (1999) Biophys J 77:553
99. Taylor DL, Woo ES, Giuliano KA (2001) Curr Opin Biotechnol 12:75
100. Yoshida N, Tamura M, Kinjo M (2000) Single Mol 1:279
101. Kok JW, ter Beest M, Scherphof G, Hoekstra D (1990) Eur J Cell Biol 53:173
102. Pagano RE, Puri V, Dominguez M, Marks DL (2000) Traffic 1:807
103. Chen CS, Petterson MC, Wheatley CL, O'Brien JF, Pagano RE (1999) Lancet 354:901
104. Gadella TWJ, Arndt-Jovin DJ, Jovin TM (1994) J Fluoresc 4:295
105. Auer M, Moore KJ, Meyer-Almes FJ, Guenther R, Pope AJ, Stoeckli KA (1998) Drug Discov Today 3:457
106. Bacia K, Majoul IV, Schwille P (2002) Biophys J 83:1184
107. Balla T, Bondeva T, Varnai P (2000) TiPS 21:238
108. Hoekstra D, de Boer T, Klappe K, Wilschut J (1984) Biochemistry 23:5675
109. Monkkonen J, Urtti A (1998) Adv Drug Deliv Rev 34:37
110. Hui SW, Langner M, Zhao YL, Ross P, Hurley E, Chan K (1996) Biophys J 71:590
111. Edidin M, Zagyansky Y, Lardner TJ (1976) Science 191:466
112. Samiotaki M, Kwiatkowski M, Ylitalo N, Landegren U (1997) Anal Biochem 253:156
113. Tsien RY (1998) Annu Rev Biochem 67:509
114. Verkman AS (2002) Trends Biochem Sci 27:27
115. Ohki S, Flanagan TD, Hoekstra D (1998) Biochemistry 37:7496
116. Chen CS, Martin OC, Pagano RE (1997) Biophys J 72:37
117. Gadella TW, Jovin TM (1995) J Cell Biol 129:1543
118. Ahlberg C (1999) Drug Discov Today 4:370
119. Ishida T, Iden DL, Allen TM (1999) FEBS Lett 460:129

Part 5

Fluorescence Spectroscopy in cells: FCS and Quantum Dots

14 Fluorescence Correlation Spectroscopy in Cell Biology

R. Brock

Keywords: Fluorescence correlation spectroscopy; Time-correlated FCS; Cellular FCS; Intracellular autocorrelation measurements

Abbreviations

EGFP	Enhanced green fluorescent protein
FCS	Fluorescence correlation spectroscopy
fpm	Fluorescence per molecule
FRAP	Fluorescence recovery after photobleaching
PCH	Photon counting histograms

14.1 Introduction

Over the past decade, fluorescence correlation spectroscopy (FCS) has been established as one of the most versatile fluorescence-based techniques for acquiring information on the dynamics of molecular processes. The popularity of FCS among other methods for single molecule detection is based on historical as well as on practical considerations. The analysis of molecule diffusion and the kinetics of molecular interactions by FCS was already presented more than 30 years ago [1, 2]. With the implementation of confocal optics and single-photon detectors in microscopy and spectroscopy, single molecule detection emerged in the early 1990s [3, 4]. At that moment, FCS had a head start for the analysis of single molecule fluorescence. In addition, in the early 1990s the use of autocorrelation analysis for signal processing also followed very practical considerations. Computer memory was still too costly to record megabytes worth of fluorescence signals for performing off-line data analysis in other ways. Instead, dedicated autocorrelator boards enabled on-line calculation of autocorrelation functions and thereby data reduction.

The relative ease by which an FCS setup can be built, the availability of robust and turn-key commercial instruments, and straightforward methodology for the analysis of autocorrelation functions contributed to establishing FCS in life science laboratories. A broad literature base provides solutions for a large number of different biological problems. Even though former restrictions on computer hardware do not hold any longer, other methods for single molecule detection such as photon counting histograms [5] and time-resolved single molecule detection are moving only slowly beyond the doors of dedicated biophysical research facilities.

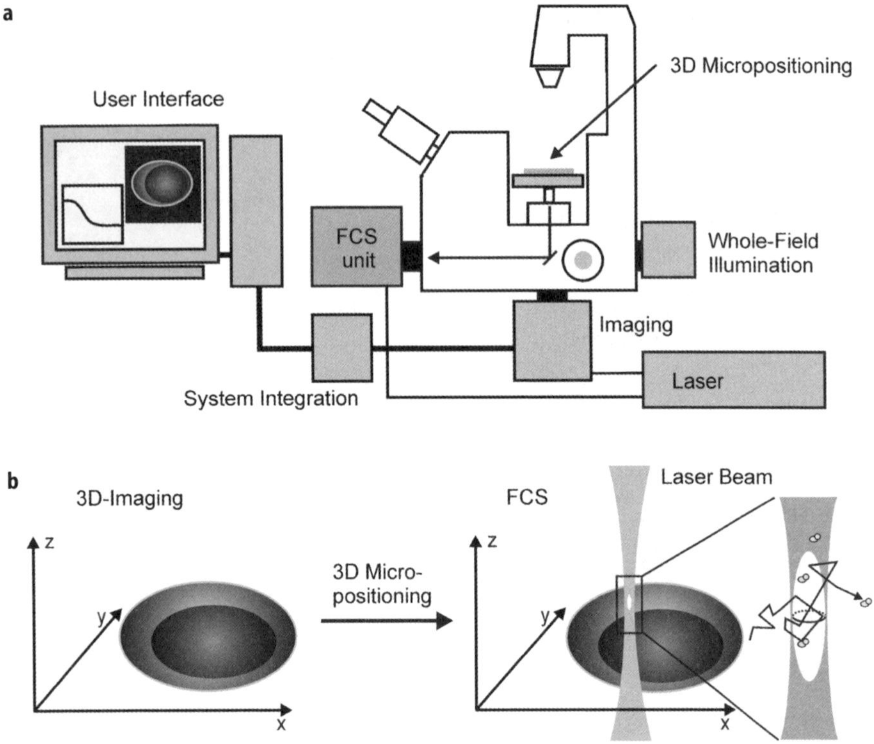

Fig. 14.1a, b. Fluorescence correlation spectroscopy in cell biology. In order to apply FCS to cell biology, the confocal and time-resolved detection of fluorescence is integrated in a fluorescence microscope. Time-correlated FCS differs in two main aspects from imaging microscopy and time-lapse imaging microscopy: (1) fluorescence is recorded with high temporal resolution, usually in the submicrosecond timescale; and (2) for one individual FCS measurement, fluorescence is only recorded from one spot of the sample. In most FCS setups realized so far, a separate unit for confocal laser scanning microscopy or a CCD camera is used for imaging (**a**). Here, a configuration with confocal imaging is shown. High-resolution imaging and high-precision positioning in 3D are required to position the confocal detection volume for FCS measurements (**b**)

FCS provides information on reaction kinetics from temporal fluctuations of the intensity of a fluorescence signal impinging on a highly sensitive detector. The information contained in the recorded signal depends on the molecular and/or photophysical nature of the processes giving rise to the fluctuations. Such processes are: (1) triplet transitions of fluorophores from a fluorescence-active state into a relatively long-lasting triplet state [6]; (2) protonation equilibria, in which the fluorescence quantum yield depends on the presence or absence of a proton [7]; (3) conformational transitions of a fluorophore affecting the fluorescence quantum yield [8]; (4) intramolecular and intermolecular dynamics where in one state of the molecule(s) the fluorophore is in the vicinity of a quencher [9]; and (5) transit of fluorescent molecules through the detection volume by diffusion or directed transport [4].

In contrast to imaging microscopy, an individual FCS measurement is performed by recording fluorescence only at a single spot of a sample (Fig. 14.1). Even when implemented into an imaging microscope, FCS is therefore not an imaging technique. However, by performing a series of FCS measurements at different locations within a sample, imaging information can be supplemented by information on the dynamics of fluorescent molecules at these specific locations. The optics used for excitation and detection of fluorescence are very similar to those used in laser scanning confocal microscopes. High numerical aperture lenses and diffraction-limited illumination in combination with confocal optics yield a volume element for the detection of fluorescence with lateral and axial dimensions of about 0.3 and 2 μm, respectively.

From the very beginning of confocal FCS, the minuteness of the detection element has motivated researchers to use this technique for addressing the dynamics of molecules inside living cells. The incorporation of FCS as an additional measurement modality into microscopes enabled a precise positioning of the measurement volume inside a cell. With such fluorescence correlation microscopes, the recording of intracellular autocorrelation functions proved to be straightforward [10]. In addition, formalisms for the analysis of cellular autocorrelation functions that account for the complexity of the intracellular environment have been presented. In contrast to other fluorescence techniques, it is a characteristic of FCS to enable concentration measurements of fluorophores that are robust to environment-dependent changes of fluorescence quantum yield. This chapter will provide an application-oriented introduction to the theoretical concepts of FCS and then give an overview of applications of FCS with isolated molecules and in intact cells. The text will strictly focus on time-correlated FCS, i.e. the analysis of temporal fluctuations of fluorescence. In similar ways as described here for time-correlated FCS, fluorescence correlation spectroscopy may be performed spatially [11]. Special attention will be paid to the requirements and limitations of intracellular concentration measurements. The number of formulas was kept at a minimum. The interested reader is referred to the cited literature.

14.2 Fluorescence Correlation Spectroscopy Step by Step

14.2.1 Theoretical Background

FCS derives the time constants of reactions and transport processes from spontaneous fluctuations of a system in thermodynamic equilibrium. The generation of information on reaction kinetics in FCS is therefore related to that of the classical methods of perturbation kinetics such as temperature jump and pressure jump. The major difference between the classical methods and FCS is that the perturbation is not caused from the outside but occurs spontaneously as a statistical fluctuation [12].

A limiting factor in detecting such statistical fluctuations is the fluctuation amplitude. This fluctuation amplitude depends on the average number of molecules in the volume element from which fluorescence is detected. If, for example, either only one or no molecule is in the detection volume at any given point in time, the fluctuation amplitude between both states will be 100%. If, in contrast, on average one thousand molecules are in the detection volume, any single molecule will cause a fluctuation in the fluorescence of only 0.1%. At too high a concentration, the fluorescence fluctuations will be indistinguishable from noise. Due to the requirement for high fluctuation amplitudes, low molecule numbers in the detection volume are mandatory. These are achieved through the combination of confocal detection optics and probe concentrations in the low to medium nanomolar range.

Even though the ability of FCS to operate at low concentrations has been promoted as one of the strengths of the technique for limiting sample consumption, the sampling of only a few molecular events at any given point in time comes at the expense of relatively long measurement times required to obtain information on kinetic parameters. In the classical perturbation techniques the whole ensemble is perturbed. The relaxation kinetics of the system back into equilibrium represent the average of an enormous number of individual molecular events. For this reason, kinetic parameters may be derived from only one single cycle of perturbation and relaxation. In FCS, instead, the statistical fluctuations of a large number of molecules are recorded sequentially over time. The ensemble average is derived by calculating the autocorrelation function of the fluorescence signal.

The reliance on fluctuations about equilibrium necessitates that the system remains constant with respect to molecule concentrations and average mobilities, at least for the time of the autocorrelation measurement. This requirement for equilibrium conditions severely limits the scope of analytical questions in cellular FCS. For slowly diffusing molecules such as transmembrane receptors, measurement times of one to a few minutes are required. During such long periods of time, the cell or at least the plasma membrane may move and thereby change the measurement conditions. In contrast, for molecules diffusing freely in solution, measurement times of a few seconds are usually sufficient to acquire an autocorrelation function. In this case, the kinetics of reactions that gradually change the concentration of reactants, such as the binding of a ligand to a receptor [13] or endonuclease digestion of oligonucleotides [14], can be investigated by a series of sequential FCS measurements [15]. Each individual FCS measurement is short enough to locally approximate the system by an equilibrium. Alternatively, and in accordance with the principle of FCS, reaction kinetics have been derived from fluctuations about equilibrium. However, the prerequisite for such measurements is that reactants and products differ in their fluorescent properties.

14.2.2 Calculation of the Autocorrelation Function

In FCS, the autocorrelation function describes the average persistence time of fluctuations in the recorded signal. The calculation of the autocorrelation function $G(\tau)$ does not require information on the molecular processes giving rise to

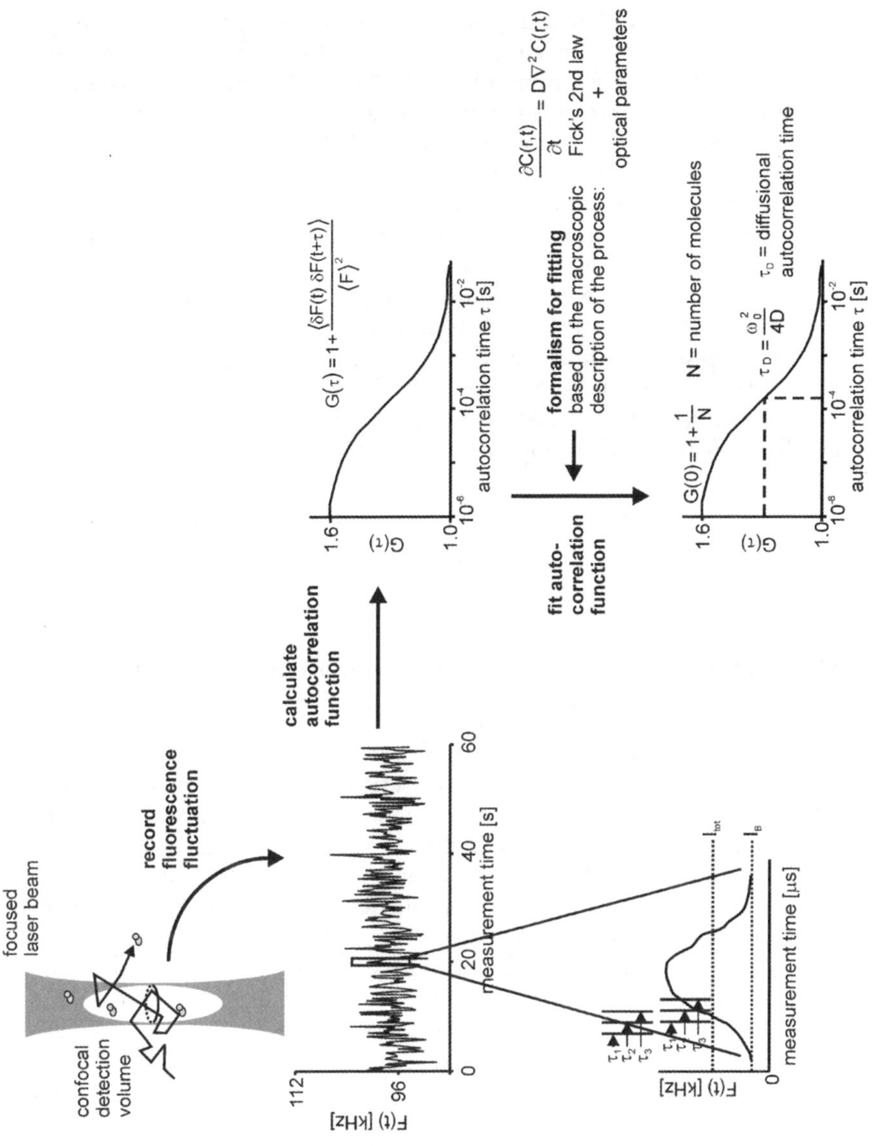

Fig. 14.2. Analysis of fluorescence fluctuations in fluorescence correlation spectroscopy exemplified for Brownian translational diffusion. At low concentrations, translational diffusion of molecules results in fluctuations of the signal recorded from the confocal detection volume. A temporal autocorrelation analysis relates the signal at any point in time t to the signal at any later point in time, given by the increment τ. The autocorrelation function is fitted by a formalism that implements the kinetic formalism of the molecular process giving rise to the fluorescence fluctuations, i.e. Fick's second law of diffusion, and the characteristics of the instrument optics, thereby yielding the diffusional autocorrelation time τ_D

the fluorescence fluctuations (Eq. 14.1). Instead, for any given time-dependent fluorescence signal $F(t)$, the autocorrelation function provides information on the extent to which, at any point in time t, the fluorescence F is related to the signal at any later point in time $t+\tau$. Information on the molecular mechanism is only implemented when a formalism for fitting the autocorrelation function is derived (Fig. 14.2). Due to the averaging over all time points t, only those molecular processes that possess a characteristic relaxation time finally contribute to the autocorrelation function, e.g. transport processes such as Brownian diffusion and chemical reactions. Fluctuations with varying time constants are averaged out and do not contribute to the autocorrelation function. However, in contrast to the analysis of signals by Fourier transformation, each individual event may occur at an arbitrary point in time.

In FCS, the fluorescence autocorrelation function is usually normalized by the square of the mean fluorescence intensity $\langle F \rangle$ [16]. Instead of the total fluorescence, the deviations $\delta F(t)$ about the mean are considered.

$$G(\tau) = \frac{\langle F(t)\,F(t+\tau)\rangle}{\langle F\rangle^2} = \frac{\langle F\rangle^2 + \langle \delta F(t)\,\delta F(t+\tau)\rangle}{\langle F\rangle^2} = 1 + \frac{\langle \delta F(t)\,\delta F(t+\tau)\rangle}{\langle F\rangle^2} \tag{14.1}$$

14.2.3
Implementation of an Analytical Formalism for Describing an Autocorrelation Function: Translational Diffusion

In perturbation kinetics, the kinetic parameters of a reaction are derived from the relaxation kinetics of the perturbed system back into equilibrium. The formalism for describing the macroscopic reaction is the starting point for deriving a formalism to describe an autocorrelation function [12]. In the case of Brownian translational diffusion this is Fick's second law of diffusion. In addition, the optical characteristics for the excitation and detection of fluorescence need to be considered. The most commonly used analytical solution for the autocorrelation function assumes a three-dimensional Gaussian distribution of the detection efficiency of fluorescence from the confocal volume element [4, 17]. For this reason, the intensity is a function of the position of each molecule in the detection volume. The calculation of the intensity has to integrate over the intensity distribution in the whole volume element. As a consequence, the relaxation of occupation number fluctuations in the detection volume does not manifest itself as an exponential decay in the fluorescence (Eq. 14.2).

The amplitude of the autocorrelation function is inversely related to the number of fluorescent particles N in the detection volume. The half maximum of the autocorrelation amplitude corresponds approximately to the autocorrelation time of the process giving rise to the fluctuation (due to the square root term (Eq. 14.2) the half maximum is shifted to slightly smaller values of τ). For transport processes that deviate from Brownian diffusion, the shape of the autocorrelation function provides information on the mechanism of transport. Steeper decays arise for example from directed transport [18, 19]. In the presence of so-called anomalous diffusion the slope decreases [20]. For one fluorescent species undergoing three-dimensional translational diffusion, the diffusional autocorrelation function is described by Eq. 14.2.

$$G(\tau) = 1 + \frac{1}{N}\left(\frac{1}{1+\tau/\tau_D}\right)\cdot\left(\frac{1}{1+(\omega_0^2/z_0^2)\,(\tau/\tau_D)}\right)^{\frac{1}{2}} \tag{14.2}$$

where ω_0 and z_0 are the $1/e^2$ beam waist radii along the x/y and z axes, respectively. The ratio z_0/ω_0 is frequently denoted as the structure parameter S; τ_D is the translational diffusion autocorrelation time, related to the translational diffusion constant D (Eq. 14.3). Practically, D may either be derived by comparison of τ_D with a diffusional autocorrelation time of a molecule with known diffusion constant or by directly determining ω_0. It is important to realize that τ_D sensitively responds to any deviation of the detection volume from the ideal situation [21].

$$\tau_D = \frac{\omega_0^2}{4D} \tag{14.3}$$

The amplitude of the diffusional autocorrelation function is inversely related to the number of molecules in the detection volume. With the size of the detection volume known, FCS provides direct access to molecular concentrations. Division of the total fluorescence by the number of molecules yields the fluorescence per molecule (fpm). The fpm carries information on environment-dependent changes of the quantum yield of a fluorophore and aggregation of fluorescently labelled molecules. Under conditions in which such environment-dependent changes of the fpm can be excluded, a lower than expected fpm is indicative of misalignment of the instrument or the presence of background fluorescence [22]. For this reason, at the beginning of each experiment, the optical alignment of the instrument is validated by acquiring autocorrelation functions for a calibration sample with known photophysical and diffusional characteristics. Moreover, misalignments will immediately manifest themselves in larger structural parameters and distortions of the shape of the autocorrelation function. The latter effect is highly relevant for the analysis of multi-component systems and cellular analyses. Distortions in the shape of the autocorrelation function may be misinterpreted as the presence of multiple components with slightly different diffusional autocorrelation times or the presence of anomalous diffusion.

14.2.4 Autocorrelation Functions Containing Several Components

The autocorrelation function reflects all reactions and molecular transport processes that give rise to temporal fluctuations in the fluorescence and that have a characteristic relaxation time. The relative amplitudes of these individual processes in the autocorrelation function provide quantitative information on the fraction of molecules contributing to each of the individual processes. However, for extracting such quantitative information, two general cases need to be distinguished: (1) those in which all fluctuations have the same amplitude per molecule and (2) those in which different processes result in different fluctuation amplitudes. The latter processes are exemplified by the diffusion of mixtures of molecules that contain monomers as well as aggregates of fluorescently labelled molecules, or by reactions in which molecules change between states of different molecular brightness, rather than between an on and an off state [23].

14.2.4.1
All Fluctuations Having the Same Amplitude per Molecule

For the analysis of receptor–ligand interactions using a fluorescently labelled ligand of low molecular weight, the free ligand will be represented in the autocorrelation function by a short autocorrelation time, and the receptor-bound ligand by a longer autocorrelation time (Fig. 14.3). The molar fractions of bound versus free ligand can be derived from the relative amplitudes of both contributions, given that the fluorescence properties remain unaffected upon binding of the ligand to the receptor. Equation 14.4 is appropriate for samples consisting of molecules with different diffusion constants but identical molecular brightness, a triplet transition with time constant τ_T and fraction of molecules in the triplet state T, and in which a constant background I_B is present in the total signal I_{tot}.

$$G_{tot}(\tau) = 1 + \frac{\overbrace{\left(1 + \frac{Te^{-\tau/\tau_T}}{1-T}\right)}^{\text{triplet}} \overbrace{\left(1 - \frac{I_B}{I_{tot}}\right)^2}^{\text{back-ground}}}{N_{tot}} \cdot \overbrace{\sum_j \varphi_j \cdot \left(\frac{1}{1+\tau/\tau_{Dj}}\right) \cdot \left(\frac{1}{1+(\omega_0^2/z_0^2)\,(\tau/\tau_{Dj})}\right)^{\frac{1}{2}}}^{\text{Brownian translational diffusion}} \tag{14.4}$$

Uncorrelated background I_B decreases the fluctuation amplitude, i.e. leads to a higher apparent number of molecules in the detection volume [29]. For each species j, φ_j is the fractional weighting factor for the jth contribution to the autocorrelation function. In the case of equal molecular brightness for all

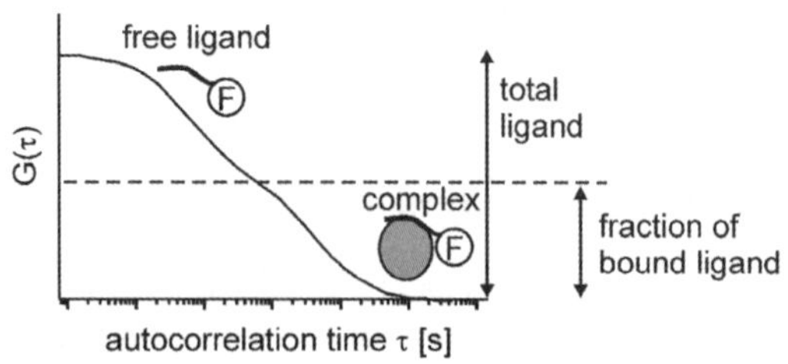

Fig. 14.3. Detection of ligand receptor interactions. Binding of a fluorescently labelled low molecular weight ligand to a receptor results in the presence of a second diffusional component in the autocorrelation function with a longer diffusional autocorrelation time. The fraction of bound ligand is derived from the contribution of the slow-moving fraction to the autocorrelation amplitude

molecular species, i.e. all fluctuations having the same amplitude per molecule, φ_j equals N_j/N_{tot}.

The part of the autocorrelation function that describes molecular diffusion is a sum of the individual contributions. In order to distinguish two different diffusional components, a considerable difference in molecular weight of both molecular species is required. The diffusion coefficient of a molecule is inversely related to its Stokes radius, which for globular molecules is related to the cubic root of the molecular weight. Under favourable experimental conditions, i.e. comparable fpm and high fluorescence signal, a difference in diffusion autocorrelation times by more than a factor of 1.6, corresponding to a difference in molecular weight by a factor of 4, is required to decompose the autocorrelation function into its two components [24]. For the detection of molecular interactions it is mandatory that the smaller component carries the fluorophore in order to distinguish the free molecular species from the complex. In some cases, conformational changes induced by ligand binding, rather than an increase in molecular weight, may result in a shift in diffusion times large enough to be resolved by the autocorrelation analysis.

The constraints on molecular weight differences for the detection of molecular interactions can be overcome by fluorescence cross-correlation spectroscopy [25]. In cross-correlation spectroscopy both constituents of a complex carry a spectrally distinct fluorophore with little or no overlap of the fluorescence emission spectra. The fluorescence of both species is separated by dichroic mirrors and detected simultaneously by two different detectors. A molecular interaction manifests itself by the concerted recording of fluorescence fluctuations in both fluorescence channels.

If, in addition to the diffusion of fluorescent molecules, photophysical processes or molecular reactions contribute to the autocorrelation function, these fluctuations will be present at smaller relaxation times. The residence time of the molecules in the detection volume sets the lower limit for rate constants of reaction kinetics to be investigated by FCS [23]. Only if a sufficient number of transitions of any such process occur while the molecule is in the detection volume can the rate constant of that process constitute itself in the autocorrelation function. Energy transfer in conjunction with cross-correlation spectroscopy may be used to extend the range of time constants to be observed by FCS [23].

If the molecule is diffusing too fast, buffers of higher viscosity can be employed in order to slow down diffusion and enable the detection of the reaction. One needs to make sure, however, that the more viscous solvent will have no impact on the relaxation of the photophysical transition or the reaction. Alternatively, the fluorescent molecules can be immobilized [26]. In the implementation of analytical formalisms for describing photophysical processes the intensity profile can be neglected. The autocorrelation function for photophysical processes decays exponentially [6], as exemplified for the triplet term in Eq. 14.4.

In addition to transitions into the triplet state, the analysis of photophysical processes by FCS has been exemplified by investigations of protonation-dependent fluorescence fluctuations of the green fluorescent protein (GFP) [7] and the light-driven *trans–cis* isomerization of indocyanine dyes [8]. Moreover, the binding of ethidium bromide to DNA is one very early example in which environment-dependent changes of the photophysical characteristics of a fluorophore were employed to analyse intermolecular interactions [2]. The detection of conformational changes of DNA hairpins [9] and RNA three-helix junctions [26] was based on fluctuations caused by conformation-dependent fluorescence resonance energy transfer.

14.2.4.2
Fluctuations Having Different Amplitudes

Instead of all components having the same fpm, in many cases molecular species with different molecular brightness are present. Aggregation equilibria of fluorescently labelled molecules, molecular interactions in which one receptor binds to two fluorescently labelled ligands, and changes of molecular brightness induced by the molecular environment and affecting only a fraction of all fluorescent particles exemplify this situation. In these cases, the contribution of a molecular component to the autocorrelation function depends on the square of the relative molecular brightness ($rfpm_j$) of this component [12] (Eq. 14.5). The molar fraction of a component j is given by $Y_j = N_j / N_{tot}$.

$$\varphi_j = \frac{(Y_j rfpm_j{}^2)}{\left(\sum_j Y_j rfpm_j\right)^2} \qquad (14.5)$$

This characteristic of FCS may affect the determination of binding constants of fluorescently labelled peptides to antibodies. If antibodies complexed to two peptides comprise 10% of the fluorescent particles in the sample, these complexes will contribute 33% to the autocorrelation amplitude, instead of only 10% [27]. In vitro, this complication can be circumvented by adding the bivalent receptor in excess. However, in cellular experiments it may be impossible to control the molecular composition of the sample to this level. For heterogeneous populations of molecules in different states of aggregation the amplitude of the autocorrelation function will therefore not provide a valid measure of the number of particles in the detection volume. In this case, photon counting histograms [5, 28] provide an independent means for analysing the presence and relative amounts of molecular aggregates in a sample.

14.2.5
Noise

The treatment of noise in fitting of the autocorrelation function deserves special attention. The signal-to-noise ratio of the autocorrelation function depends on the number of detected photons per fluctuation and per fluorescent molecule [29]. Thus, for an autocorrelation function the statistical accuracy is not a constant for all autocorrelation times. Instead, for short autocorrelation times, the autocorrelation function is less well defined. The higher standard deviation at shorter autocorrelation times is a result of the fact that the more rapidly a fluctuation occurs, the fewer photons will be detected for the respective event. Long autocorrelation times are subject to a systematic error – referred to as bias – that can only be decreased by longer sampling times [30]. A pragmatic approach to considering the dependence of noise on the autocorrelation time is by recording a series of autocorrelation functions for each sample, and weighting the contribution of each data point to the fit with the inverse of the standard deviation of the average over all curves [31].

14.3 Cellular FCS

14.3.1 Molecular Dynamics

Under diffraction-limited conditions the confocal detection volume measures about 0.3 fl, three to four orders of magnitude smaller than a eukaryotic cell. For molecules that are freely diffusive inside the cytoplasm and/or the nucleus one may readily record and analyse autocorrelation functions using the same protocols and algorithms described above. Autocorrelation functions recorded for free GFP inside the cytoplasm and the nucleus can be fitted with algorithms containing one diffusive component only [32]. The diffusional autocorrelation time differs by a factor only attributed to cytoplasmic viscosity [33]. The detection of freely diffusive molecules has been valuable for comparing the efficiency of import of molecules into the cell cytoplasm [22]. Moreover, FCS offers specific advantages when using fluorescent probes with environment-dependent photophysical properties [34]. Given the same fluorescence per molecule for all fluorophores, the number of molecules derived from the autocorrelation amplitude will be unaffected by changes in molecular brightness. Therefore, in contrast to other fluorescence techniques, FCS provides a concentration-normalized measure of environment-dependent changes in fluorescence intensity.

The first applications of cellular FCS addressed the mobility of molecules introduced into the cells by microinjection and considered to be inert with respect to the interaction with intracellular components [35, 36]. In the past few years, however, an increasing number of applications have used the technology for addressing the dynamics of molecules with respect to their function inside the cell. These applications have been greatly promoted by the use of fusion proteins of fluorescent proteins. The analysis of diffusion and transport of molecules in the cytoplasm, nucleus and cellular membranes provides a wealth of information on the cellular fine structure.

Deviations from the ideal case of non-interacting molecules and free Brownian diffusion constitute themselves in the shape of the autocorrelation function. Hindrance of free Brownian diffusion by the presence of obstacles [37] or by residence of a molecule within membrane domains [38] was analysed by applying algorithms implementing anomalous diffusion. In contrast to free Brownian diffusion, anomalous diffusion is characterized by a time dependence of the mean square displacement of a particle [39].

Further publications on cellular FCS that accounted for the cellular fine structure addressed (1) active transport through plastid tubules [40], (2) the measurement of diffusion of molecules next to a fluctuating membrane [41], (3) diffusion on curved membranes [42] and (4) detection volumes that are partially delimited by the plasma membrane [43]. However, especially for the models that describe diffusion in or next to membranes, detailed information on the cell morphology and position of the detection volume is required in order to extract the correct molecular parameters. Misplacement of the detection volume relative to the plane of the membrane leads to the recording of autocorrelation functions from which too small diffusion constants will be derived. Recently, a protocol for measuring diffusion of membrane-associated proteins was presented that solved the positioning problem [44].

14.3.2
Intracellular Concentration Measurements

The ability to determine molecule concentrations independently of fluorescence calibration samples is a unique feature of FCS. For proteins involved in signal transduction, the concentrations of the protein may determine the dose-response characteristics to an external ligand, the kinetics of signalling and finally the functional response of a cell. Through analysis of the interdependence of protein concentration and a cellular response, FCS should therefore contribute significantly to model building in systems biology [45].

In contrast to in vitro measurements using microlitre sample volumes, the volume of a cell is a few picolitres only. For this reason, photobleaching will lead to a continuous depletion of fluorophores. A solution to this problem was developed in the context of measurements of the concentration of EGFP (enhanced green fluorescent protein)-tubulin in cells of budding yeast (Fig. 14.4) [46]. From biochemical experiments it was estimated that tubulin was present at a total concentration of about 500 nM, 20 to 40% of which was EGFP-tubulin, corresponding to an EGFP-tubulin concentration of 100 to 200 nM. The diameter of a yeast cell is about 5 μm, corresponding to a volume of 200 fl. While one fraction of the tubulin is sequestered into microtubules, the other fraction is freely diffusive inside the cytoplasm. Positioning of the detection volume was based on images taken with a CCD camera and scans along the optical axis [32]. For each yeast cell, a series of ten autocorrelation measurements over 15 s each was performed. Autocorrelation functions could readily be fitted with a term containing one diffusive component and one triplet term. At the beginning of each series of measurements, a rapid decrease of fluorescence was observed that was attributed to the photobleaching of EGFP-tubulin molecules present in microtubules. In order to describe the decay of fluorescence, two more exponential components were required. These were attributed to: (1) a local depletion of fluorophores inside the detection volume, resulting in a concentration gradient towards the detection volume; and (2) a depletion of fluorophores inside the finite cellular volume. In vitro, high laser powers lead to a local decrease in the number of fluorophores as

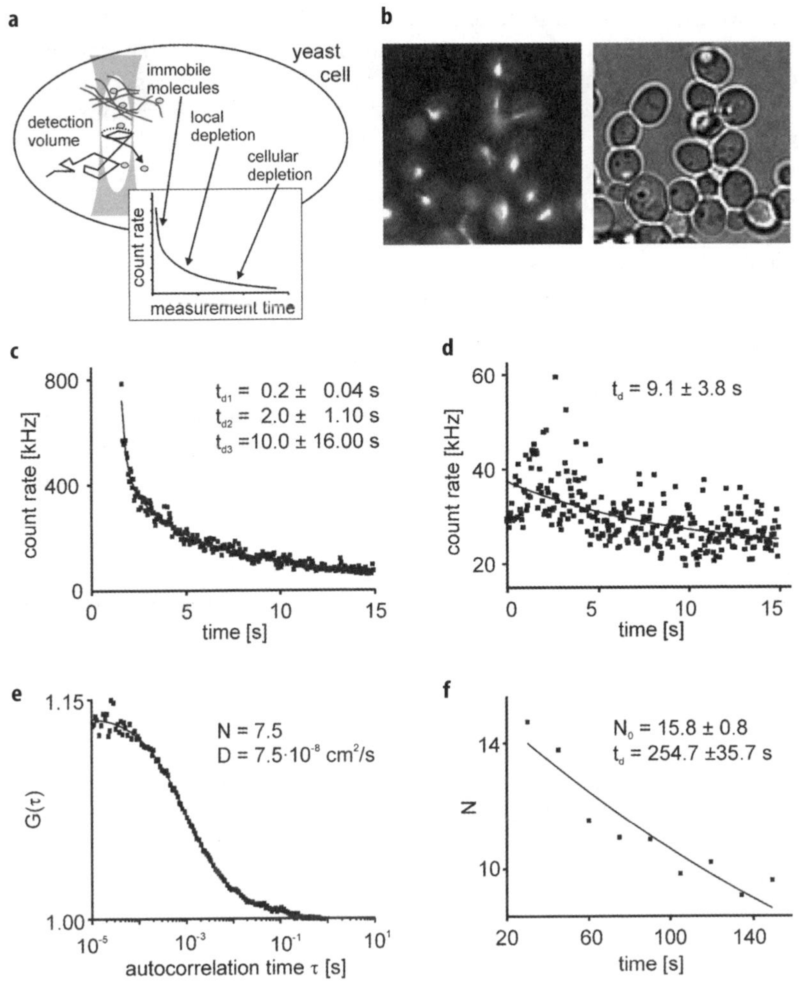

Fig. 14.4a–f. Measurements of free EGFP-tubulin in budding yeast. **a** Schematic representation of processes contributing to a decay in fluorescence. Immobile molecules are bleached rapidly. Local depletion relates to a local reduction in the number of fluorophores due to photobleaching during the residence time of molecules in the focus. The limited cellular volume leads to a photodepletion of molecules in the cell. **b** Fluorescence (*left*) and transmission (*right*) images of budding yeast cells, expressing EGFP-tubulin. **c, d** Count traces of cellular GFP fluorescence **c** at the beginning of excitation and **d** over a subsequent 15 s interval. In **c** the decay was fitted with three exponentials with decay rates t_{d1} to t_{d3}, in **d** with one exponential t_d only. Once immobile fluorescence has been bleached and the local depletion has occurred, the decay reflects the cellular photodepletion only. **e** Cellular autocorrelation function, recorded over a measurement time of 15 s. **f** Number of molecules N derived from a series of autocorrelation measurements, fitted with a monoexponential. The number of molecules N_0 at the zero time point was extrapolated from the fit. Measurements were performed on a ConfoCor2 fluorescence correlation spectroscope (Carl Zeiss, Jena, Germany) equipped with a cooled CCD camera (PCO Computer Optics GmbH, Kelheim, Germany). The yeast cells were provided by P. Maddox, Laboratory of E. D. Salmon, University of North Carolina at Chapel Hill

well. However, in contrast to cellular measurements, the practically infinite size of the reservoir of fluorophores results in the formation of a stable concentration gradient, yielding a constant fluorescence once this gradient has formed.

In the experiments depicted in Fig. 14.4, intracellular autocorrelation measurements were performed at a laser power lower than the one used for the generation of the bleaching curves. Still, the initial photobleaching and depletion of fluorophores could not be avoided. When plotting the number of molecules derived from the individual measurements versus the total measurement time, this depletion was apparent from a monoexponential decay in the molecule number. At this lower laser power, each individual measurement was unaffected by the depletion, illustrating the ability of FCS to approximate a non-equilibrium state by local equilibria at sufficiently short measurement times. Extrapolation to the zero time point yielded the number N_0 of free EGFP-tubulin molecules before the measurements were started. A direct determination of this molecule number was compromised by the initial rapid photobleaching of immobile molecules. In addition, extrapolation with a monoexponential overcame the inability to calculate an average molecule number. In spite of the decay, nine out of ten of the individual measurements contributed to the determination of the molecule number.

Partial overlap of the detection volume with the vacuole very likely contributed to deviations between individual measurements. The results obtained by FCS compared favourably with the biochemical estimates. A molecule number N_0 of 18 EGFP-tubulin molecules was obtained as a mean for several cells. For a detection volume of 0.15 fl, this figure corresponds to a concentration of 180 nM. Comparison of this concentration with the one obtained for yeast cells treated with the microtubule depolymerizing agent, nacodazole, should enable the determination of the fraction of tubulin sequestered into microtubules. The determination of microtubule-associated EGFP-tubulin by performing FCS measurements in the absence and presence of nacodazole represents a paradigmatic application of intracellular concentration measurements by FCS.

14.3.3 Limiting Factors in Cellular FCS

When considering FCS for the analysis of molecular organization inside the cell, three key factors limit the suitability of the technique (Fig. 14.5). These are:

1. Aggregation of fluorescent molecules resulting in low diffusion constants and in high fluorescence per aggregate, the latter compromising concentration measurements (Eq. 14.5).
2. The association of molecules with immobile cellular structures such as the cytoskeleton. Upon laser excitation, these molecules will be bleached rapidly.
3. Molecular diffusion with a diffusion autocorrelation time longer than the time in which changes in the cellular distribution of molecules or cellular morphology occur.

The measurement time for recording an autocorrelation function must be longer than the time constant of the slowest process to be investigated. For photophysical transitions and the diffusion of small molecules such as labelled peptides

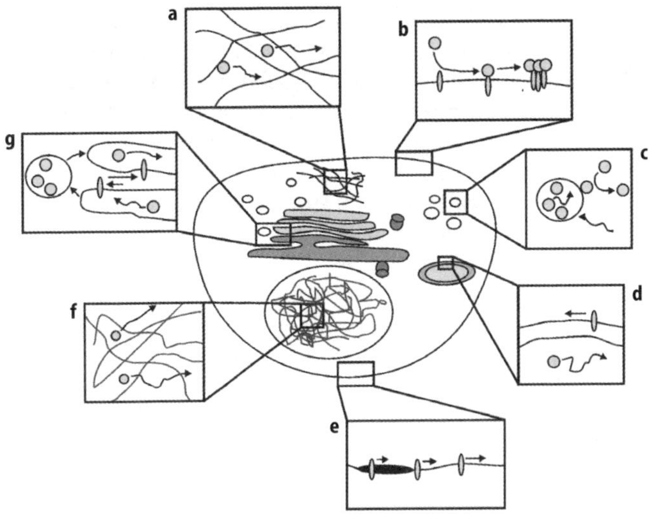

Fig. 14.5 a–g. Accessibility of cellular molecular processes to analysis by FCS. **a** Diffusion of molecules. Obstruction by cellular components leads to anomalous diffusion. **b** Binding of ligands to cell surface receptors. The quantification of receptor-bound ligand is compromised by rapid aggregation and subsequent internalization of receptor–ligand complexes. **c** Vesicle transport and association of molecules with vesicles. Due to the accumulation of fluorophores, the vesicle corresponds to a molecular entity with a very high fpm. The vesicle movement will dominate the autocorrelation function. Diffusion of molecules inside the vesicles or fluctuations due to transient association of molecules with a vesicle will not be detectable. Two-colour cross-correlation spectroscopy can be used as a sensitive method for detecting the co-localization of molecules in or co-associated with vesicles [47]. **d** Molecular mobility and concentration measurements in micrometre-sized compartments. For cylinder- or worm-shaped organelles such as mitochondria, positioning of the optical axis perpendicular to the axis of the organelle will in principle enable the acquisition of autocorrelation functions [48]. However, such measurements are confronted with the rapid depletion of fluorophores. **e** Membrane organization. Transient association of molecules with membrane microdomains, i.e. anomalous diffusion, affects the shape of the autocorrelation function. **f** Diffusion inside the nucleus. The analysis of diffusional modes of molecules inside the nucleus [37] and association of oligonucleotides with molecular complexes in the nucleus [49] have been presented. **g** Diffusion of molecules in the secretory pathway. Such measurements are very much related to measurements of molecules at the plasma membrane. Molecules trapped inside vesicles will be bleached

and globular proteins, with residence times in the detection volume of tens to hundreds of microseconds, this requirement can easily be fulfilled. However, for transmembrane receptors with residence times in the upper millisecond to second time range, measurement times of several minutes will be required in order to record an autocorrelation function. In this case, aggregation and internalization evoked by stimulation with ligand will be too fast for acquiring an autocorrelation function that describes the state of the system at any given point in time. Changes in the cellular morphology that affect the position of the detection volume relative to the compartment of interest by themselves cause fluctuations that may mask those resulting from the motion of molecules.

Even in situations in which the molecules of interest are mobile but molecular states with very different diffusion constants exist, FCS measurements may still fail to provide a full description of the molecular dynamics. For rapidly diffusing molecules, higher laser powers are required in order to record a sufficient number of photons per residence time in the detection volume than for slowly diffusing molecules. At these higher laser powers, however, slowly diffusing particles may be bleached. These counteracting requirements of photons needed to describe a process and avoidance of photobleaching define the dynamic range for autocorrelation measurements. The width of this dynamic range depends on the photostability of the fluorophore. For GFP fusion proteins, free cytosolic proteins and membrane-associated proteins may be detected simultaneously. In contrast, for fluorescein one may fail to simultaneously detect a free fluorescein-labelled ligand and a receptor-bound ligand.

14.4 Perspectives

In FCS considerable progress in the analysis of molecular processes in cells is to be expected through the integration of different modalities of data acquisition and analysis. In particular, dedicated methods for the analysis of fluorescence that account for the cellular morphology will be implemented.

14.4.1 Combinations of Detection Modalities

Cellular single molecule spectroscopy will follow the trend of in vitro single molecule spectroscopy to replace or complement the temporal autocorrelation analysis by photon counting histograms (PCH) [5, 50]. FCS, PCH and imaging were combined for the analysis of the membrane association, cellular distribution and aggregation of isoforms of adenylate kinase [51]. In these experiments, FCS provided information on the mobility of EGFP-tagged proteins, localized in the cytosol and at the plasma membrane by imaging microscopy. PCH was employed to address whether a diffusion constant that was smaller than expected for the protein of the respective molecular weight was due to homo-aggregation of the protein. Still, in spite of the growing significance of these modes of data analysis, the on-line calculation of autocorrelation functions will retain its significance as the primary measurement modality for validating the measurement position and adjustment of the instrument.

For the analysis of molecular mobilities, the combination of FCS and photobleaching techniques such as FRAP (fluorescence recovery after photobleaching) and continuous photobleaching will gain significance [52]. FCS fails to provide information on mobilities and transport of molecules occurring over minutes to hours. In addition, high local densities of proteins may compromise the acquisition of autocorrelation functions.

Imaging microscopy and FCS will complement each other in providing information on concentrations of immobile and mobile molecules. In cellular compartments in which molecules are mobile the number of molecules determined by FCS may be correlated with the image intensity, thereby providing an internal calibration for the quantitation of fluorescence in those parts of the image that represent immobile molecules [53].

14.4.2
Alternative Methods for Analysing Diffusional Modes

In autocorrelation measurements, the shape of the autocorrelation function carries information on the mode of diffusion of the fluorescent particles. However, a number of recent publications have demonstrated that the shape of the autocorrelation function is affected in very similar ways by a number of different conditions. In particular, a flattening of the autocorrelation function, as occurs for anomalous diffusion, may instead be caused by the presence of two components with similar diffusional autocorrelation times [37], obstruction of the detection volume by the plasma membrane [43], and diffusion of particles on curved membranes [42]. In cellular measurements, however, it is very difficult to define the geometry of the detection volume and the molecular composition of the sample to the point that any of the aforementioned conditions can be fully excluded. Very recently, Weiss et al. demonstrated the detection of anomalous diffusion based on the fractal dimensions of the recorded fluorescence fluctuations [54]. This method is unaffected by all of the factors listed above.

Acknowledgements
The author wishes to thank Oda Stoevesandt for critical reading of the manuscript and assistance in preparation of figures.

References

1. Madge D, Elson EL, Webb WW (1972) Phys Rev Lett 29:705
2. Madge D, Elson EL, Webb WW (1974) Biopolymers 13:29
3. Eigen M, Rigler R (1994) Proc Natl Acad Sci USA 91:5740
4. Rigler R, Mets Ü, Widengren J, Kask P (1993) Eur Biophys J 22:169
5. Chen Y, Müller JD, So PTC, Gratton E (1999) Biophys J 77:553
6. Widengren J, Mets Ü, Rigler R (1995) J Phys Chem 99:13368
7. Haupts U, Maiti S, Schwille P, Webb WW (1998) Proc Natl Acad Sci USA 95:13573
8. Widengren J, Schwille P (2000) J Phys Chem A 104:6416
9. Bonnet G, Krichevsky O, Libchaber A (1998) Proc Natl Acad Sci USA 95:8602
10. Brock R, Jovin TM (1998) Fluorescence correlation microscopy (FCM): fluorescence correlation spectroscopy (FCS) in cell biology. In: Elson EL, Rigler R (eds) Fluorescence correlation spectroscopy: theory and applications. Springer, Berlin Heidelberg New York
11. Wiseman PW, Hodedelius P, Peterson NO, Magnusson KE (1997) FEBS Lett 401:43
12. Elson EL, Madge D (1974) Biopolymers 13:1
13. Rauer B, Neumann E, Widengren J, Rigler R (1996) Biophys Chem 58:3
14. Kettling U, Koltermann A, Schwille P, Eigen M (1998) Proc Natl Acad Sci USA 95:1416
15. Schwille P, Bieschke J, Oehlenschläger F (1997) Biophys Chem 66:211
16. Thompson NL (1991) Fluorescence correlation spectroscopy. In: Lakowicz JR (ed) Topics in fluorescence spectroscopy. Plenum, New York, p 337
17. Aragón SR, Pecora R (1976) J Chem Phys 64:1791
18. Kunst BH, Schots A, Visser AJWG (2002) Anal Chem 74:5350
19. Madge D, Webb WW, Elson EL (1978) Biopolymers 17:361

20. Schwille P, Korlach J, Webb WW (1999) Cytometry 36:176
21. Hess ST, Webb WW (2002) Biophys J 83:2300
22. Waizenegger T, Fischer R, Brock R (2002) Biol Chem 383:291
23. Horn EFY, Verkman AS (2002) Biophys J 83:533
24. Meseth U, Wohland T, Rigler R, Vogel H (1999) Biophys J 76:1619
25. Schwille P, Meyer-Almes F-J, Rigler R (1997) Biophys J 72:1878
26. Kim HD, Nienhaus U, Ha T, Orr JW, Williamson JR, Chu S (2002) Proc Acad Natl Sci USA 99:4284
27. Chen Y, Müller JD, Tetin SY, Tyner JD, Gratton E (2000) Biophys J 79:1074
28. Müller JD, Chen Y, Gratton E (2000) Biophys J 78:474
29. Koppel DE (1974) Phys Rev A 10:1938
30. Saffarian S, Elson EL (2003) Biophys J 84:2030
31. Wohland T, Rigler R, Vogel H (2001) Biophys J 80:2987
32. Brock R, Vàmosi G, Vereb G, Jovin TM (1999) Proc Natl Acad Sci USA 96:10123
33. Chen Y, Müller JD, Ruan Q, Gratton E (2002) Biophys J 82:133
34. Widengren J, Terry B, Rigler R (1999) Chem Phys 249:259
35. Berland KM, So PTC, Gratton E (1995) Biophys J 68:694
36. Brock R, Hink MA, Jovin TM (1998) Biophys J 75:2547
37. Wachsmuth M, Waldeck W, Langowski J (2000) J Mol Biol 298:677
38. Schütz GJ, Kada G, Pastushenko VPh, Schindler H (2000) EMBO J 19:892
39. Feder TJ, Brust-Mascher I, Slattery JP, Baird B, Webb WW (1996) Biophys J 70:2767
40. Köhler RH, Schwille P, Webb WW, Hanson MR (2000) J Cell Sci 113:3921
41. Fradin C, Abu-Arish A, Granek R, Elbaum M (2003) Biophys J 84:2005
42. Milon S, Hovius R, Vogel H, Wohland T (2003) Chem Phys 288:171
43. Gennerich A, Schild D (2000) Biophys J 79:3294
44. Benda A, Beneš M, Marecek V, Lhotský A, Hermens WTh, Hof M (2003) Langmuir 19:4120
45. Kitano H (2002) Nature 420:206
46. Maddox PS, Bloom KS, Salmon ED (2000) Nat Cell Biol 2:36
47. Bacia K, Majoul IV, Schwille P (2002) Biophys J 83:1184
48. Cluzel P, Surette M, Leibler S (2000) Science 287:1652
49. Politz JC, Browne ES, Wolf DE, Pederson T (1998) Proc Natl Acad Sci USA 95:6043
50. Palo K, Brand L, Eggeling C, Jager S, Kask P, Gall K (2002) Biophys J 83:605
51. Ruan Q, Chen Y, Gratton E, Glaser M, Mantulin WW (2002) Biophys J 83:3177
52. Wachsmuth M, Weidemann T, Muller G, Hoffmann-Rohrer UW, Knoch TA, Waldeck W, Langowski J (2003) Biophys J 84:3353
53. Weidemann T, Wachsmuth M, Knoch TA, Müller G, Waldeck W, Langowski J (2003) J Mol Biol 334:229
54. Weiss M, Hashimoto H, Nilsson T (2003) Biophys J 84:4043

15 Fluorescent Quantum Dots: Properties and Applications

M. R. Warnement and S. J. Rosenthal

Keywords: Nanometer-sized fluorescent semiconductor crystallites; Quantum dots; Serotonin transporter proteins

Abbreviations

ANA	Anti-nuclear antigen
FLIM	Fluorescence lifetime microscopy
FRET	Fluorescence resonance energy transfer
GlyRs	Glycine receptors
IgG	Immunoglobulin G
PEG	Polyethylene glycol
QDs	Quantum dots
SERT	Serotonin transporter protein

15.1 Introduction

Labeling biological molecules using fluorescent tags is a common and very useful practice in biological science. Advances in techniques such as fluorescence resonance energy transfer (FRET), fluorescence lifetime microscopy (FLIM), and two-photon fluorescence spectroscopy continue to provide additional information concerning cellular processes. All of these techniques, however, traditionally have suffered several limitations due to the use of organic dyes as fluorophores. Most notably, organic dyes have a narrow spectral window for efficient excitation, broad emission spectra, and are readily susceptible to quenching. Nanometer-sized fluorescent semiconductor crystallites (nanocrystals or quantum dots, QDs) have the potential to revolutionize and expand fluorescence imaging and screening applications in biology and medicine.

Unlike organic dyes and fluorescent proteins, nanocrystals have size-tunable, narrow, Gaussian emission spectra which can be simultaneously excited with a single excitation source, enabling several receptors, cellular components, or processes to be visualized simultaneously without bleed-through or crosstalk. This multiplexed fluorescence detection capability also lends itself to applications in high-throughput screening and fluorescence-based assays. All sizes of nanocrystals can be conveniently photoexcited at wavelengths longer than the ultraviolet, extending cell life. As the nanocrystals are inorganic, they are robust, with photobleaching occurring on a timescale of hours to days as opposed to minutes. This will enable visualization of dynamic processes with continuous illumination. Finally, the surface chemistry of the nanocrystals has been developed such that they are both water soluble and compatible with conjugation to biologically active small molecules, peptides, and antibodies.

The development and application of semiconductor nanocrystals as fluorescent probes is one of the most exciting areas of research which spans the fields of chemistry, physics, and biology. There are numerous examples of their use as biological labels in the literature [1–20], only a few of which have been selected for review in this chapter to illustrate current applications. Initially, we will present a brief quantum mechanical description of the photophysical properties of semiconductor nanocrystals. Then we will review several applications of these nanocrystals illustrating their use as fluorescent probes.

15.2 Photophysical properties of Quantum Dots

An explanation of the photophysical properties of quantum dots begins with an understanding of the semiconductor band gap. When a nanocrystal absorbs a photon, an electron from the valance band is promoted to the conduction band, leaving a hole in the valance band. The unique physical properties of quantum dots originate from the quantum confinement of the electron–hole pair. Quantum confinement arises when one of the dimensions of the object is of the order of the exciton Bohr radius [21]. In the case of CdSe nanocrystals, the Bohr exciton radius is 56 Å. As such, CdSe semiconductor nanocrystals vary in diameter from 20 to 70 Å and contain from 200 to 10,000 atoms, as illustrated in Fig. 15.1.

This quantum confinement leads to the emergence of discrete electronic states in the nanocrystal and size-dependent electronic and optical properties. Just as in the quantum mechanical "particle-in-a-box", these energy levels vary inversely with the square of the confinement length. Therefore, a smaller nanocrystal leads to a larger difference in energy levels and, consequently, a higher energy (shorter wavelength) emission. Conversely, a larger nanocrystal will result in a lower energy (longer wavelength) emission. As a result the color of emission is dictated by the size of the nanocrystal, as illustrated in Fig. 15.2.

In 1998 two revolutionary papers appeared in which it was demonstrated that a special type of nanocrystal, a "core/shell", can be used for fluorescent tags in biological applications [1, 2]. Individual semiconductor nanocrystals have dangling bonds on their surface which act to trap electrons and holes, preventing recombination and fluorescence. This acts to drastically reduce the quantum yields of these nanocrystals. In order to design an improved fluorescent probe with enhanced quantum yields, a core of one semiconductor nanocrystal can be wrapped in a shell of a second semiconductor material [22]. As long as the second material has a wider band gap, it will be energetically favorable for the electron and hole to remain in the core, in close proximity to each other, such that they will recombine to give off fluorescence. The shell also acts to passivate the dangling bonds of the core, eliminating the traps of electrons and holes. However, because of the offsets in the crystal

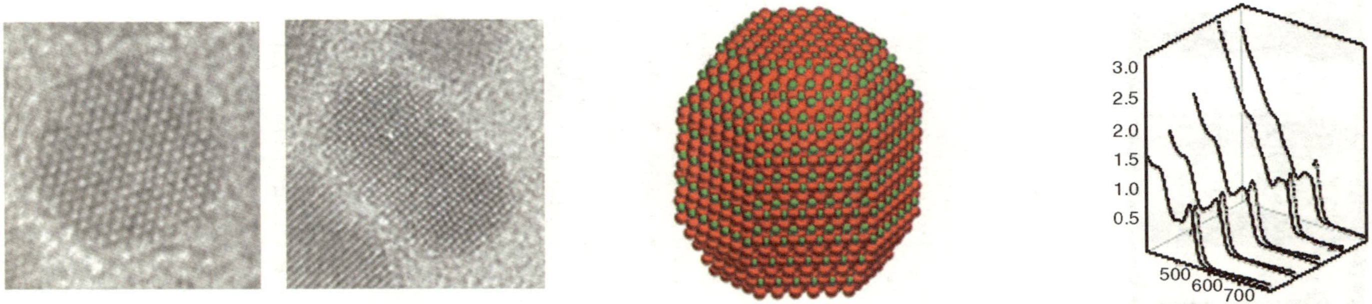

Fig. 15.1. *Left*: High-resolution transmission electron microscopy (HR-TEM) images of CdSe nanocrystals obtained at the Vanderbilt Microscopy Facility. *Center*: Model of a CdSe nanocrystal obtained from analysis of HR-TEM images. *Right*: Absorption and emission spectra of a series of CdSe nanocrystals. As the nanocrystal size increases the energy of the first absorption feature decreases. The emission spectrum is narrow and shifts with size in a like manner. The absorption spectrum is continuous above the band gap so any standard excitation source can be used to excite the nanocrystal

Fig. 15.2. The color of the emission is dictated by the size of the core, as illustrated for a series of highly fluorescent CdSe/ZnS core/shell nanocrystals from the Quantum Dot Corporation. The CdSe nanocrystal core is smallest (18 Å) for blue emitting dots and largest (60 Å) for red emitting dots

lattices of the two materials, the passivation is imperfect and different trap states are introduced at or near the interface. Overall the fluorescence quantum yield of the core/shell is enhanced over that of the core, but is less than unity due to imperfect passivation and/or crystal defects at or near the interface.

However, recent advancements by Quantum Dot Corporation in the synthesis of core/shell nanocrystals have produced quantum dots with significantly improved quantum yields. These nanocrystals have emission spectra with full width at half maximum height (FWHM) as narrow as 20 nm, and reproducible fluorescence quantum yields as high as 85%. The key to the high quantum yield is matching the Zn and S atoms to the Cd and Se atoms at the interface between the core and the shell, so that there are no defects which can act as traps for the photoexcited electron or hole and prevent them from recombining to yield fluorescence. Quantum Dot Corporation chemists have optimized the lattice matching between the core and shell by adding a dopant during shell growth.

Core/shell nanocrystals have several advantages over organic molecules as fluorescent labels, including resistance to photodegradation, improved brightness, nontoxicity, and size-dependent, narrow emission spectra that enable the monitoring of several processes simultaneously. Additionally their absorption spectrum is continuous above the band gap (Fig. 15.1) so that any standard excitation source can be used to excite the nanocrystal, and a single wavelength can excite all sizes of nanocrystal, and hence all colors. The superior spectral characteristics of fluorescent nanocrystals are illustrated in Fig. 15.3, which compares a series of Alexa dye molecules with CdSe/ZnS core/shell nanocrystals. Both the absorption and emission lineshapes of organic dye molecules are log normal, leading to the necessity of multiple excitation sources to simultaneously excite the dyes and spectral overlap in the emission. In contrast the emission spectrum of an ensemble of core/shell nanocrystals is narrow, as small as 20 nm FWHM, and the emission lineshape is Gaussian, yielding spectrally distinct signatures. The width of the emission spectrum is dictated by the size distribution of the cores, which can be exquisitely controlled to ±2 Å.

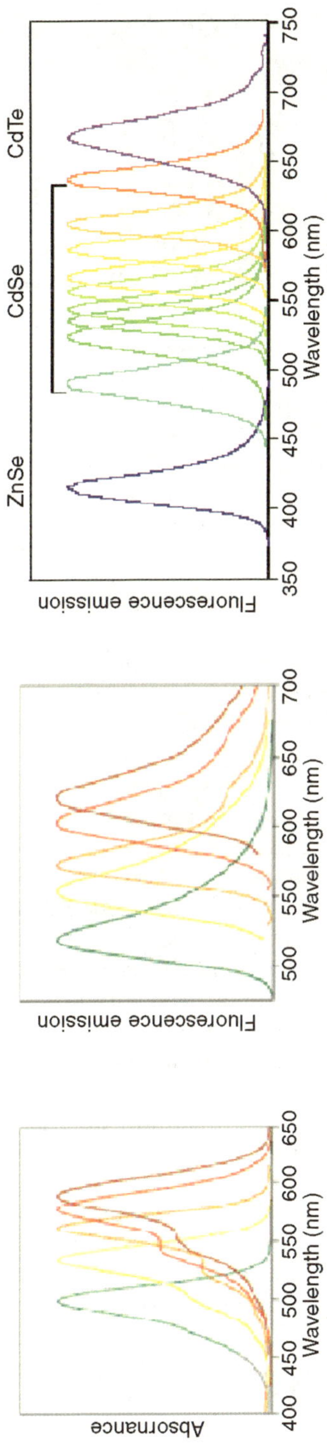

Fig. 15.3. Absorption (*left*) and emission (*center*) spectra of a series of Alexa dye molecules. Notice multiple excitation wavelengths are required to excite this series of molecules and the emission spectra overlap. In contrast, a single wavelength can be used to excite all sizes of nanocrystals and the emission spectra of quantum dots (*right*) are Gaussian in nature and spectrally isolable when using several fluorophores, as illustrated by this series of emission spectra from Quantum Dot Corporation

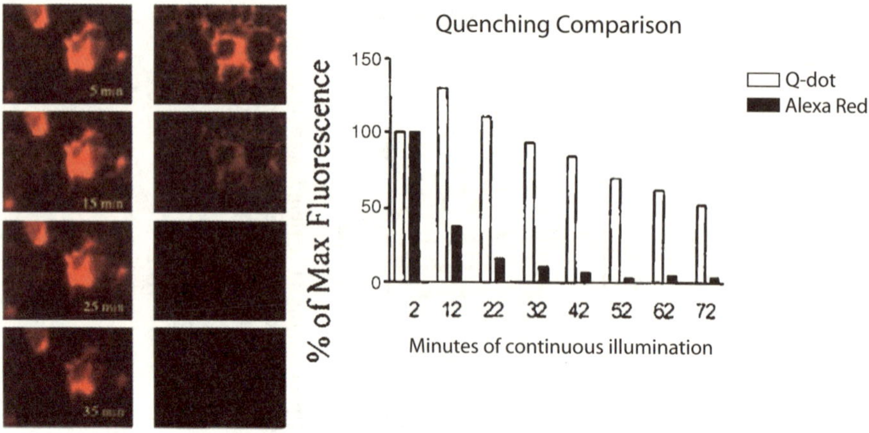

Fig. 15.4. *Left*: Live primary hippocampal cells labeled with monoclonal anti-LAMP (lysosomal-associated membrane protein) followed first by biotinylated anti-mouse IgG and then by (PEG)SA-QDs (*far left*) or streptavidin-labeled Alexa Red (*near left*) [23]. *Right*: Time dependence of integrated fluorescence intensity illustrating rapid photobleaching of Alexa Red

High-quality core/shell nanocrystals also show a remarkable resistance to photodegradation, as illustrated in Fig. 15.4. In this experiment live hippocampal cells are first labeled with monoclonal anti-LAMP (lysosomal-associated membrane protein), followed by biotinylated anti-mouse IgG and then by either streptavidin-conjugated quantum dots or streptavidin-labeled Alexa Red [23]. After 15 min of continuous illumination the Alexa Red has faded while the nanocrystals remain bright for more than an hour. These are modest results; new generations of core/shell nanocrystals remain bright after many hours of continuous illumination. It is precisely this resistance to photobleaching that will eventually enable dynamic imaging in living tissue. With fluorescence quantum yields of 85% and enormous extinction coefficients of 800,000 M^{-1} cm^{-1}, quantum dots can be used with lower intensities of excitation radiation, extending cell life during long-exposure experiments. It is also possible to excite all sizes of nanocrystals without using UV light, again extending cell life.

An interesting property of single quantum dots is the phenomenon of blinking, first described by Nirmal et al. [24] and later described in terms of Auger ionization by Efros and Rosen [25]. This property is observed when investigating individual quantum dots. Single nanocrystals alternate between an emitting state ("on" or "bright" state) and a nonemitting state ("off" or "dark" state). This result is cancelled out when an aggregate of several quantum dots is observed due to simple statistical considerations. As such, the average (observed) ensemble quantum yield is in fact underestimating the actual quantum yields that would be observed if all quantum dots were emitting [21]. This, however, is not as problematic as it may seem. Even with blinking occurring, researchers have been able to synthesize nanocrystals with quantum yields in excess of 85%. In addition, the blinking serves as a very useful signature for single molecule studies.

It is evident that quantum dots possess the photophysical properties necessary to overcome the limitations of conventional fluorophores. Their broad excitation spectra, size-tunable narrow emission spectra, high quantum yields, and resistance to photodegradation make them superior to conventional fluorophores in several ways. In order to be effective probes, however, they must also be made to be biologically active. It is in this area that most of the current research is directed. An effective probe must bind specifically to whatever target is being investigated. The challenge is to modify the surfaces of these relatively large inorganic molecules in such a way that will make them hydrophilic as well as site specific. We will now review several examples of current research that have been undertaken in order to advance the use of quantum dots as fluorescent probes.

15.3 Applications of Quantum Dots as Fluorescent Probes

Two seminal papers demonstrating the use of quantum dots as fluorescent probes were first published in *Science* in 1998 [1, 2]. Alivistatos and coworkers first enclosed two sizes of CdSe/CdS core/shell nanocrystals in a silica shell. The smaller nanocrystal (2-nm core) emitted green fluorescence at 550 nm with a 15% quantum yield. The larger nanocrystal (4-nm core) emitted red fluorescence at 630 nm with a 6% quantum yield. The surface of the green nanocrystals was then coated with trimethoxysilylpropyl urea and acetate groups, which were shown to bind with high affinity to the cell nucleus. The surface of the red nanocrystals was coated with biotin in order to label streptavidin-conjugated F-actin filaments. Effective dual-emission, single-excitation labeling of two different nanoconjugates on mouse fibroblasts was demonstrated, after incubation with both size probes, which allowed for the clearly resolved imaging of both red and green nanocrystal labels. There was, however, nonspecific labeling of the nuclear envelope by both the red and the green probes (resulting in a yellow color), but the actin filaments were specifically stained with red nanocrystals [1]. In complementary work, Nie and coworkers demonstrated that nanocrystal-labeled transferrin could undergo endocytosis, with the internalized nanoconjugate readily observed within cells [2]. CdSe/ZnS core/shell nanocrystals were first coated in mercaptoacetic acid to make them hydrophilic and then transferrin proteins were covalently bound to the carboxylic acid functionality by cross-linking to reactive amine groups. Cells were then incubated with mercapto-QDs (without the protein) as a control and no endocytosis of QDs was observed. In contrast, the transferrin-QD bioconjugates were transported into the cell, demonstrating that receptor-mediated endocytosis had occurred. This illustrated that the attached transferrin proteins were still active and able to be recognized by the receptor. In comparing core/shell nanocrystal fluorescence to dyes such as rhodamine, it was also shown that the QDs were 20 times as bright, 100 times as stable against photobleaching, and one third as wide in spectral linewidth.

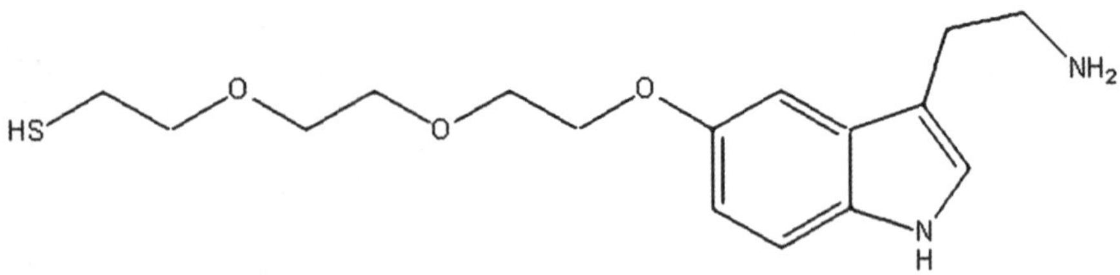

Fig. 15.5. Modified serotonin ligand with PEG linker arm. The thiol terminus allows covalent attachment to Zn on the surface of the core/shell nanocrystal

While these results were promising, it was still unclear whether QDs could specifically label molecular targets at a subcellular level. To overcome this nonspecificity we employed a strategy involving ligand-conjugated core/shell nanocrystals designed to specifically label designated targets. The first nanoconjugate probe synthesized was a serotonin-conjugated nanocrystal [3]. We first developed a strategy to attach the serotonin directly to the nanocrystal through the 5-hydroxyl functionality. Direct conjugation of the serotonin ligand, however, quenched the fluorescence of the nanocrystals, presumably through charge-transfer interactions, thereby reducing the valuable signal for imaging. Attachment of serotonin through a polyethylene glycol (PEG) linker arm, as shown in Fig. 15.5, helps both to defeat charge transfer and to provide additional degrees of freedom for interaction with cellular targets. The thiol anchor allows attachment to CdSe/ZnS core/shell nanocrystals via ligand exchange. This probe was then used to visualize serotonin transporter proteins (SERT) expressed in HEK cells. These ligand-conjugated nanocrystals were found to label SERT-transfected cells, but did not label either nontransfected cells or transfected cells co-incubated with the high-affinity SERT antagonist paroxetine, indicating the specificity of the nanoconjugate–SERT interactions. This approach to drug conjugation also has the additional advantage of being highly modular, allowing a great deal of flexibility to overcome experimental difficulties as they are encountered.

Bruchez and coworkers more recently employed quantum dots conjugated to immunoglobulin G (IgG) and streptavidin to specifically label different types of target proteins at different subcellular locations in fixed cells [4]. The QD-IgG probes successfully labeled Her2 on the surface of human SK-BR-3 breast cancer cells after the cells were incubated with a monoclonal anti-Her2 antibody, which binds to the external domain of Her2. This study also investigated the possibility of labeling intracellular proteins with quantum dot probes. Even though these probes are larger than organic dye molecules, steric factors did not limit their access to intracellular targets. This was demonstrated by labeling microtubules in fixed mouse 3T3 fibroblast cells using QD-streptavidin as the secondary label reagent. Specific labeling of antigens inside the nucleus was also demonstrated by incubating fixed human epithelial cells with human anti-nuclear antigen (ANA) antibodies, followed by biotinylated anti-human IgG and QD-streptavidin. To demonstrate the specificity of these probes, two different targets in the same cell were imaged simultaneously with different colors of QDs or combinations of QDs and organic dyes. Labeling was shown to be specific for the intended targets, brighter, and significantly more photostable than comparable organic dyes. In fact, all of the tested quantum dot probes were at least two to four times brighter than comparable Alexa dyes, which are reportedly brighter than any other known organic dye [26].

Quantum dots have also been used for *in vivo* applications. Peptides that specifically target lung blood vessel endothelial cells, tumor blood vessels, and tumor cell lymphatic vessels were conjugated to quantum dots and intravenously injected into mice [5]. Each of the peptides directed the quantum dots to the appropriate target within the mice, showing that they can be targeted *in vivo* with specificity. Quantum dots conjugated to lung-targeting peptides were observed in the lungs 5 min after intravenous injection, and no acute toxicity was observed, even after 24 h of circulation. Similarly, quantum dots coated with peptides specific to tumor vasculature accumulated in the tumors. Since the peptides used target tumor vessels, the probes actually targeted distinct structures within the tumors. The regional specificity of quantum dot delivery within a tumor demonstrates the feasibility of targeting functionally distinct components of a tumor. However, these probes were found to nonspecifically label organs with a prominent reticuloendothelial component, which mediates uptake of circulating particulates. Co-adsorption of PEG onto the quantum dots nearly eliminated these nonspecific interactions, without noticeably altering the probes' accumulation in tumor tissue.

Work published by Yamamoto and coworkers demonstrated that quantum dot probes can be useful bioimaging tools for tracing target cells over a period of a week *in vivo* [6]. In this study, EL-4 cells were incubated for 15 min with albumin-conjugated quantum dots, at a concentration of 0.1 mg/ml, so that uptake could occur via endocytotic pathways. The labeling concentration was set at 0.1 mg/ml because no significant cytotoxicity was observed at this level. After ten days in culture, approximately 10% of the cells still held observable quantum dots, which by this point were highly concentrated in the endosomes. These labeled cells were then injected into mice in order to assess their survival *in vivo*. The cells injected into mice were observed in the peripheral blood for 5 days after injection by fluorescence microscopy and flow cytometry. However, approximately 70% of the quantum dot-labeled cells were eliminated from the blood circulation within 2 h. Roughly 20% of these cells were detected in the kidneys, liver, lungs, and spleen and could still be observed 7 days after injection. In addition, no damage or toxicity due to injection of these quantum dot-labeled cells was observed, but cytotoxicity was observed at high incubation concentrations of quantum dots. These results indicate that quantum dots can be used as a cell-tracing marker, especially for that of the transplanted target cells, and that these probes can also be applied to flow cytometric analysis.

Quantum dots have been shown to be enabling in the area of multiphoton microscopy as well, having the largest reported two-photon cross sections of any probe used in multiphoton microscopy [7]. Multiphoton microscopy enables deep imaging of a variety of biological samples with less overall photobleaching than wide-field or confocal microscopy, and has become the primary fluorescence imaging technique for thick samples. Webb and coworkers investigated CdSe/ZnS core/shell nanocrystals coated in an amphiphilic polymer. These probes were found to have an action cross section two to three orders of magnitude larger than conventional fluorescent probes. This means that the probability of two-photon excitation occurring with quantum dots is as much as three orders of magnitude higher than with conventional probes, and is six orders of magnitude higher than with intrinsic molecules which add to background effects. Increasing the probe excitation probability relative to that of intrinsic species increases both the signal-to-noise ratio and the sample viability. To demonstrate multiphoton imaging *in vivo*, mice were first intravenously injected with the amphiphilic quantum dots. Imaging through the intact skin of a live mouse to a depth of 900 nm, the quantum dot-containing vasculature was clearly visible. In a comparison to conventional fluorophores, the same imaging experiment using fluorescein isothiocyanate-dextran as the probe shows considerably less detail even with five times the excitation power. Due to their large cross sections, quantum dots enable imaging at greater depths than conventional probes allow, using less average power. In addition, the mice showed no noticeable ill effects, and are being maintained as part of an investigation of long-term quantum dot toxicity.

Quantum dots can also serve as ideal probes in single molecule detection studies. Recently, Dahan and coworkers employed quantum dots to track individual glycine receptors (GlyRs) in order to analyze their diffusional dynamics in the neuronal membrane of living cells [8]. Receptors labeled with quantum dots could be visualized for time periods exceeding 20 min, while receptors labeled with the organic dye Cy3 could only be visualized for approximately 5 s. This high resistance to photobleaching is of paramount importance in single molecule tracking experiments simply due to the timescale required. Conventional dyes may bleach in the time it takes to focus the microscope, while quantum dots remain robust for extended periods of time. In this particular publication, biotinylated primary antibodies were first used to specifically label glycine receptors of fixed neuronal cells, and streptavidin-conjugated quantum dots were then employed as a secondary label. GlyRs were detected within the synaptic and extrasynaptic domains. Single quantum dots were identified by their blinking properties, another useful signature in single molecule detection studies. The spots were detected with a signal-to-noise ratio of about 50 using an integration time of 75 ms, almost an order of magnitude larger than the signal obtained with the Cy3 dye. Single molecule tracking deconvolution of the fluorescent spots resulted in a spatial resolution of approximately 10 nm. In this experiment, quantum dots were never observed intracellularly, indicating that the quantum dot-GlyRs were not internalized during the course of the experiments. This work demonstrates that quantum dots can be employed to record the mobility of individual molecules, even in confined cellular compartments.

15.4 Summary

Fluorescent quantum dots have experienced a rapid evolution in both the quality of their photophysical properties and the range of applications exploiting these properties. Indeed quantum dots are now commercially available which have emission spectra tunable throughout the visible, are photostable, and are exceptionally bright due to large absorption extinction coefficients and high fluorescence quantum yields. Various surface modifications for these commercial quantum dots are also available, for example streptavidin-conjugated dots, protein A nanoconjugates, and anti-mouse IgG antibody-conjugated dots to name a few. The list of demonstrated applications spans several fields in the biological sciences and has grown from simple staining applications of fixed cells to *in vivo* imaging applications.

The future for fluorescent quantum dots is bright. Improvements will continue to be made in quantum dot materials and surface chemistry. Next-generation materials will extend the spectral coverage to include the near-IR and IR spectral regions [27], enhancing *in vivo* and diagnostic applications. In addition to protein- and antibody-conjugated dots, the field will see the emergence of peptide- and small molecule-conjugated dots, further extending the flexibility of this platform. The range of demonstrated applications will continue to grow and will soon include, for example, multiplexed dynamic imaging of cellular processes in living cells. Finally, in addition to further demonstrations of novel applications enabled by quantum dots, we will soon also see a transition to the emergence of studies in which the properties of fluorescent quantum dots are exploited to advance discovery in the biological sciences.

References

1. Bruchez MP, Moronne M, Gin P, Weiss S, Alivisatos AP (1998) Science 281:2013
2. Chan WCW, Nie S (1998) Science 281:2016
3. Rosenthal SJ, Tomilson ID, Adkins EM, Schroeter S, Adams S, Swafford L, McBride J, Wang Y, DeFelice LJ, Blakely RD (2002) J Am Chem Soc 124:4586
4. Wu X, Liu H, Liu J, Haley KN, Treadway JA, Larson JP, Ge N, Peale F, Bruchez MP (2003) Nat Biotechnol 21:41
5. Akerman ME, Chan WCW, Laakkonen P, Bhatia SN, Ruoslahti E (2002) Proc Natl Acad Sci USA 99:12617
6. Hoshino A, Hanaki K, Suzuki K, Yamamoto K (2004) Biochem Biophys Res Commun 314:46
7. Larson DR, Zipfel WR, Williams RM, Clark SW, Bruchez MP, Wise FW, Webb WW (2003) Science 300:1434
8. Dahan M, Levi S, Luccardini C, Rostaing P, Riveau B, Triller A (2003) Science 302:442
9. Jaiswal JK, Mattoussi H, Mauro JM, Simon SM (2003) Nat Biotechnol 21:47
10. Pathak S, Choi S, Arnheim N, Thompson ME (2001) J Am Chem Soc 123:4103
11. Dahan M, Laurence T, Pinaud F, Chemla DS, Alivisatos AP, Sauer M, Weiss S (2001) Opt Lett 26:825
12. Byassee TA, Chan WCW, Nie S (2000) Anal Chem 72:5606
13. Han M, Gao X, Su JZ, Nie S (2001) Nat Biotechnol 19:631
14. Parak WJ, Boudreau R, Le Gros M, Gerion D, Zanchet D, Micheel CM, Williams SC, Alivisatos AP, Larabell C (2002) Adv Mater 14:882

15. Winter JO, Liu TY, Korgel BA, Schmidt CE (2001) Adv Mater 13:1673
16. Dubertret B, Skourides P, Norris DJ, Noireaux V, Brivanlou AH, Libchaber A (2002) Science 298:1759
17. Chan WCW, Maxwell DJ, Gao X, Bailey RE, Han M, Nie S (2002) Curr Opin Biotechnol 13:40
18. Sutherland AJ (2002) Curr Opin Solid State Mater Sci 6:365
19. Mitchell P (2001) Nat Biotechnol 19:1013
20. Klarreich E (2001) Nature 413:450
21. Michalet X, Pinaud F, Lacoste TD, Dahan M, Bruchez MP, Alivisatos AP, Weiss S (2001) Single Mol 4:261
22. Hines MA, Guyot-Sionnest P (1996) J Phys Chem US 100:468
23. Mason J, Tomilson ID, Rosenthal SJ, Blakely RD (2004) Bionanotechnology protocols, Humana, New York (accepted)
24. Nirmal M, Dabbousi BO, Bawendi MG, Macklin JJ, Trautman JK, Harris TD, Brus LE (1996) Nature 383:802
25. Efros AL, Rosen M (1997) Phys Rev Lett 78:1110
26. Panchuk-Voloshina N et al (1999) J Histochem Cytochem 47:1179
27. Bailey RE, Nie S (2003) J Am Chem Soc 125:7100

16 Heat Stress of Cancer Cells: Fluorescence Imaging of Structural Changes with Quantum Dots 605 and Alexa 488

O. Minet, C. Dressler and J. Beuthan

Keywords: Biomedical fluorescence; Heat stress; Quantum dots

Abbreviations

ConA	Concanavalin A
CLSM	Confocal laser scan microscope
PBS	Phosphate-buffered saline
Qdots	Luminescent quantum dots
SNOM	Scanning near-field optical microscopy

16.1 Introduction

In life sciences the probing of viable cells in full color over extended periods of time is necessary in order to visualize the rather complicated processes that occur during embryogenesis, carcinogenesis, and heat stress, for example. Existing fluorescence techniques use molecules like organic dyes, proteins or synthesized markers for optical molecular imaging [1]. In this case light will be emitted over a wide spectral range, which means that the spectra will overlap and a differentiation of the dyes becomes harder the more dyes are involved. Inorganic semiconductor nanocrystals – called quantum dots – promise alternatives to the chemical fluorophores and visible fluorescent proteins.

On the cellular level stress conditions can be induced by different factors like malnutrition, toxic agents, genetic defects or temperature mismatch [2]. Medical applications of hyperthermia are based on thermal tissue effects. Depending on the medical indication, thermal effects have to be generated by exact radiation parameters and appropriate laser or radio-frequency source specifications. Local tissue coagulations are induced with temperatures between 50 and 100 °C, while surgical tissue ablations and hemostasis need to be performed at temperatures higher than 100 °C [3]. Especially in the temperature range beneath 60 °C, therapeutic successes are predominantly influenced by the properties of the target tissue. At subcoagulating temperatures specific stress response mechanisms, including the complex reaction cascades during apoptosis or necrosis [4], might be activated in single cells or confined tissue areas. However, the outcome will always be either cell survival or cell death [5].

The molecular genetics of thermal cell stress responses have already been investigated in detail, mainly focusing on the so-called heat-shock proteins [5, 6] first described in 1962 by Ritossa [7]. Detailed knowledge of the phenomenology and pathophysiology of tissue-specific heat-induced stress reactions is not yet available. Only a few studies on selected tissue models have been published so far [e.g., 8–11]. Especially in the context of thermotherapies of tumors, like laser-induced interstitial thermotherapy [12–14], systematic studies exactly describing the dose–effect relationships are necessary for the microdosimetric optimization. Active and passive cellular heat-stress responses are dictated by the applied stress dose in a functional and phenomenological manner [15, 16].

A very suitable approach to study the phenomenon of heat stress in single cells is fluorescence microscopy, as one of the best established standard techniques in biomedical analytics and clinical diagnostics today. Fluorescence detection of specific structures and functions in cellular systems is benefited and limited, simultaneously, by the tremendous variability of fluorescent probes and markers concerning target specificities as well as spectral properties [17–19]. Many different organic fluorescent dyes and GFP-based biosensors have been routinely used for fluorescent imaging of cells, and now luminescent quantum dots (Qdots) have recently entered the field of bioanalytics.

Qdots are semiconductor nanocrystals of small extension with special optical properties [20]. A structure consisting of some thousands of GaAs/AlGaAs atoms is as small as 4 nm in diameter, and an In/P/AlGaAs structure is even 3 nm. Therefore, such structures are frequently called zero-dimensional objects.

The order of magnitude of the electrons' de Broglie wavelength is similar to the size of quantum dot. The electron–hole pairs are located in the three spatial dimensions because the core is contained within a protective shell. This band gap has a higher electronic level which confines the electron–hole pairs. The fluorescence emission is confined to a narrow band of typically 20–30 nm full width at half maximum height. Emission wavelengths are directly correlated with particle diameters [21]. For example, a Qdot made from Cd/Se radiates green light of 520 nm or red light of 630 nm depending on whether the diameter is 3 or 5.5 nm. Qdots can be excited over a broad spectrum, which means that a mixture of different kinds of Qdots for multiple color imaging can be excited by a light source of a single wavelength between UV and red [22]. The photoluminescence yield strongly depends on the environmental temperature [23, 24] exhibiting a memory effect (Fig. 16.1).

The inorganic shell of the Qdots is of hydrophobic nature. When putting Qdots to use in biology, they must be prevented from coming into direct contact with water and must be prevented from quenching. Despite their name, Qdots are about ten times bigger than conventional organic dyes, which might limit cellular uptake. Various modifications of the outer surface have been successful, always accompanied by increased size and a compromise in the colloidal and fluorescent properties. A courageous step toward these constraints was done by encapsulating the Qdots in micelles – a simple chemical aggregate made from two kinds of phospholipids [25]. The hydrophobic tails move away from the water, while the polar heads immerse themselves in aqueous environments. Qdots stay protected at the center of the micelles. A wavelength in the infrared spectrum and the relatively high quantum efficiency is a bonus when the light has to travel through bulk tissue. Besides

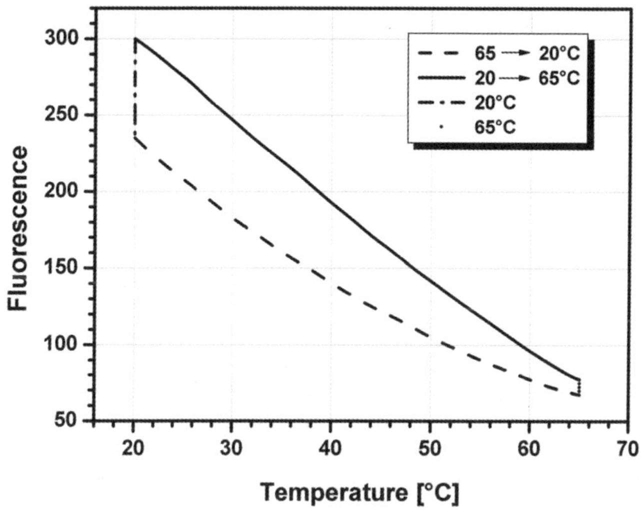

Fig. 16.1. Fluorescence emission intensities of Qdots are influenced by the surrounding temperature exhibiting hysteresis (according to [24])

many other biological observations, they have no detrimental effect on embryonic development and were photostable in vivo, for example in embryos of an African frog, for days (e.g., 14 h by a 50-mW, 488-nm laser [26]). Also Qdot targeting of human tumors in vivo worked in mice when the Qdots were conjugated to a peptide or antibody to recognize a specific cancer cell [25]. Finally, after targeting Qdots must be not more than weak toxic, should recognize selectively a specific target in a cellular structure, and must not interfere with normal physiology.

16.2 Experiments

16.2.1 Cell Cultivation and Heat Stressing

Human undifferentiated breast cancer cells of the line MX1 (Deutsches Krebsforschungszentrum, Germany) were employed as tissue model. Cells were maintained in RPMI 1640 medium supplemented with 20 mM HEPES and 10% (v/v) heat-inactivated fetal calf serum, at 37 °C and 5% CO_2 in a humidified atmosphere. Cultures were dissociated with 0.05% trypsin –0.02% EDTA. A 1% antibiotic–antimycotic solution (100 U/mL penicillin, 100 µg/mL streptomycin, 0.25 µg/mL amphotericin B; GIBCO, Invitrogen GmbH, Germany) was also added to the medium. All other culture medium components and solutions were purchased from Biochrom KG Seromed, Germany. Experimental cells were grown in chamber slides (Nunc GmbH & Co. KG, Germany) until subconfluent cell densities were achieved. Heat stress treatments were performed in a temperature-regulated water bath at the temperatures given with the results (40, 42, 45, 50, or 56 °C) for 30 min each. Growth controls were carried out in every experiment. Immediately after heat stressing cells were submitted to further experimental processing.

16.2.2
Cell Viability Screening via Colorimetric Microassay

Prior to fluorescence labeling experiments the impact of heat stressing on cell viability was screened with a colorimetric microassay. AlamarBlue (BioSource International Inc., USA) was used as metabolic reduction–oxidation indicator exhibiting a blue color in the native oxidized form or a red color after metabolic reduction (Fig. 16.2).

The cells were heat stressed as described before and submitted to the AlamarBlue cell activity assay according to the manufacturer's recommendations [27]. After 4 h reincubation under physiological conditions the reduction reaction was stopped by refrigerating the microplates (4 °C) until measurement. Absorption photometric evaluations of the plates were performed using a Dynex MRX II microplate reader combined with the Revelation G 3.2 software (Dynex Technologies, USA) at 570 nm (measurement wavelength) and 630 nm (reference wavelength). Temperature-dependent cell activities were quantified in correlation with control cells (100% cell viability) and negative controls (no cells).

16.2.3
Fluorescence Imaging of Cytoskeletal F-Actin in Cells

Visualization of heat-induced intracellular alterations was approached by fluorescence labeling of cytoskeletons. Generally, the cytoskeleton is a highly sen-

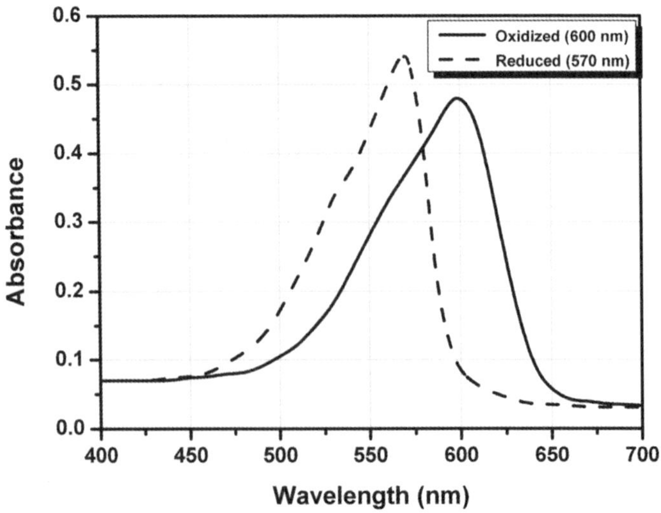

Fig. 16.2. Absorption spectra of oxidized and reduced AlamarBlue (Biosource Intl., Inc.)

sitive indicator of cellular stress reactions exhibiting distinct rearrangements under mild stress conditions. When exposed to severe stress conditions the cytoskeletal network will be disrupted and show fragmentations. Here the filamentous actin (F-actin) component of the cytoskeleton was selected as the target structure. F-actin was targeted by actin-binding phalloidin conjugated with Alexa Fluor 488 (Molecular Probes, NL). Fluorescence labeling was performed according to the manufacturer's recommendations [19]. Heat-stressed and control cells were rinsed twice in PBS and incubated with 100 μg/mL DL-α-lysophosphatidylcholine palmitoyl (ICN Biomedicals Inc., USA) in 3.7% formaldehyde added by 10 units/mL fluorescent phalloidin in methanol for 20 min at 4 °C. Then, cells were washed three times with PBS, mounted with cover slips, and examined under a microscope.

Alexa 488-labeled F-actin networks in cells were viewed by an Axioplan2 fluorescence light microscope combined with an AxioCam MRc digital camera and the AxioVision 3.1 software. All microscope components were purchased from Carl Zeiss, Germany. Fluorescence was excited with blue light (band pass 470±20 nm) and green emissions were detected (long pass >520 nm).

16.2.4 Quantum Dot Labeling of Cells

In general, the surface of the Qdots can be conjugated with ligand-specific molecules like antibodies, peptides or streptavidin for driving the biological activities [28, 29]. In particular, the streptavidin–biotin system is a widely used bridging tool in many labeling technologies. Since avidin and its derivatives selectively bind biotin and biotinylated proteins with high binding affinities (e.g., 10^{15} mol/L for streptavidin:biotin), they offer very versatile secondary fluorescent labels now extended by Qdot–streptavidin conjugates [19, 20, 30]. The general structure of a Qdot–streptavidin conjugate is illustrated in Fig. 16.3.

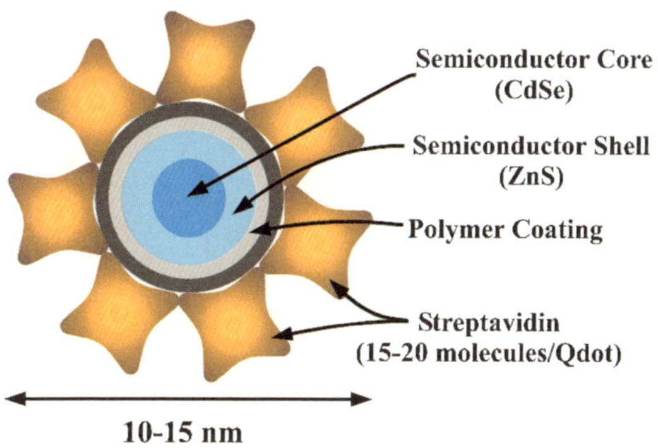

Fig. 16.3. Schematic overall structure of a Qdot-streptavidin conjugate. The core consists of semiconductor material (CdSe) which has been coated with a semiconductor shell (ZnS). The polymer coating provides a biochemical surface modification of the Qdot particle by direct coupling of the polymer to streptavidin (according to [20])

This study was a methodical approach and intended to test the Qdot–streptavidin conjugate for investigating the microdosimetry of heat stress using mammary carcinoma cells (species MX1) as tumor model. Red fluorescent Qdots 605 were selected, because autofluorescence of cells in the red spectral range is very low. The extinction coefficient of Qdot 605–streptavidin conjugate is 650,000 M^{-1} cm^{-1} at 600 nm. Besides the complex network of metabolic processes, the plasma membrane is the structure most prone to heat stress as the structural interphase between intracellular metabolism and the environment [31]. Therefore we chose glycoproteins and glycolipids as universal cellular plasma membrane targets. Biotinylated concanavalin A (ConA) served as the targeting probe. ConA selectively binds to α-mannopyranosyl and α-glucopyranosyl residues in glycoproteins and glycolipids abundantly located in the glycocalyx of cellular plasma membranes [19, 31]. This sandwich system was tested for its ability to support the characterization of heat-stress-induced cellular changes using a confocal laser scan microscope (CLSM).

Cells were rinsed in cold phosphate-buffered saline (PBS) and internal biotin was blocked by incubating cells in 50 µg/mL avidin (Sigma Aldrich Chem., Germany) in PBS (30 min, 37 °C). Cells were triple rinsed in cold PBS before application of 250 µg/mL concanavalin A-biotin conjugate (Molecular Probes, NL) in PBS (30 min, 37 °C). Streptavidin-conjugated Qdot 605 (Quantum Dot Corp., USA) nanocrystals were used to label cell-bound ConA mediated by biotin: streptavidin bridging. Qdot-conjugate stock solution was diluted 1:10 in incubation buffer supplied by the manufacturer 20 min prior to application. The final concentration of 0.1 µM was obtained by mixing equal volumes of PBS and Qdot solution. After 30 min incubation the cells were rinsed three times and micrographed in PBS. This labeling protocol is mainly based on the manufacturer's recommendations [20, 30]; alterations have been mentioned above.

Qdot fluorescence emissions were measured with an LSM 410 confocal microscope (Zeiss, Germany). The wavelengths 488 and 568 nm of an external argon-krypton laser (Spectra Physics, Germany) were used for exciting fluorescence signals. Emission signals were detected using a dichroic beam splitter FT 580 and a long-pass filter LP 590. A scan resolution of 3×8 bits/0.5 s (512×512 pixels) was applied for fluorescence imaging. All measurements were performed under confocal conditions (pinhole 20) at room temperature.

16.3 Results

16.3.1 Cell Viability Screening

Heat stresses at temperatures of 40 or 42 °C for 30 min did not impair cellular activities compared with the unstressed control cells, as shown by the AlamarBlue-based microassay (Fig. 16.4). On the other hand, cells stressed at 40 °C were activated and exhibited an approximately 20% higher viability than the control. Only stress doses above 42 °C decreased cellular viabilities with increasing tem-

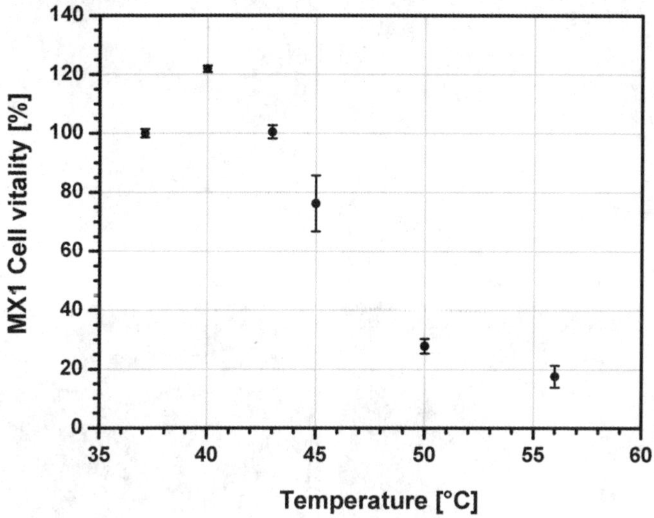

Fig. 16.4. Screening the metabolic activities of breast cancer cells (MX1) after heat stress using the AlamarBlue microassay. Forty-eight measurements were averaged for every experimental group. Control cells were incubated at 37 °C. The distributions of single measurements are indicated. Measurement wavelength: 570 nm, reference wavelength: 630 nm

peratures, as expected. After treating cells at 56 °C a minor metabolic rest activity below 20% was detected.

16.3.2 Fluorescence Microscopic Investigations

The temperature-dependent dynamics of cytoskeletal changes in MX1 tumor cells exposed to heat stress are illustrated in Fig. 16.5. While Alexa Fluor 488-labeled F-actin networks were intact in control cells (37 °C), the heat-stressed cells (40, 42, 45, 50 or 56 °C, 30 min each) exhibited increasing disintegration and disruption of actin fibers with rising temperatures. Intracellular restructuring was accompanied by drastic changes of cell shapes (rounding) and increasing disintegration of tissue integrity. These observations indicated that cell activities should be impaired after a 45 °C stress temperature, which corresponded well with the viability data (Fig. 16.4). Higher stress temperatures at 50 and 56 °C induced increasing fragmentation of actin fibers.

16.3.3 Quantum Dots

As expected, control cells (37 °C) and cells stressed at 40 or 42 °C exhibited regular growth conformations with intact tissue integrities and tissue-specific cell morphologies. The Qdot fluorescence was mainly restricted to the external leaflets of plasma membranes and intercellular areas (Fig. 16.6a–c). These data

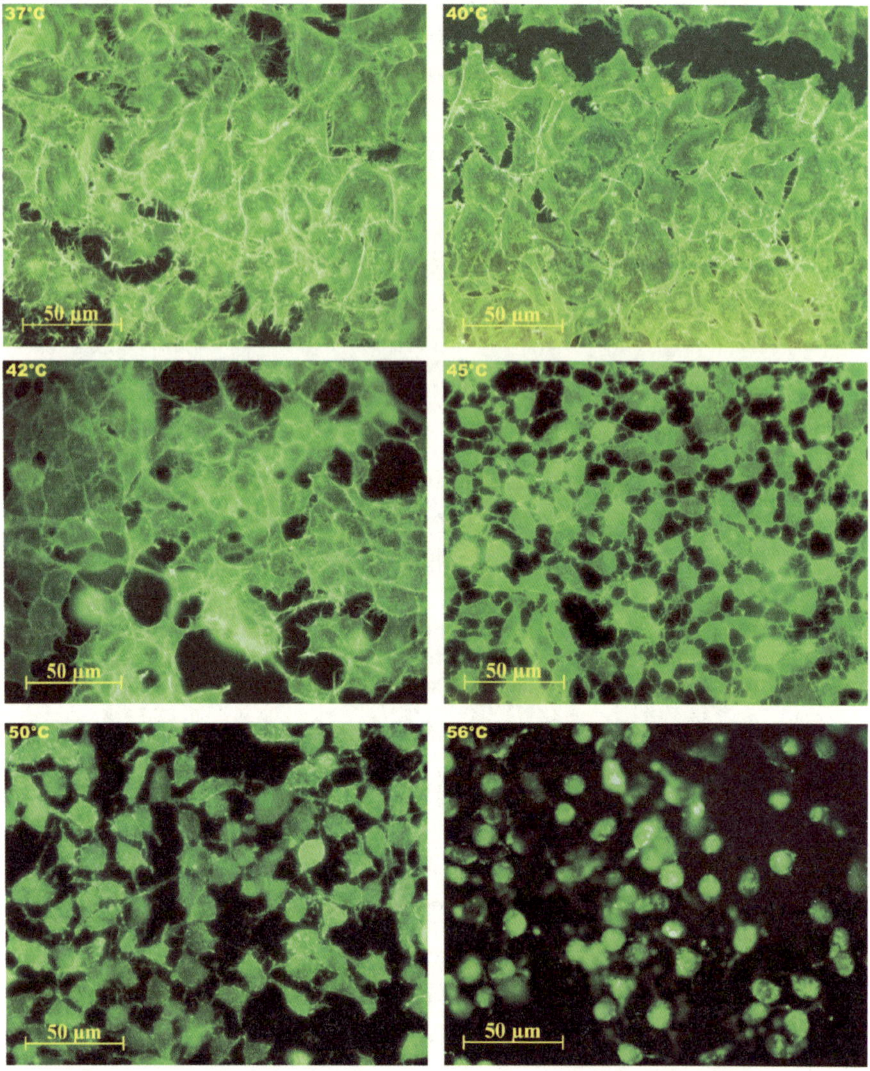

Fig. 16.5. Fluorescence imaging of F-actin in the cytoskeletons of tumor cells labeled with Alexa 488-conjugated phalloidin (Molecular Probes, NL). The cells were cultivated under physiological conditions and then heat stressed at 40, 42, 45, 50 or 56 °C for 30 min. Control cells were incubated at 37 °C. Excitation: 470±20 nm, detection: >520 nm

Fig. 16.6. Laser scan microscopy of MX1 cancer cells labeled with quantum dots. The first row shows control cells; heat-stressed cells (30 min each) are shown in the next five rows for 40 to 56 °C. All rows include the superposition of fluorescence micrograph and transmission (*left*), 3D depth coding in the colors of the rainbow (blue: 0 µm; red: 250 µm) with an actual range between 0 and 150 µm, approx. (*center*), and the 3D signal intensity profile (*right*). Fluorescence emissions were excited with the laser wavelengths 488 and 568 nm and emission signals were detected in the spectral range above 590 nm (FT 580, LP 590). The brightness of the 3D depth coding is the product between the maximum of brightness and the intensity plus offset. For the 3D signal intensity profile the integration time was 8 s/scan

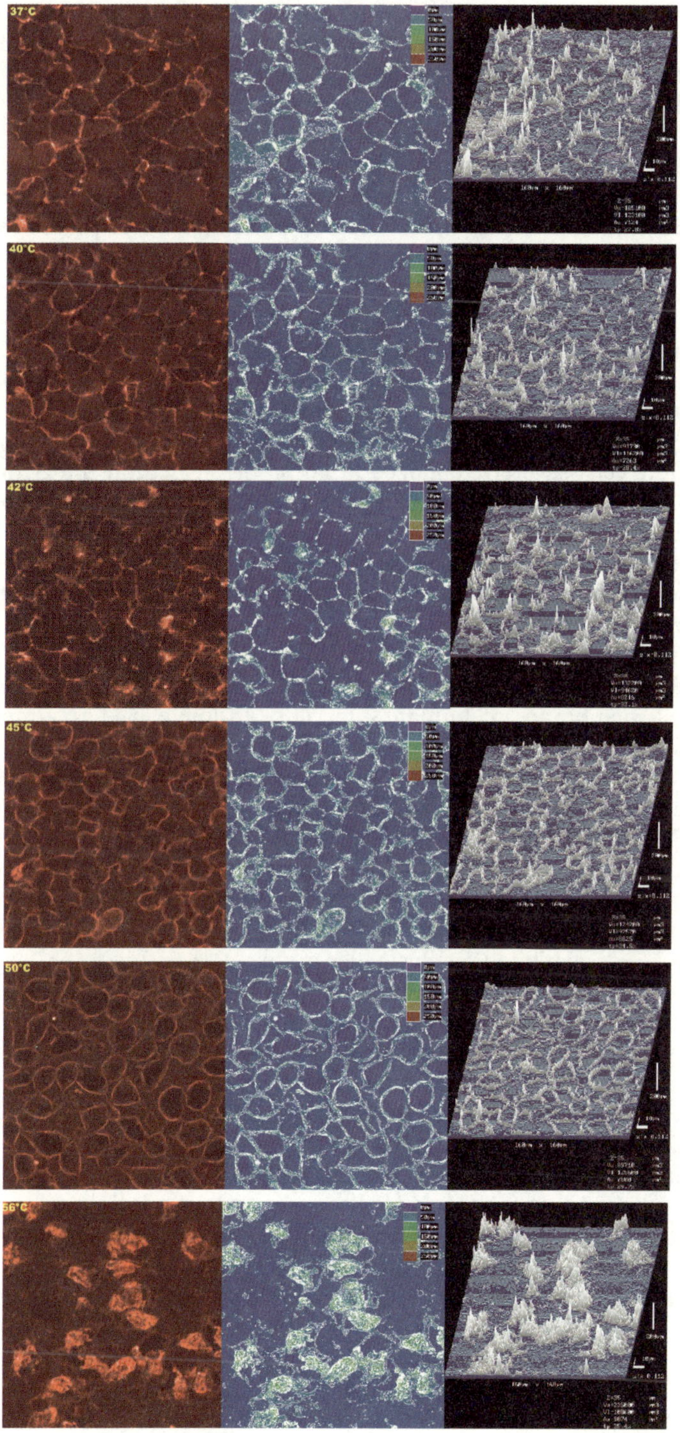
37°C
40°C
42°C
45°C
50°C
56°C

correlated well with the fluorescence micrographs measured by LSM. The superposition of fluorescence and transmission micrographs (left column in Fig. 16.6) shows functional and structural information together. It was done by a linear weighted addition in the sense of image processing algorithms.

Nevertheless the fluorescence intensity distributions were irregular, as illustrated by the 3D intensity profiles. When stressing was performed at 45 or 50 °C (Fig. 16.6d,e) the cells were smaller and rounder than in the control group (Fig. 16.6a). In addition, the tissue integrities were clearly dispersed indicating active cellular stress responses during heat treatment. This effect was more pronounced in the 50 °C group than in the 45 °C group. Here the 3D profiles exhibited more regular fluorescence intensity distributions than under minor stress conditions (40 and 42 °C). At the same time Qdot fluorescence was also detected at the cytoplasmic leaflets of plasma membranes, supporting the presumption that plasma membranes were reorganized and permitted Qdot conjugates to enter the cells. Nuclei in these cells were still intact and exhibited no fluorescence, while in the cells severely damaged at 56 °C the structural integrities of the plasma membranes and nuclear membranes were destroyed and fluorescence accumulated intracellularly penetrating the nuclear compartments (Fig. 16.6f). At 56 °C a more or less spontaneous necrosis was induced since the cells did not exhibit round shapes, indicating that heat stress at 56 °C was too severe to allow active stress responses.

Our results correspond well with the data published as an internet presentation by Hymer et al. [11], where temperatures higher than 45 °C were found to injure cell membranes in human skin cells. Although heat treatments were performed with a time regimen (1–300 s) different from the one used in our study, it can generally be concluded that the highly conserved membrane structure of animal cells is destabilized when faced with temperatures of 45 °C or higher. Cell-specific and individual differences in heat sensitivities obviously are most pronounced in this temperature range. This statement is supported by the comparatively large mean square deviations (m.s.d.) of the cell viability at 45 °C, as shown in Fig. 16.4, while the viabilities in the cell groups stressed at 40, 42, 50, and 56 °C exhibited much smaller m.s.d.

Qdots themselves are not toxic. The toxicity of the buffer was measured (data not shown), but did not induce any structural changes in cells. Functional defects were not investigated. Since Qdot labeling and microscopic viewing of specimens were performed after heat stress interventions, the measured fluorescence intensities did not need to be corrected with respect to the temperature effects shown in Fig. 16.1.

The experimental protocol recommended by the manufacturer [20] has been modified in the way that the Qdot labeling reaction was performed in a 50% PBS and 50% incubation buffer (provided by manufacturer) solution and not in 100% incubation buffer. In a previous experiment when Qdot labeling was performed in 100% incubation buffer, cells obviously were intoxicated and the formation of vacuoles was induced as shown in Fig. 16.7. Since the original incubation buffer is basic (pH 8.3) it employs nonphysiological conditions during incubations. Therefore the experimental conditions should be optimized for every cell type prior to Qdot labeling procedures. Valuable information on the preconditioning of labeling experiments using Qdots are given in refs. [20, 26, 28, 29, 32].

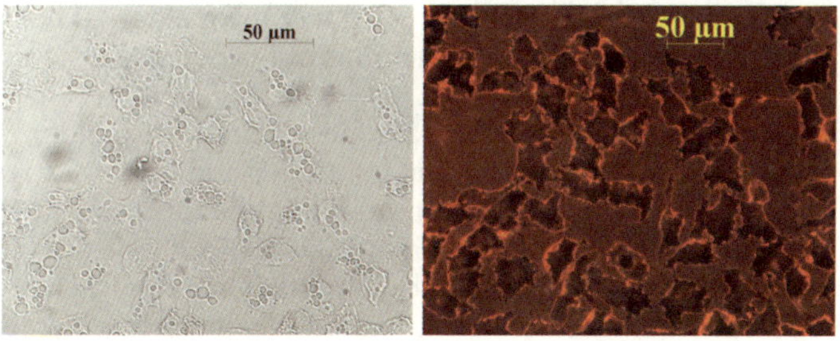

Fig. 16.7. Light (*left*) and fluorescence (*right*) microscopy (Axioplan 2, Carl Zeiss, Germany) of MX1 cancer cells labeled with quantum dots in 100% incubation buffer provided by the manufacturer [20]. After a 30 min incubation period the cells exhibited vacuoles

16.4 Discussion

Under the experimental conditions applied in this study the phenomenology of cellular response to heat stress was outlined. General cytomorphological features, mainly cell shape, tissue conformation, membrane integrity, and the intracellular organization of nuclei and F-actin skeletons, were visualized after fluorescence detection of Qdots or fluorescently labeled phalloidin. Investigations on specific subcellular features, especially protein functions and their molecular genetic basis, will have to use target-specific probes and appropriate detection methods, respectively. In general, Qdots show a much brighter fluorescence light due to higher quantum yields in comparison to organic fluorescent dyes. But future availability of different Qdot-conjugated probes will benefit bioanalytical studies on the dynamics of stress responses induced by physiological or environmental stress factors. The benefits of Qdot labeling will be accomplished by minimal spectral overlaps in multicolor labeling applications and long-term photostability. This will be associated with reduced technological requirements for fluorescence detection systems. The employment of Qdot labeling of specific targets will be advantageous for real-time monitoring of various intracellular processes. In particular, the use of Qdot-encoded multiplexed assays and high-throughput analyses of genes and proteins are of interest [22, 33]. Different Qdot-conjugated probes will also provide minimal spectral overlaps in multicolor labeling applications.

Also in the field of single molecule detection and spectroscopy Qdots are promising alternatives to fluorescent biomolecules because of their high photoluminescence yields [34]. Highly sensitive techniques like scanning near-field optical microscopy (SNOM), providing high lateral resolutions, could detect single Qdots and might favor the localization and quantification of specific targets or low-abundance biomolecules on cell membranes or various subcellular organelles or selected structures.

Our study only focused on spontaneous stress responses while long-term impacts on cell survival, apoptotic pathways or genetic overexpression rates were not included. Cellular genetics and phenomenologies of apoptotic and necrotic processes are complementary [4], not only for external noxious heat, but also they might be interrelated end points of the defense systems of living organisms.

Cancer cells especially are in a general state of genetic instability, eventually causing various subclones to be resistant to internal and external stress factors [35–37]. This implies the necessity for studying phenotypes and genotypes of heat stress responses in different tissue species in detail. Further results dealing with these subjects will enhance hyperthermia treatments.

With the results presented here the employment of luminescent Qdots was shown to be a reasonable alternative to organic fluorescent labels on the cellular level. But the high quality of modern fluorescent dyes like Alexa Fluor 488 and the successful application of standardized labeling methods will not entirely be substituted by Qdots today. Particularly in clinical diagnostics, standard labeling protocols cannot easily be adapted to Qdots as fluorescent markers. Therefore, the use in the near future of Qdots in biomedical sciences will mainly focus on different topics in experimental and basic research, e.g. [38].

References

1. Tung CH, Mahmood U, Bredow S, Weissleder R (2000) Cancer Research 60:4953
2. Csermely P (ed) (1998) Stress of life – from molecules to man. Ann NY Acad Sci Vol 851
3. Minet O, Dörschel K, Müller G (2004) Lasers in biology and medicine. In: Poprawe R, Weber H, Herziger G (eds) Laser applications. Landolt-Börnstein vol VIII/1c. Springer, Berlin Heidelberg New York (in press)
4. Proskuryakov SY, Konoplyannikov AG, Gabai VL (2003) Exp Cell Res 283:1
5. Vaux DL (2002) Toxicology 181–182:3
6. Rutherford SL (2003) Nat Rev Genet 4:263
7. Ritossa FM (1962) Experientia 18:571
8. Macouillard-Poulletier de Gannes F, Leducq N, Diolez P, Belloc F, Merle M, Canioni P, Voisin PJ (2000) Neurochem Int 36:233
9. Yagui-Beltran A, Graig AL, Lawrie L, Thompson D, Pospisilova S, Johnston D, Kernohan N, Hopwood D, Dillon JF, Hupp TR (2001) Eur J Biochem 268:5343
10. Jakubowicz-Gil J, Rzymowska J, Gawron A (2002) Biochem Pharmacol 64:1591
11. http://www.ntar3.sr.unh.edu/Download/191/paper191.DOC
12. Jacques S, Newman C, He XY (1991) Am Soc Mech Eng 189:71
13. Gewiese B, Beuthan J, Fobbe F, Stiller D, Müller G, Böse-Landgraf J, Wolf KJ, Deimling M (1994) Invest Radiol 29:345
14. Diaz SH, Nelson JS, Wong BJF (2002) Phys Med Biol 48:19
15. Welch AJ, van Germert MJC (1995) Optical-thermal response of laser-irradiated tissue. Plenum, New York
16. Chato JC, Lee, RC (1998) Ann NY Acad Sci 858:1
17. Oksvold MP, Skarpen E, Widerberg J, Huitfeldt HS (2002) J Histochem Cytochem 50:289
18. Stephens DJ, Allen VJ (2003) Science 300:82
19. http://www.probes.com/handbook
20. http://www.qdots.com
21. Woggon U (1997) Optical properties of semiconductor quantum dots. Springer, Berlin Heidelberg New York
22. Chan WCW, Maxwell DJ, Gao X, Bailey RE, Han M, Nie M (2002) Curr Opin Biotechnol 13:40
23. Braskén M, Lindberg M, Sopanen M., Lipsanen H, Tulkki J (1998) Phys Rev B 58:15993
24. Knight A (2003) 8th conference on methods and applications of fluorescence: spectroscopy, imaging and probes, Prague, 24–27 August 2003 (private communication)

25. Dubertret B, Skourides P, Norris DJ, Noireaux V, Brivanlou AH, Libchaber A (2002) Science 298:1759
26. Jaiswal JK, Mattoussi H, Mauro JM, Simon SM (2003) Nat Biotechnol 21:47
27. http://www.biosource.com/content/literatureContent/PDFs/alamarbluebooklet.pdf
28. Goldman ER, Anderson GP, Tran PT, Matoussi H, Charles PT, Mauro JM (2002) Anal Chem 74:841
29. Akerman ME, Chan WCW, Laakkonen P, Bhatia SN, Ruoslahti E (2002) Proc Natl Acad Sci USA 99:12617
30. Stayton PS, Nelson KE, McDevitt TC, Bulmus V, Shimooji T, Ding Z, Hoffman AS (1999) Biomol Eng 16:93
31. Rüdiger H, Gabius HJ (2001) Glycoconj J 18:589
32. Wu X, Liu H, Liu J, Haley KN, Treadway JA, Larson JP, Ge N, Peale F, Bruchez MP (2003) Nat Biotechnol 21:41
33. Xu H, Sha MY, Wong EY, Uphoff J, Xu Y, Treadway JA, Truong A, O'Brien E, Asquith S, Stubbins M, Spurr NK, Lai EH, Mahoney W (2003) Nucleic Acids Res 31:8 e43
34. Ikeda K, Matsuda K, Saito H, Nishi K, Saiki T (2001) J Microsc 202:209
35. Narita N, Noda L, Ohtsubo T, Fujeda S, Tokuriki M, Saito T, Saito H (2002) Int J Radiat Oncol Biol Phys 53:190
36. Zhang X, Li Y, Huang Q, Wang H, Yan B, Dewhirst MW, Li CY (2003) Clin Cancer Res 9:1155
37. Zeise E, Rensing L (2002) Int J Hyperthermia 18:344
38. Minet O, Beuthan J, Licha K, Mahnke C (2002) The biomedical use of rescaling procedures in optical biopsy and optical molecular imaging. In: Kraayenhof R, Visser AJWG, Gerritsen HC (eds) Fluorescence Spectroscopy, Imaging and Probes (Springer Series on Fluorescence Methods and Applications, Vol. 2) Springer, Berlin Heidelberg New York, pp 349

Sachverzeichnis

E

G

Q